Algorithms for Intelligent Systems

This book series publishes research on the analysis and development of algorithms for intelligent systems with their applications to various real world problems. It covers research related to autonomous agents, multi-agent systems, behavioral modeling, reinforcement learning, game theory, mechanism design, machine learning, meta-heuristic search, optimization, planning and scheduling, artificial neural networks, evolutionary computation, swarm intelligence and other algorithms for intelligent systems.

The book series includes recent advancements, modification and applications of the artificial neural networks, evolutionary computation, swarm intelligence, artificial immune systems, fuzzy system, autonomous and multi agent systems, machine learning and other intelligent systems related areas. The material will be beneficial for the graduate students, post-graduate students as well as the researchers who want a broader view of advances in algorithms for intelligent systems. The contents will also be useful to the researchers from other fields who have no knowledge of the power of intelligent systems, e.g. the researchers in the field of bioinformatics, biochemists, mechanical and chemical engineers, economists, musicians and medical practitioners.

The series publishes monographs, edited volumes, advanced textbooks and selected proceedings.

More information about this series at http://www.springer.com/series/16171

V. K. Giri · Nishchal K. Verma ·
R. K. Patel · V. P. Singh
Editors

Computing Algorithms with Applications in Engineering

Proceedings of ICCAEEE 2019

Springer

Editors
V. K. Giri
Department of Electrical Engineering
Madan Mohan Malaviya University
of Technology
Gorakhpur, Uttar Pradesh, India

R. K. Patel
Department of Electrical Engineering
Rajkiya Engineering College Sonbhadra
Churk, India

Nishchal K. Verma
Department of Electrical Engineering
Indian Institute of Technology Kanpur
Kanpur, India

V. P. Singh
Department of Electrical Engineering
Rajkiya Engineering College Sonbhadra
Churk, India

ISSN 2524-7565 ISSN 2524-7573 (electronic)
Algorithms for Intelligent Systems
ISBN 978-981-15-2368-7 ISBN 978-981-15-2369-4 (eBook)
https://doi.org/10.1007/978-981-15-2369-4

This Springer imprint is published by the registered company Springer Nature Singapore Pte Ltd.
The registered company address is: 152 Beach Road, #21-01/04 Gateway East, Singapore 189721, Singapore

Preface

Computing applications in engineering are a vibrant field which grow in importance from many years. It is also a subject that students and research scholars enjoy at all levels from graduation to post-graduate college. The main objective of learning computing applications in engineering domain varies with the age group. For graduation students, computing applications in a particular field are an educational toy; for students in the mid of their graduation, computing application in engineering can increase the motivation of the students to study computer, electrical, electronics, mathematics (CEEM) at the introductory college level, and students can learn how the electrical engineering, electronics engineering, and computer engineering that they study can be applied to practical engineering projects; finally, upper-level graduate and post-graduate students and research scholars prepare for carriers for higher studies. This book is aimed for engineering students at the middle of the age range such as students in undergraduate/post-graduate course and for research scholars. We focus on computing applications in different engineering fields. We believe that nowadays, due to advent of modern devices and due to million number of devices which has been already deployed in field requires some smart computing technology in order to share and exchange the information among others. The presentation of the computing algorithms without advanced mathematics and engineering is necessarily simplified, but we believe that the concepts and algorithms of computing technology can be learned and appreciated at this level and can serve as a bridge to the study of computing applications in engineering at the advanced undergraduate and graduate levels.

This book introduces research papers presented at the "International Conference on Computing Applications in Electrical and Electronics Engineering (ICCAEEE 2019)," a two-day conference which provides a platform that brings together academicians/ research scholars involved in research activities to share their ideas and findings on all aspects of computational applications in engineering and present their research papers.

ICCAEEE 2019 received 94 papers from 246 authors (240 are local authors and 6 are foreign authors from 6 countries), and 42 papers were finally accepted for presentation, and finally, 39 papers are included in this book. In fact, this book presents the novel contributions in areas of computational intelligence and it serves as a reference material for advanced research. This book is organized into two parts. Part 1 deals with the computational intelligence and smart computing technology which consists of 20 research papers, and part 2 deals with the intelligent algorithms for engineering system which consists of 19 research papers.

This book covers the research in the areas of big data analytics, IoT and smart infrastructures, machine learning, artificial intelligence and deep learning, crowd-sourcing and social intelligence, natural language processing, expert systems and business intelligence, pervasive and high-performance computing, distributed, cloud and P2P computing cluster, grid and fog computing, wireless, mobile and green communications, ad hoc, sensor and mesh networks, SDN and network virtualization, cognitive systems, swarm intelligence, human–computer interaction, computer vision and virtual reality, computational biology and bioinformatics, algorithms and programming languages, code generation and optimization, network and information security, control communication and monitoring of smart grid, intelligent control, soft computing, neuro-control, fuzzy control and their applications, networked control systems, power system monitoring, control and protection, micro-grids and distributed generation, renewable energy sources and technology, power generation, transmission and distribution, adaptive communication systems and networks, intelligent protocols and designing for communication systems and networks, device modeling and process simulation, advanced VLSI and embedded systems, advanced signal and image processing techniques, pattern recognition and object tracking, cognitive and software-defined radio, MEMs, sensor devices and applications, smart microwave imaging and remote sensing and robotics, etc.

Gorakhpur, India — Prof. V. K. Giri
Kanpur, India — Prof. Nishchal K. Verma
Churk, India — Dr. R. K. Patel
Churk, India — Dr. V. P. Singh

Acknowledgements

This book arose from the selected papers from the first "International Conference in Computing Applications in Electrical and Electronics Engineering" (ICCAEEE 2019) held at Rajkiya Engineering College, Sonbhadra, from August 30 to 31, 2019. We would like to thank all the authors, reviewers, expert members, TEQIP-III of AKTU, Lucknow, IEEE computational intelligent society and organizing team of the conference without whose efforts this book could not have been written.

We would like to thank the staff at Springer, in particular Aninda Bose, Jagdish Chandra Bansal, and Ashok Kumar for their continuous help and support.

Contents

About the Editors

Dr. V. K. Giri obtained his B.E. (Electrical) degree from REC (presently SVNIT), Surat, Gujrat, in 1988, M.E. (Measurement and Instrumentation) Hons. degree from University of Roorkee in 1997 and Ph.D. degree from Indian Institute of Technology Roorkee, in 2003. He joined the Electrical Engineering Department of Madan Mohan Malviya University of Technology, Gorakhpur, U.P. (formerly, M.M.M Engineering College, Gorakhpur,) in 1989 as lecturer. He holds the position of Professor in the same department since 2008 and currently working as Director in Rajkiya Engineering College, Sonbhadra, since 2016. He has published more than 110 research papers, guided 21 PG students, and supervised 03 Ph.D. & is supervising 04 Ph.D. book. He has received many awards including "The Corps of Engineers Prize", i.e. best paper award from the Institution of Engineers (India) in 23rd Indian Engineering Congress in year 2008. He was elected as Fellow of the Institution of Engineers (I), Institution of Electronics and Telecommunication Engineers (IETE), and is a member of many professional bodies such as life member ISTE, member IEE and member CSI. He is reviewer of several international and national journals. He has been the member of advisory and technical committee of many international and national conferences. He has been the Member of Board of Governor, MMMEC and Chairman/Member of BOS of different Institutes/Universities. Apart from the academics, he had been holding almost all the possible positions of any technical Institutes/Universities. He has also undertaken large number of consultancy, testing & sponsored projects from industries and government departments. His research interests include digital signal processing, measurement and instrumentation, biomedical instrumentation, ECG data compression, telemedicine and health monitoring of rotating machines.

Dr. Nishchal K. Verma (SM'13) is Professor, Department of Electrical Engineering and Interdisciplinary Program in Cognitive Science, Indian Institute of Technology Kanpur, India. He obtained his Ph.D. in Electrical Engineering from Indian Institute of Technology Delhi, India. He is an awardee of Devendra Shukla Young Faculty Research Fellowship by Indian Institute of Technology Kanpur, India, for year 2013–16. His research interests include big data analysis, deep learning of neural and fuzzy networks, machine learning algorithms, computational intelligence, computer vision, brain–computer/machine interface, intelligent informatics, soft computing in modelling and control, Internet of Things/ cyber-physical systems, cognitive science and intelligent fault diagnosis systems, prognosis and health management. He has authored more than 200 research papers.

Dr. Verma is an IETE Fellow. He is currently serving as Guest Editor of the IEEE Access Special Section "Advance in Prognostics and System Health Management", an Editor of the IETE Technical Review Journal, an Associate Editor of the IEEE Transactions on Neural Networks and Learning Systems, an Associate Editor of the IEEE Computational Intelligence Magazine, an Associate Editor of the Transactions of the Institute of Measurement and Control, U.K., and Editorial Board Member for several journals and conferences.

Dr. R. K. Patel is working as an Assistant Professor, Department of Electrical Engineering, Rajkiya Engineering College, Sonbhadra. He obtained his M.Tech. from NIT Hamirpur (H.P.) in Condition Monitoring Control and Protection of Electrical Apparatus and received Director's Medal for M.Tech. degree in year 2013. Dr. Patel obtained his Ph.D. from AKTU, Lucknow, in Electrical Engineering. He received "SHRESTHA SHIKSHAK PURASKAR-2007" on Teachers' Day, by Vice-Chairman of Sharda Group Institution. He also received best paper award in IEEE Students' Conference on Engineering and Systems (SCES -2012) MNNIT Allahabad, Prayagraj, and best paper award on Research Scholar Day 30th March, 2015, at MMMUT, Gorakhpur. His research interests include condition monitoring and fault diagnosis of electrical machines, structure health monitoring, prognosis and health management. He has authored 12 research papers and 7 conference papers. He is having total 10 years teaching experience at UG level.

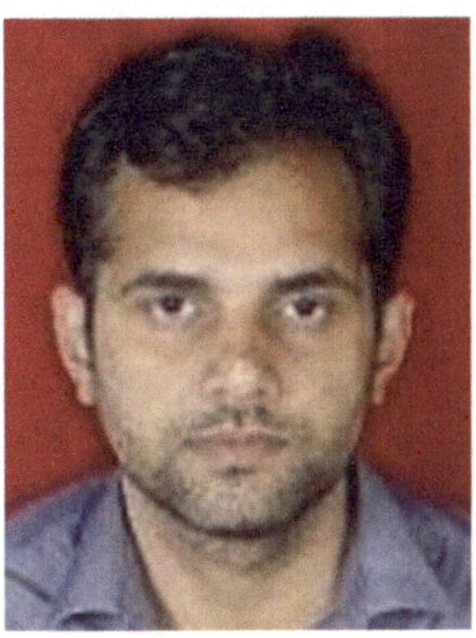

Dr. V. P. Singh is working as an Assistant Professor, Department of Electrical Engineering, Rajkiya Engineering College, Sonbhadra. He obtained his M.Tech. and Ph.D. in Electrical Engineering from MNNIT Allahabad, Prayagraj, India. He has total teaching experience 08 years at UG level and research experience 03 years. His research interests include smart grid, renewable energy resources, distributed generation system, multi-agent system, power system and load frequency control. He has authored 09 SCI research papers published in high impact factor journals and 10 international conference papers. Presently, Dr. Singh is guiding 2 Ph.D. students. He chaired a session in IEEE ISGT ASIA-2015, Thailand, Bangkok. He is regular reviewer of IEEE Transaction on Energy Conversion, IEEE Transaction on Industrial Informatics, IEEE Systems Journals, IET, Generation Transmission and Distribution, Electric Power System and Research and Electrical Energy and Power System. Dr. Singh has delivered many expert talks in various reputed institutions in Faculty development program and conference in India.

Part I
Computational Intelligence and Smart Computing Technology

Chapter 1
Reduction of Discrete Interval System Using Mixed Approach

V. Singh, A. P. Padhy and V. P. Singh

1 Introduction

Model order reduction (MOR) is an innovative concept in the areas of engineering and science. It approximates higher-order system into ROM, by retaining the important properties of original system. Thus, by replacing the original system to a ROM, analysis of higher-order system becomes easier. Furthermore, for the analysis and synthesis purpose, mathematical modeling of complex system faces problem. Therefore, the MOR technique efficiently analyzes the complex systems.

From last few decades, many methods have been proposed [1–6] for model order reduction of fixed-coefficient systems. Now, the systems under consideration have modified to the interval structure resulting from parameter variation and other unmodeled dynamics. Thus, a system with coefficients having uncertainty within certain ranges is described as an interval system. In case of interval systems [7–11], the numerator and denominator coefficients vary within the range of defined intervals. For this, interval arithmetic [7, 12–14] is used to deal with interval coefficients. Also, in [7], authors proposed Routh-Pade approximation for MOR of CIS. In [15–17], the new techniques like gamma-delta Routh approximations are proposed for computation of TM and Markov parameters of CIS. There are many techniques proposed in [18–21] for order reduction of DIS. However, these techniques are not so familiar because of their complex computational methods.

This paper aims at developing a new approach for MOR of DIS. In the first step, DIS is converted to CIS using linear transformation. The denominator and numerator of the ROM are computed by direct truncation and matching of TM, respectively. In

V. Singh · A. P. Padhy (✉)
Department of Electrical Engineering, NIT, Raipur, India

V. Singh
e-mail: vsingh.ele@nitrr.ac.in

V. P. Singh
Department of Electrical Engineering, MNIT, Jaipur, India

V. K. Giri et al. (eds.), *Computing Algorithms with Applications in Engineering*, Algorithms for Intelligent Systems, https://doi.org/10.1007/978-981-15-2369-4_1

the next step, the ROM of CIS is reconverted to discrete domain using inverse linear transformation.

In this research, the problem statement is described in Sect. 2. Section 3 depicts interval systems along with their TM. Section 4 describes the direct truncation method. Validation of this method is carried out by considering a two-tank system which has been described in Sect. 5. At the end, Sect. 6 concludes the entire article.

2 Conceptualization Towards the Problem

Consider a SISO system with higher-order DIS given as:

$$G_{\mathrm{nh}}(z) = \frac{N(z)}{D(z)} = \frac{c_0 + c_1 z + \cdots + c_{\mathrm{nh}-1} z^{\mathrm{nh}-1}}{d_0 + d_1 z + \cdots + d_{\mathrm{nh}} z^{\mathrm{nh}}} \tag{1}$$

where $c_x = [c_x^-, c_x^+]$ for $x = 0, 1, \ldots, \mathrm{nh}-1$ and $d_x = [d_x^-, d_x^+]$ for $x = 0, 1, \ldots, \mathrm{nh}$ are the interval coefficients of numerator and denominator, respectively, in the discrete domain.

By using linear transformation, i.e., $z = p + 1$, the discrete transfer function (1) is converted into continuous transfer function as given in (2). Now, the modified Eq. (1) in p-domain becomes

$$H_{\mathrm{nh}}(p) = \frac{N(p)}{D(p)} = \frac{e_0 + e_1 p + \cdots + e_{\mathrm{nh}-1} p^{\mathrm{nh}-1}}{f_0 + f_1 p + \cdots + f_{\mathrm{nh}} p^{\mathrm{nh}}} \tag{2}$$

where $e_x = [e_x^-, e_x^+]$ for $x = 0, 1, \ldots, (\mathrm{nh} - 1)$ and $f_x = [f_x^-, f_x^+]$ for $x = 0, 1, \ldots, \mathrm{nh}$ are the interval coefficients of the original system in the transformed continuous domain (TCD).

Let us consider a ROM that has a continuous interval transfer function which is represented as:

$$\hat{H}_{\mathrm{rl}}(p) = \frac{\hat{N}(p)}{\hat{D}(p)} = \frac{\hat{e}_0 + \hat{e}_1 p + \cdots + \hat{e}_{\mathrm{rl}-1} p^{\mathrm{rl}-1}}{\hat{f}_0 + \hat{f}_1 p + \cdots + \hat{f}_{\mathrm{rl}} p^{\mathrm{rl}}} \tag{3}$$

where $\hat{e}_y = [\hat{e}_y^-, \hat{e}_y^+]$ for $y = 0, 1, \ldots, (\mathrm{rl} - 1)$ and $\hat{f}_y = \left[\hat{f}_y^-, \hat{f}_y^+\right]$ for $y = 0, 1, \ldots, \mathrm{rl}$ are the interval coefficients of the model in the continuous domain.

The above reduced order continuous interval model is reconverted into discrete interval model using inverse linear transformation, i.e., $p = z - 1$; now, the modified form of Eq. (3) is given by

$$\hat{G}_{\mathrm{rl}}(z) = \frac{\hat{N}(z)}{\hat{D}(z)} = \frac{\hat{c}_0 + \hat{c}_1 z + \cdots + \hat{c}_{\mathrm{rl}-1} z^{\mathrm{rl}-1}}{\hat{d}_0 + \hat{d}_1 z + \cdots + \hat{d}_{\mathrm{rl}} z^{\mathrm{rl}}} \tag{4}$$

where $\hat{c}_y = \left[\hat{c}_y^-, \hat{c}_y\right]$ for $y = 0, 1, \ldots, \text{rl} - 1$ and $\hat{d}_y = \left[\hat{d}_y^-, \hat{d}_y^+\right]$ for $y = 0, 1, \ldots, \text{rl}$ are the interval coefficients of the ROM in the transformed discrete domain (TDD).

3 Time Moments of the Interval Systems

Let us consider a CIS denoted by the transfer function

$$G_{\text{nh}}(z) = \frac{N(z)}{D(z)} = \frac{c_0 + c_1 z + \cdots + c_{\text{nh}-1} z^{\text{nh}-1}}{d_0 + d_1 z + \cdots + d_{\text{nh}} z^{\text{nh}}} \tag{5}$$

By applying linear transformation, Eq. (5) becomes

$$H_{\text{nh}}(p) = \frac{N(p)}{D(p)} = \frac{e_0 + e_1 p + \cdots + e_{\text{nh}-1} p^{\text{nh}-1}}{f_0 + f_1 p + \cdots + f_{\text{nh}} p^{\text{nh}}} \tag{6}$$

For the system, the power series expansion around $p = 0$ can be written as

$$H_{\text{nh}}(p) = T_0 + T_1 p + \cdots + T_k p^k + \cdots \quad \text{(expansion around } p = 0) \tag{7}$$

where $T_i = \left[T_i^-, T_i^+\right]$ for $i = 0, 1, \ldots, k, \ldots$ are the TM of the interval system.

The TM of the system can be calculated from the equations as given below

$$\left.\begin{array}{l}
\left[e_0^-, e_0^+\right] = \left[f_0^-, f_0^+\right]\left[T_0^-, T_0^+\right] \\
\left[e_1^-, e_1^+\right] = \left[f_0^-, f_0^+\right]\left[T_1^-, T_1^+\right] + \left[f_1^-, f_1^+\right]\left[T_0^-, T_0^+\right] \\
\left[e_2^-, e_2^+\right] = \left[f_0^-, f_0^+\right]\left[T_2^-, T_2^+\right] + \left[f_1^-, f_1^+\right]\left[T_1^-, T_1^+\right] + \left[f_2^-, f_2^+\right]\left[T_0^-, T_0^+\right] \\
\quad \vdots
\end{array}\right\} \tag{8}$$

Consider a stable rlth order model described by the transfer function

$$\hat{H}_{\text{rl}}(p) = \frac{\hat{N}_{\text{rl}}(p)}{\hat{D}_{\text{rl}}(p)} = \frac{\hat{e}_0 + \hat{e}_1 p + \cdots + \hat{e}_{\text{rl}-1} p^{\text{rl}-1}}{\hat{f}_0 + \hat{f}_1 p + \cdots + \hat{f}_{\text{rl}} p^{\text{rl}}} \tag{9}$$

where $\text{rl} < \text{nh}$.

The parameters $\hat{e}_y = \left[\hat{e}_y^-, \hat{e}_y^+\right]$ for $y = 0, 1, \ldots, \text{rl} - 1$ and $\hat{f}_y = \left[\hat{f}_y^-, \hat{f}_y^+\right]$ for $y = 0, 1, \ldots, \text{rl}$ denote the interval coefficients of model.

The model, (9), in terms of TMs can be expanded as

$$G_{\text{rl}}(p) = \hat{T}_0 + \hat{T}_1 p + \cdots + \hat{T}_k p^k + \cdots \tag{10}$$

where $\hat{T}_i$ for $i = 0, 1, \ldots, k, \ldots$ denotes TM. Hence, similar expressions as obtained in Eq. (8) can be derived for the model given in Eq. (9). The expressions for Eq. (9) became,

$$\left.\begin{aligned}
\left[\hat{e}_0^-, \hat{e}_0^+\right] &= \left[\hat{f}_0^-, \hat{f}_0^+\right]\left[\hat{T}_0^-, \hat{T}_0^+\right] \\
\left[\hat{e}_1^-, \hat{e}_1^+\right] &= \left[\hat{f}_0^-, \hat{f}_0^+\right]\left[\hat{T}_1^-, \hat{T}_1^+\right] + \left[\hat{f}_1^-, \hat{f}_1^+\right]\left[\hat{T}_0^-, \hat{T}_0^+\right] \\
\left[\hat{e}_2^-, \hat{e}_2^+\right] &= \left[\hat{f}_0^-, \hat{f}_0^+\right]\left[\hat{T}_2^-, \hat{T}_2^+\right] + \left[\hat{f}_1^-, \hat{f}_1^+\right]\left[T_1^-, T_1^+\right] + \left[\hat{f}_2^-, \hat{f}_2^+\right]\left[\hat{T}_0^-, \hat{T}_0^+\right] \\
&\vdots
\end{aligned}\right\} \tag{11}$$

The numerator of the model is computed by matching rl TM of the system and model as given below.

$$T_i = \hat{T}_i \tag{12}$$

The numerator coefficients $\hat{e}_y = \left[\hat{e}_y^-, \hat{e}_y^+\right]$ of the model can be computed by using Eq. (12). Finally, the numerator $\hat{N}_{\mathrm{rl}}(p)$ became,

$$\hat{N}_{\mathrm{rl}}(p) = \left[\hat{e}_0^-, \hat{e}_0^+\right] + \left[\hat{e}_1^-, \hat{e}_1^+\right]p + \left[\hat{e}_2^-, \hat{e}_2^+\right]p^2 + \cdots + \left[\hat{e}_{\mathrm{rl}-1}^-, \hat{e}_{\mathrm{rl}-1}^+\right]p^{\mathrm{rl}-1} \tag{13}$$

where $\hat{N}_{\mathrm{rl}}(p)$ is the numerator of the model and $\left[\hat{e}_0^-, \hat{e}_0^+\right], \left[\hat{e}_1^-, \hat{e}_1^+\right] \cdots$ denote the coefficients of numerator.

4 Direct Truncation Method

The denominator can be constructed by applying direct truncation method [22], and this method was proposed by Shamash for fixed-coefficient systems. Further, this technique is extended to interval system in this investigation. This method is very simple. The denominator of hlth order system is given in Eq. (2) which is

$$D(p) = f_0 + f_1 p + \cdots + f_{\mathrm{nh}} p^{\mathrm{nh}} \tag{14}$$

The denominator of rlth order model derived from (14) becomes

$$\hat{D}_{\mathrm{rl}}(p) = f_0 + f_1 p + \cdots + f_{\mathrm{rl}} p^{\mathrm{rl}} \tag{15}$$

where $f_0, f_1, \ldots, f_{\mathrm{rl}}$ symbolize the coefficients of the ROM.

5 Test Case

Let us consider a two-tank system whose second-order discrete interval transfer function is given as

$$G_{\mathrm{nh}}(z) = \frac{[6, 6]}{[8, 8.5]z^2 + [-11, -9.8]z + [2.8, 3.5]} \tag{16}$$

The above discrete system can be converted into the continuous system by replacing z with $1 + p$. After conversion, the following transfer function is obtained.

$$H_{\mathrm{nh}}(p) = \frac{[6, 6]}{[8, 8.5]p^2 + [6, 6.2]p + [1, 1]} \tag{17}$$

By applying direct truncation method, the denominator of the above transfer function is expressed as,

$$\hat{D}(p) = [1, 1] + [6, 6.2]p$$

By matching the TM of system and model, $T_0 = \hat{T}_0$, the first-order model is given by

$$\hat{H}_{\mathrm{rl}}(p) = \frac{[6, 6]}{[6, 6.2]p + [1, 1]} \tag{18}$$

In z-domain, the above transfer function is,

$$\hat{G}_{\mathrm{rl}}(z) = \frac{[6, 6]}{[-5.2, -5]z + [1, 1]} \tag{19}$$

The first-order model calculated using method proposed in [16] and matching of TM is

$$\hat{H}_{\mathrm{rl}}(p) = \frac{[6, 6]}{[6, 6.2]p + [0.9677, 1.033]} \tag{20}$$

which is expressed in z-domain as,

$$\hat{G}_{\mathrm{rl}}(z) = \frac{[6, 6]}{[6, 6.2]z + [6.9677, 7.233]} \tag{21}$$

The first-order model calculated using matching of TM and the technique proposed in [23] is

$$\hat{H}_{\mathrm{rl}}(p) = \frac{[36, 37.2]}{[16, 17]p + [6, 6.2]} \tag{22}$$

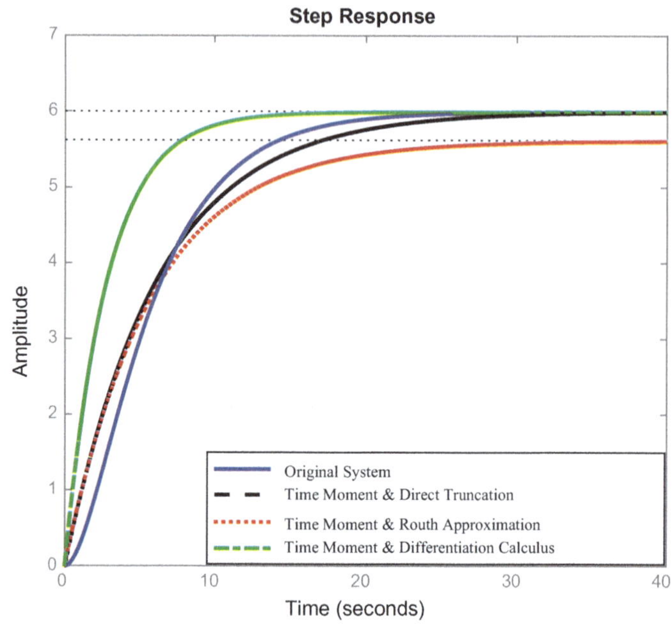

Fig. 1 Step responses of the system and models

By applying inverse linear transformation, the above transfer function is given as,

$$\hat{H}_{\text{rl}}(z) = \frac{[36, 37.2]}{[16, 17]z + [-11, -9.8]} \tag{23}$$

The step responses of rational systems obtained using Kharitonov's polynomials [24, 25] of numerator and denominator of interval system, (16), and different models, (19), (21) and (23), are plotted in Fig. 1.

From Fig. 1, the steady state of step responses of suggested model, (19), are matching with (16). Figures 2 and 3 depict the frequency responses of the reduced models for the lower and upper limits of transfer functions, respectively. This shows that proposed method produces a better reduced order models of original higher-order interval systems.

The proposed MOR technique provides acceptable results. The merits of the proposed method are:

(1) The stability of the original system in the ROM is ensured.
(2) The steady-state value of the original system is preserved in the ROM.

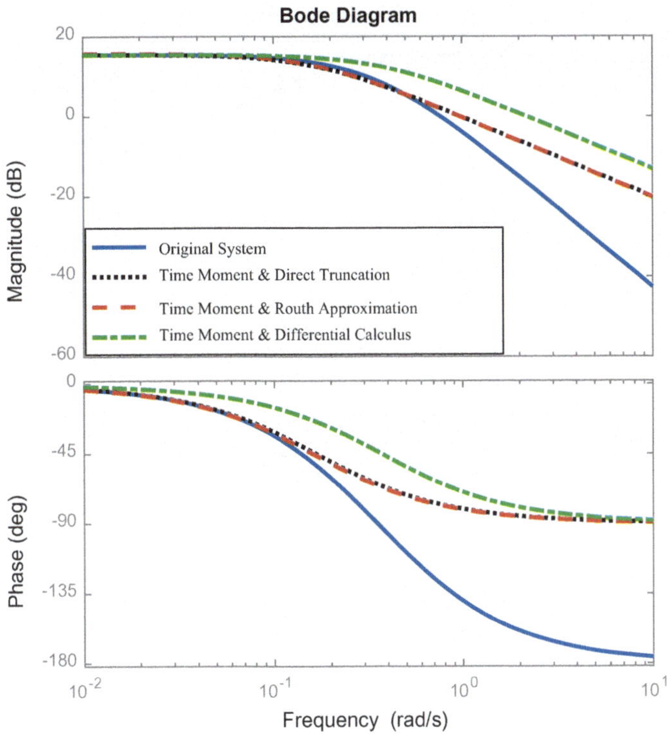

Fig. 2 Frequency responses of system and models (lower limit)

6 Conclusions

A new order reduction technique for discrete interval system using time moments matching in continuous domain is presented in this paper. A simple linear transformation approach is used to convert discrete time interval transfer function into continuous interval system. Then, the denominator of the ROM is formed by using direct truncation method and the numerator is calculated by matching time moments in the continuous domain. Further, the obtained ROM is reconverted to the discrete domain using inverse linear transformation. This method provides sufficient degree of approximation.

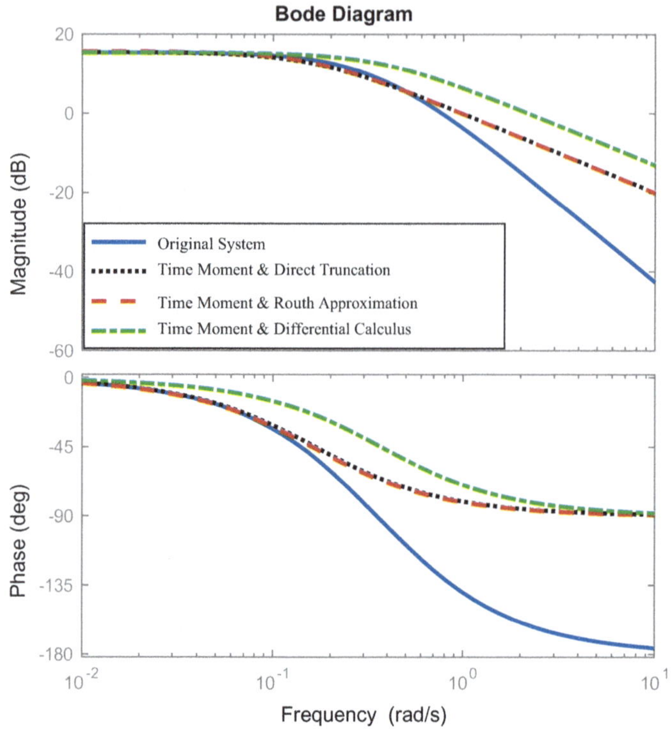

Fig. 3 Frequency responses of system and models (upper limit)

References

1. Fortuna L, Nunnari G, Gallo A (2012) Model order reduction techniques with applications in electrical engineering. Springer, London
2. Singh J, Vishwakarma C, Chattterjee K (2016) Biased reduction method by combining improved modified pole clustering and improved Pade approximations. Appl Math Model 40(2):1418–1426
3. Prajapati AK, Prasad R (2019) Reduced-order modelling of LTI systems by using Routh approximation and factor division methods. Circuits Syst Signal Process 38:3340–3355
4. Sastry G, Krishnamurthy V (1987) Biased model reduction by simplified Routh approximation method. Electron Lett 23(20):1045–1047
5. Chen T, Chang C, Han K (1980) Stable reduced-order Padé approximants using stability-equation method. Electron Lett 16(9):345–346
6. Panda S et al (2009) Reduction of linear time-invariant systems using Routh-approximation and PSO. Int J Appl Math Comput Sci 5(2):82–89
7. Bandyopadhyay B, Ismail O, Gorez R (1994) Routh-Pade approximation for interval systems. IEEE Trans Autom Control 39(12):2454–2456
8. Dolgin Y, Zeheb E (2003) On Routh-Pade model reduction of interval systems. IEEE Trans Autom Control 48(9):1610–1612
9. Ismail O, Bandyopadhyay B (1995) Model reduction of linear interval systems using Pade approximation. In: 1995 IEEE international symposium on circuits and systems, 1995. ISCAS'95. IEEE

10. Kumar DK, Nagar S, Tiwari J (2011) Model order reduction of interval systems using criterion and factor division method. Int J Comput Appl 28(11):4–8
11. Sastry G, Rao GR (2003) Simplified polynomial derivative technique for the reduction of large-scale interval systems. IETE J Res 49(6):405–409
12. Choudhary AK, Nagar SK (2016) Revisiting approximation techniques to reduce order of interval system. IFAC-PapersOnLine 49(1):241–246
13. Potturu SR, Prasad R (2018) Order reduction of interval systems using Kharitonov's theorem and stability equation method. In: 2018 annual American control conference (ACC). IEEE
14. Sastry G, Rao RR, Rao PM (2000) Large scale interval system modelling using Routh approximants. Electron Lett 36(8):768–769
15. Bandyopadhyay B, Upadhye A, Ismail O (1997) /spl gamma/-/spl delta/Routh approximation for interval systems. IEEE Trans Autom Control 42(8):1127–1130
16. Singh V et al (2017) On time moments and Markov parameters of continuous interval systems. J Circuits Syst Comput 26(03):1750038
17. Choudhary AK, Nagar SK (2018) Order reduction techniques via Routh approximation: a critical survey. IETE J Res 65:365–379
18. Choudhary AK, Nagar SK (2017) Novel arrangement of Routh array for order reduction of z-domain uncertain system. Syst Sci Control Eng 5(1):232–242
19. Choudhary AK, Nagar SK (2018) Order reduction in z-domain for interval system using an arithmetic operator. Circuits Syst Signal Process 38:1023–1038
20. Choudhary AK, Nagar SK (2013) Gamma Delta approximation for reduction of discrete interval system. In: international joint conferences on ARTCom and ARTEE, Bangalore. Elsevier
21. Choudhary AK, Nagar SK (2018) Model order reduction of discrete-time interval system based on Mikhailov stability criterion. Int J Dyn Control 6:1558–1566
22. Smamash Y (1981) Truncation method of reduction: a viable alternative. Electron Lett 17(2):97–99
23. Choudhary AK, Nagar SK (2018) Model order reduction of discrete-time interval systems by differentiation calculus. Autom Control Comput Sci 52(5):402–411
24. Padhy AP, Singh VP, Pattnaik S (2018) Model reduction of multi-input-multi-output discrete interval systems using gain adjustment. Int J Pure Appl Math 119(12):12721–12739
25. Singh V, Chandra D (2012) Reduction of discrete interval system using clustering of poles with Padé approximation: a computer-aided approach. Int J Eng Sci Technol 4(1):97–105

Chapter 2
Techno-economic Analysis of Partially Shaded PV Plant: An Application of Artificial Intelligence-Based Software

Rachit Srivastava, A. N. Tiwari and V. K. Giri

1 Introduction

Solar energy is one of the most promising sources of energy. Photovoltaic process is one of the most important ways to convert solar energy into electricity [1–3]. Performance evaluation is very important to determine various aspects of PV plant such as annual electricity production, performance ratio, and the payback period of the plant [4, 5]. In this study, performance evaluation of 100-kW parking integrated grid-connected PV system has been carried out. The number of artificial intelligence-based software is available for PV plant performance evaluation, such as PVSyst [6, 7], PVWatts, PVGis, RETScreen [8], SAM [9], HOMER [10] and TRNSYS. Some trees are also present nearby the conceded plant. Hence, shading is also required to be addressed. Few softwares are available which are able to simulate shading effect on the plant. PV*SOL premium is one of the software that is able to address the shading effect.

El Gindi et al. have used PV*SOL premium software for retrofitting of a building integrated PV system in Egypt [11]. Firat has also used the same software for performance estimation of hybrid PV plant [12]. Simulation and performance evaluation of a grid-connected PV system have also been carried out in Jourdan using PV*SOL premium software [13]. Malara et al. have carried out a shading effect on PV plant using PV*SOL premium software [14]. In all these cases, PV*SOL premium software shows good results. Hence in this case study, techno-economic analysis of 100-kW parking integrated grid-connected PV plant has been carried out using PV*SOL premium software. In this study, PV*SOL premium 2019 version has been used. Details of the PV*SOL premium can be seen from [15].

The paper is divided into four sections. Second section deals with various specifications of the plant. The third section is dedicated to various outcomes from the

R. Srivastava (✉) · A. N. Tiwari · V. K. Giri
Department of Electrical Engineering, Madan Mohan Malaviya University of Technology, Gorakhpur, Uttar Pradesh 273010, India

V. K. Giri et al. (eds.), *Computing Algorithms with Applications in Engineering*, Algorithms for Intelligent Systems, https://doi.org/10.1007/978-981-15-2369-4_2

Table 1 PV module specifications

Material	Polycrystalline
Manufacturer	Su-Kam
Module weight	17.30 kg
Solar surface area	16,094.98 cm^2
Rated power output	250 W
Rated voltage	24 V
Open-circuit voltage	44.66 V
Short-circuit current	7.24 A
MPP voltage	35.96 A
MPP current	6.96 A
Efficiency	15.53%

software results such as annual electricity generation from the grid, the performance ratio of the plant effects of shading on plant outcomes, and financial results. The fourth and the last section is dedicated to concussions of the paper.

2 Plant Description

Grid-connected PV plant is located in the Madan Mohan Malaviya University of Technology (MMMUT), Gorakhpur, which is situated in the northern region of India. The plant is located at 26°43′50.41″ north and longitude of 83°26′2.8″ east. PV models are placed on the parking stands. Total 7 parking stands are employed. 400 PV models of 250 W were employed in this plant. Six parking stands consist of 360 PV module, and 7th stand consists of 40 modules. Two inverters of 50 kW were used in this plant. The specifications of PV module and inverter are presented in Tables 1 and 2, respectively [16, 17]. There are some trees nearby the plant. The picture view of the actual plant is shown in Fig. 1.

3 Software Result and Discussion

The 3D environment of the actual system has been simulated in the software to evaluate actual plant performance by applying. For this, first of all, site location has been selected in software; then, actual data of PV module and inverter have been inserted into the software. A real 3D system has been created into the software. For this, real dimensions have been adopted. The software uses inbuilt artificial intelligence-based algorithm to calculate various results. Figure 2 shows a 3D diagram of the plant created into the software. Figure 3 represents the overall circuit diagram of the plant. Figure 4 shows the shading effect on the plant. The software uses inbuilt artificial

Table 2 Inverter specifications

Model	RPI M50A
Manufacturer	Delta energy system
DC nominal output	52.00 kW
Maximum DC power	58.00 kW
AC rated power	50.00 kW
Maximum AC power	55.00 kW
Night consumption	2.50 W
Maximum input voltage	1100.00 V
Nom. DC voltage	600.00 V
No. of input tracker	2 Nos.
Maximum input current per MPP tracker	50 A
Minimum MPP voltage	250 V
Maximum MPP voltage	800 V

Fig. 1 Picture view of actual PV plant

intelligence-based algorithm to calculate various results. Some important outcomes from the algorithm are presented in this section.

3.1 Electricity Generation

From the simulation, it has been found that plant is generating 136.621 MWh electricity in a year. The specific annual yield of the plant is 136.62 kWh/kW. Figure 5 shows the energy generation graph for one year. From the graph, we can observe that minimum electricity is generating in the month of January (8814.4 kWh), whereas maximum electricity is generating in the month of May (14,154.7 kWh). From the simulation, it has also been observed that we are losing 2.9% yield electricity due to

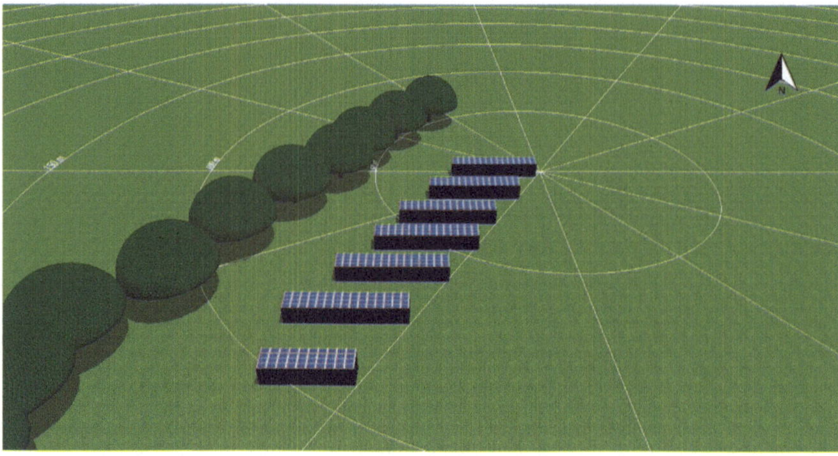

Fig. 2 Simulation of plant and trees

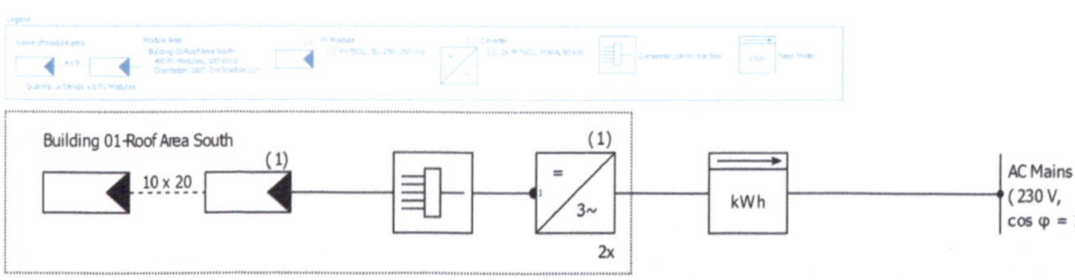

Fig. 3 Circuit diagram of plant

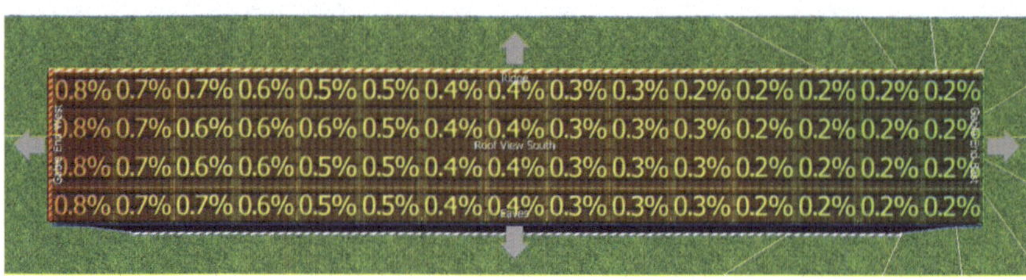

Fig. 4 Shading level (in percentage) on various modules

the shading on the plant. This plant is avoiding 81,973 kg/year CO_2 emission in the atmosphere. Energy flow between grid and plant is shown in Fig. 6.

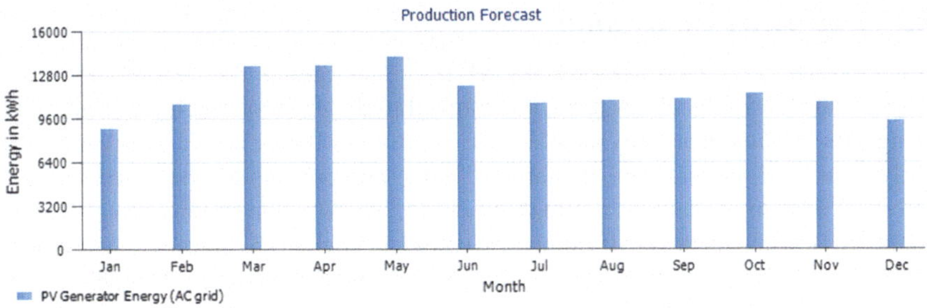

Fig. 5 Electricity production graph for one year

Fig. 6 Power flow between grid and plant in a year (values are in kWh)

3.2 *Performance Ratio*

The performance ratio is one of the most important parameters to observe the performance of any PV power plant. As high the performance ratio is, as good the plant is working on. The maximum value of the performance ratio is 100%, which is not feasible. From the simulation, it has been found that plant performance ratio is 75.9%. It has been found that the performance ratio of the plant varies from 78.61% (in the month of January) to 74.36% (in the month of April).

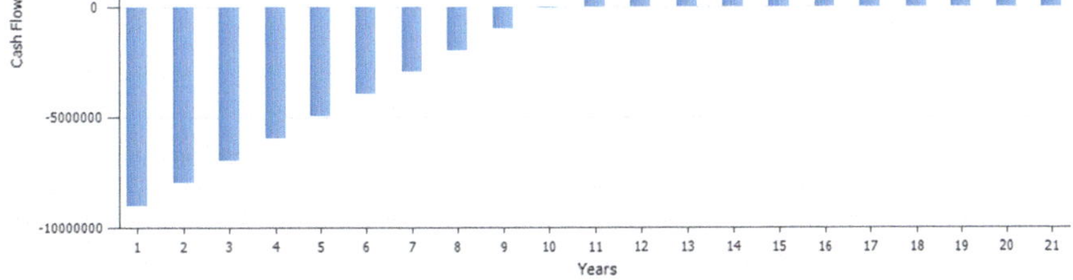

Fig. 7 Cash flow graph

3.3 *Electricity Generation*

To check the feasibility of any plant, economical study of any system is also important along with the technical analysis. PV*SOL premium software is able to do economic analysis as well. For this, the actual cost of the components has been fed into the software. System capital cost is 100,000 ₹ ($142,785.70) (conversion rate—1 US$ = 70.04 INR), whereas the tariff rate is 7.70 ₹ ($0.11) per kWh. From the simulation, it has been observed the return of assists is 8.55% and total payment from the utility is 1,051,984.51 ₹ ($15,019.76) per year. Figure 7 shows the flow of cash in 20 years. From the figure, it can be observed that the payback period of the plant is 10 years. After 10 years, the plant will deliver the returns. From the results, we can say that the plant is economical at least the life of a PV plant is 20 years [18].

4 Conclusions

In this paper, techno-economic analysis of parking integrated 100-kW grid-connected PV system has been carried out. Real system environment has been created in the PV*SOL premium software by conceding real system data and real dimensions. The real 3D system has been created for shading analysis. Real financial data were used to check financial feasibility. The following conclusions are drawn from the study:

- The monthly total energy generated varied between 136.62 kWh/kW in December and 141.5 kWh/kWp in June, while the annual total energy generated was 885.1 kWh/kW.
- The performance ratio (PR) experienced a slight variation within the range of 74.36–78.61%, with an annual average value of 75.9%.
- Payback period of the plant is 10 years.
- The carbon credits that can be earned from the plant are resulted as 81,973 kg/year.
- Loss of electricity due to shading is 2.9%.

From the observations, it has been observed that the plant is showing good results although it is facing shading due to the nearby trees. The observed experience of the PV system can be applied for future large-scale projects.

References

1. Panwar NL, Kaushik SC, Kothari S (2011) Role of renewable energy sources in environmental protection: a review. Renew Sustain Energy Rev 15:1513–1524
2. Kabir E, Kumar P, Kumar S, Adelodun AA, Kim KH (2018) Solar energy: potential and future prospects. Renew Sustain Energy Rev 82:894–900
3. Kannan N, Vakeesan D (2016) Solar energy for future world: a review. Renew Sustain Energy Rev 62:1092–1105

4. Sundaram S, Babu JSC (2015) Performance evaluation and validation of 5 MW_p grid connected solar photovoltaic plant in South India. Energy Convers Manag 100:429–439. https://doi.org/10.1016/j.enconman.2015.04.069
5. Leloux J, Narvarte L, Trebosc D (2012) Review of the performance of residential PV systems in France. Renew Sustain Energy Rev 16:1369–1376
6. Mermoud A (2012) Pvsyst: software for the study and simulation of photovoltaic systems. ISE, University of Geneva. www.pvsyst.com
7. Rachit S, Giri VK (2018) Design of grid connected PV system using PVsyst. i-manager's J Electr Eng 10:14–18. https://doi.org/10.26634/jee.10.1.8195
8. Psomopoulos CS, Ioannidis GC, Kaminaris SD, Mardikis KD, Katsikas NG (2015) A comparative evaluation of photovoltaic electricity production assessment software (PVGIS, PVWatts and RETScreen). Environ Process 2:175–189
9. Wang J, He X, Deng Y (1999) Introducing software architecture specification and analysis in SAM through an example. Inf Softw Technol 41:451–467. https://doi.org/10.1016/S0950-5849(99)00009-9
10. Srivastava R, Giri VK (2016) Techno-economical analysis of grid connected PV system for a university in India. Int J Renew Energy Res 6:535–540
11. El Gindi S, Abdin AR, Hassan A (2017) Building integrated photovoltaic retrofitting in office buildings. Energy Procedia 115:239–252
12. Firat Y (2018) Utility-scale solar photovoltaic hybrid system and performance analysis for eco-friendly electric vehicle charging and sustainable home. Energy Sources, Part A Recover Util Environ Eff 41:734–745
13. Etier I, Ababneh M, Al Tarabsheh A (2015) Design and simulation of a PV-grid connected system. Int J Comput Sci Eng. https://doi.org/10.1504/ijcse.2015.070991
14. Malara A, Marino C, Nucara A, Pietrafesa M, Scopelliti F, Streva G (2016) Energetic and economic analysis of shading effects on PV panels energy production. Int J Heat Technol. https://doi.org/10.18280/ijht.340316
15. Umar N, Bora B, Banerjee C, Panwar BS (2018) Comparison of different PV power simulation softwares: case study on performance analysis of 1 MW grid-connected PV solar power plant. Int J Eng Sci Invent 7:11–24
16. Su-kam (2016) Ecofriendly power solution for sustainable future. http://sukam-solar.com/wp-content/uploads/Solar-Panels-Catalogue-.pdf?caa478
17. Delta Electronics Inc. Delta M30A / M50A. http://www.deltaww.com/filecenter/Products/Download/05/0501/M30A_M50A.pdf. Accessed 28 Nov 2017
18. Sherwani AF, Usmani JA, Varun (2010) Life cycle assessment of solar PV based electricity generation systems: a review. Renew Sustain Energy Rev 14:540–544

Chapter 3
An Assessment of Cloud Computing and Mobile Cloud Computing in E-Learning

Bhatt Milind and Amod Kumar Tiwari

1 Introduction

To build a better society, every nation demands persons who are knowledgeable as well as skilled in their profession. To achieve this, education should reach the commonalities and make them able to learn and apply the earned knowledge into their skilled domain. The acknowledgment of on-line instructive degrees offered by a foundation has an extraordinary impulse on foreign nationals [1]. In India too, both industry and academic institutions are showing great acceptability in certification programs offered by IIT-Bombay under the banner C-DEEP and a joint effort IITs and NITs, national program on teaching enhanced learning (NPTEL) and Swayam a project of Ministry of Human Recourse and Development India. In year 2015, Ministry of Human Resource Development Department of higher education has launched a Digital India initiative [2] where a vision area 3: digital empowerment of citizens includes—universally accessibility and availability of digital resources [3] motivates to work in this field.

To impart such education to students, professionals and aging populations, the use of mobile cloud computing (MCC) and e-leaning services can become a better add-on in compared to all other traditional and modern approaches of learning. The MCC forum defines MCC as follows [4–6]. "MCC at its simplest refers to an infrastructure where both the data storage and the data processing happen outside of the mobile device." MCC applications move the computing power and data storage away from mobile phones and into the cloud, bringing applications and mobile computing to not just smartphone users but a broader range of mobile subscribers [7]. Alternatively,

B. Milind (✉)
Department of Computer Application, PSIT College of Higher Education, Kanpur, Uttar Pradesh, India

A. K. Tiwari
Department of of Computer Science, Rajkiya Engineering College, Sonbhadra, Uttar Pradesh, India

V. K. Giri et al. (eds.), *Computing Algorithms with Applications in Engineering*, Algorithms for Intelligent Systems, https://doi.org/10.1007/978-981-15-2369-4_3

"MCC can be defined as a combination of mobile Web and cloud computing which is the most popular tool for mobile users to access applications and services on the Internet" [8, 9]. In past few years, lot of effort in the direction of cloud and mobile computing and e-learning and collaborative distance learning has been done but complexity of performance issues and their challenges make it hard to design and develop a brisk working of MCC frameworks and models for collaborative e-learning.

Keeping these issues in mind, here, we have conducted a detailed survey to have a clear picture of the present days. The key objectives for this study were to study, analyze, and evaluate the merits and issues mentioned in the next section.

2 Methodology

As in a literature review, ethical approval is not a prerequisite. Hence, literature reviewed by various researchers in field of cloud computing, e-learning, mobile cloud computing has been studied to identify the broader and finer details. Search strings were concatenated using operators "AND" and "OR" to collect all relevant literature and article suggestions. MCC and e-learning is quite new to research. Therefore, in total, we found only 164 such national and international papers. Most of the papers were from IEEE, Springers, Elsevier, ACM, Science Direct, Academic Premier, and Scopus. Some major websites are also visited to have a clear view of recent products and their details. Here, out of 154, only 34 were found suitable for comparative study of the work analysis. The findings were reviewed under the following one or more issues: E-learning Cloud Architecture, MCC—Specific Requirements and Reservation, Impact and Analysis Mechanism, Mobile Cloud-Based Content Delivery Model Web2.0/Web-Based Content Management System (CMS). All these findings are categorically mentioned in Table 1.

3 Outcomes Based on the Review

From the literature review presented above, it is observed that significant work has been done in the area of assessment and improvement of MCC, e-learning, and collaborative distance learning-based systems in different communicating environments. Major work done so far mainly focuses on the various factors such as efficient installation and user-transparent execution, real-time improved management of heterogeneous computing environments, offloading techniques and methods, mobility management, connection protocols, mono- or bi-objective optimization frameworks, data portability and interoperability, motivations to collaborate, automated service promising, scalability and availability of systems, etc. Mobile cloud computing creates virtualized resources for the users which are also highly scalable [17]. No any

Table 1 Issue-wise study of parameters

Ref. No.	E-learning cloud architecture	MCC—specific requirements and reservation	Impact and analysis mechanism	Mobile cloud-based content delivery model	Web2.0/Web-based content management system (CMS)
[1]	Use of smart mobile agents build an environment which enables designing and monitoring of learning content including creating a well-placed juncture for exploring new ideas	Virtual and personal learning environment VPLE/VLE offers and monitors e-learning path-ways and confrontational instruction with on-line modules but they lack in learner centric approach. So, instructors have to take care of learners' learning			CMS provide formal learning for registered learners to access text on multimedia-based course content. Further facility of threaded discussion in required

(continued)

Table 1 (continued)

Ref. No.	E-learning cloud architecture	MCC—specific requirements and reservation	Impact and analysis mechanism	Mobile cloud-based content delivery model	Web2.0/Web-based content management system (CMS)
[10–12]	The present e-learning tools and applications require tremendous venture on the foundation regardless of the system being business or open source. Institutions are demanding/in search of low-cost solution for the same. Education at low cost is the demand of mass [10, 11]	An e-learning stage dependent on open-standards with least starting expense of venture utilizing cloud computing will ready to instruct individuals to accomplish information-based learning [12]	Cloud model is based on dynamic configuration and service-level architecture (SLA) is still applicable to static development model. But QoS is dependent on SLA. Therefore, QoS for cloud computing is still incomplete [12]	There are issues of dynamic scalability and collaboration within independent educational institute. Due to propriety software it is quite expensive to unite with other systems [10, 11]	Hardware of client machine must be compliant with Web2.0 based e-learning technology [12]
[13]		Cost and risk management decides the level of integration (IaaS, SaaS, PaaS) with e-learning and Web2.0		To exploit MCC using Web2.0 requires (a) Web2.0 technologies like—Ajax, xml, etc. (b) Web2.0-based service and applications (weblogs, wikis, podcasts, etc. (c) Web2.0-based activities as collaborative content writing, sharing, information linking, etc.	Web2.0 allows learners to personalize learning environment which places learner at the center of the activity and facilitates new form of creation, collaboration and consumption

(continued)

Table 1 (continued)

Ref. No.	E-learning cloud architecture	MCC—specific requirements and reservation	Impact and analysis mechanism	Mobile cloud-based content delivery model	Web2.0/Web-based content management system (CMS)
[6, 14]	In all e-learning infrastructures, including hardware and software computing resources, a virtualized system is build where resource management layer plays a key role to select appropriate service from PaaS, IaaS, SaaS to build an application [6]		The challenges for e-learning cloud are charges, bandwidth, security, user idea, resources and management rules [6]	The e-learning multimedia content is non-linear multimedia and it is interactive MM type which consumes much higher power consumption in compare to non-interactive MM. Similarly, network availability and intermittency are two major concerns for the multimedia content [14]	
[15]				For offloading a system named CloneCloud is build which automatically transfers mobile applications at runtime by partitioning it and thread of it is migrated to mobile device in cloud. This produces 20× execution speed. But it cannot handle unique native content remotely	

(continued)

Table 1 (continued)

Ref. No.	E-learning cloud architecture	MCC—specific requirements and reservation	Impact and analysis mechanism	Mobile cloud-based content delivery model	Web2.0/Web-based content management system (CMS)
[7]				It is not possible to access native resources in CloudClone via virtualization. With the help of Cloudlets, mobile devices can be offloaded to nearby devices and this shows real-time collaborative responses having very low latency, only one-hop distance and optimal bandwidth utilization. But trust over cloudlet is a point of concern	

(continued)

Table 1 (continued)

Ref. No.	E-learning cloud architecture	MCC—specific requirements and reservation	Impact and analysis mechanism	Mobile cloud-based content delivery model	Web2.0/Web-based content management system (CMS)
[16]			Here, learning platform issues are addressed and this shows that available courseware management systems having similarity with contentment management systems are less effective in utilization of resources. On the other hand, only 12% of mobile users know that cloud computing can also be used in learning activities	In their research model, flexible data storage requirement and carefully authentication and authorization are done. In their model, data manipulation process needs improvement and client registration is lengthy process	

(continued)

Table 1 (continued)

Ref. No.	E-learning cloud architecture	MCC—specific requirements and reservation	Impact and analysis mechanism	Mobile cloud-based content delivery model	Web2.0/Web-based content management system (CMS)
[17]	They identified that cloud for e-learning can easily be topped both horizontally and vertically. A company managing the cloud also charge the institution as per the count of active servers which are actually depending upon the enrolled students to the course			Four-layered system is built having (a) infrastructure layer for h/w, s/w, Internet, and teaching resources. (b) s/w resource layer for handling operating system and middleware. (c) Resource management layer to loosely couple the s/w with demanded h/w. (d) Service layer where SaaS is used to provide cloud computing services to customers over Internet	

(continued)

Table 1 (continued)

Ref. No.	E-learning cloud architecture	MCC—specific requirements and reservation	Impact and analysis mechanism	Mobile cloud-based content delivery model	Web2.0/Web-based content management system (CMS)
[18]				Any MCC-based model may have (a) simple APIs offer transparent accessibility to mobile services also this do not demand any other specific knowledge. (b) Can easily deploy applications over the heterogeneous carrier networks. (c) Each carrier's specific network policy, which works for flawless integration	

(continued)

Table 1 (continued)

Ref. No.	E-learning cloud architecture	MCC—specific requirements and reservation	Impact and analysis mechanism	Mobile cloud-based content delivery model	Web2.0/Web-based content management system (CMS)
[19–21]		Light weight and optimal distributed application deployment solution should incorporate optimum ways for the development, deployment and management of runtime distributed platform for MCC. The elastic feature of the old-fashioned offloading framework seems, by all accounts, to be a proper ideal answer for tending to the issues of current distributed application processing frameworks [21]	Factor affective on performance like service-level agreements, recovery, security, availability bandwidth, location, etc., and performance evaluation criteria like average response time per unit time, network capacity per second, number of I/O commands per second [20]	MCC is categorized as general-purpose mobile cloud computing, where mobile device us Internet resources on-demand to offload the execution to cloud. Another is application-specific mobile cloud computing, it is simpler in computation power like mobile service clouds, weblets of elastic application [19]	

(continued)

Table 1 (continued)

Ref. No.	E-learning cloud architecture	MCC—specific requirements and reservation	Impact and analysis mechanism	Mobile cloud-based content delivery model	Web2.0/Web-based content management system (CMS)
[7, 9, 22, 23]	The fuse of SaaS administration provisioning model and IaaS ought to likewise be researched for computational offloading in MCC [22]	Whenever single cloud doesn't play enough role for on-the-go user's demand, then the scheme should be in a state to automatically realize and constitute a service for user [7]	QoS in MCC may have problems like congestion due to limited bandwidth, network disconnection and the signal attenuation caused by mobile user's mobility [7]	Combination of femtocells and cloud computing will deliver highly economical, scalable and secure network for mobile operators, here additional resource is automatically added as required to meet demands [7] Runtime computational offloading is useful in mobile ad hoc networks and the attributes of the availability of rich resources, the scalability of services and the centralized service provisioning model in the form of IaaS, PaaS and SaaS in computational clouds; motivate accessing the pre-configured services at an on-demand basis [22]	The intelligent use of web2.0 in education will help learners to develop and conceptualize "classroom" as learning take place across the physical and cyber spaces, providing learners with an array of choices about the substance and location of their experiences
[23–25]	The heterogeneity of cloud application tools will be hitch if the personal learning environment (PLE) is not being shaped correctly, will escalate its complexity, and hampering usability. Therefore, research ought to be done to flawlessly incorporate heterogeneous cloud tools [24]	The fundamental point of the application execution structure in MCC is to enlarge the assets of portable communication devices by utilizing the assets and the cloud services [23]	There is always a great demand to deploy light weight procedure for execution in optimal way for resource hungry mobile applications, for this, MCC plays a vital role. This offloading requires complex application partitioning of different granularity levels and migration of various modules to cloud server [25]	The execution in the offloading architecture involves three phases: (a) Replicator—synchronizing the changes in phone s/w and state of the clone (b) Controller—on the mobile device starts an augmented execution and integrates the result back to mobile device (c) Augmenter—on the clone side it is responsible for local execution and returning of results [23]	

extraordinary knowledge is required to connect the PC to cloud servers hosting applications over the Internet. These servers can swap their competing periods on their own [4].

The main driving forces [26] behind CC are ubiquity of wireless networking and broadband and others are regularly falling cost of storage, improvements in Internet computing, mobile applications, and their support for CC in a form of MCC.

Theoretically, Web2.0 looks like to symbolize "knowledge" as "collective agreement" that "may combine facts with other dimensions of human experience, such as opinions, values, and spiritual beliefs." Web2.0 is both a platform on which advanced technologies have been built and a space where users are as important as the content they upload and share with others, like knowledge development through wikis (e.g., Wikipedia), creative works, such as podcasts, video-casts, blogs, and microblogs (e.g., Twitter, Blogger); content aggregation, really simple syndication (RSS) feeds and tagging tools [9].

For the proposed future work, identification of research problem is broadly categorized in three major areas.

3.1 Collaborative E-Learning Issues

- The existing e-learning platforms are not able to scale dynamically and collaborate with other educational institutions. It is expensive to integrate with other systems as it is based on proprietary software [11].
- The prevailing mobile hardwares usually are not be accommodating with e-learning technology exploiting Web2.0 [12].
- Study of "TUSK" collaborative software of Christian Medical College (CMC), India with US-based Tufts University shows that there is a positive and successful augmentation in conveying clinical training in remote and rural areas on India [27]. Also because of unreliable Internet connectivity and poor bandwidth of communication, mobile e-learning applications are being rapidly growing to teach clinical trainings especially to medical doctors practicing in rural areas of India [2].

3.2 Mobile Cloud-Based Content Delivery Model Issues

Service-level agreements (SLA) are not well defined in cloud business model [12]. Cloud model is based on dynamic configuration, but the SLA is still applicable for static deployment model [28]. Quality of Service (QoS) is dependent on the SLA. QoS is to be well defined to ensure application usability, availability, and experience of the users.

The device discovery to form mobile ad hoc cloud is big challenge to realize the optimized application execution in the scenario of application partitioning, task scheduling, application migration, authentication, authorization, etc. [25].

3.3 *Learning Platform Issues*

Open source e-learning applications like Moodle and virtual board [16] are widely adopted by educational institutions. The initial cost of S/W is very low but requires high learning curve and costly infrastructure. Open source does not ensure 99.9% uptime and its implementation in the institute takes longer time than commercial software. Web2.0-enabled social network and smart agents are the two major layers of application [13]. Currently, there is less effective utilization of learning management system comparable to content management system [16].

In academic institutions or universities, the need of applications is of two types:

a. Where university data, learning material and assessment sheets like assignment, etc. These are mostly shared by Intranet due to high-speed data transfer required.
b. Similarly, student management and reporting system used by HR and Finance.

In short, currently developing framework and applications mostly take limited advantage of MCC capabilities, both ubiquitous access, and scalability of virtually limitless computing as well as storage resources. Various current application execution framework needs adaptability characteristic and QoS support as well as experience the ill effects of profiling overhead [22]. Research focusing on the demerits of currently available models is the need of time.

Therefore, a comprehensive utilization of this low cost previously mentioned affordances should be a key focus area for researchers and if one can build a model using available open sources services and platform then there should be good solution which will be capable to provide collaborative e-learning over mobile cloud computing.

4 Expected Impacts on Academics/Industry

Indian government is promoting the low-cost or almost no-cost solutions for the masses to educate them in various fields. Mobile cloud computing can be considered as most effective solution in academic institutions, particularly, for the purpose of collaborative distance learning (a new dimension to e-learning). This will without a doubt diminishes costs and gives trustworthy information storage and retrieval and it gives almost unbounded plausibility to clients to utilize the mobile portable devices over network using the Internet. It is normal that this mobile cloud will develop in size and get changes the ICT and get radical changes the space of community-oriented e-learning.

Impact of the proposed work can be observed in a following way:

I. **Lower costs**—Utilizing on-line applications on mobiles need not required any memory space and as no product projects must be stacked and no archive documents should be spared.
II. **Improved data reliability**—One crash of hard disk will kill the data available in your PC. But in case of cloud, all your data is safe even in case of pc disk crash. Therefore, MCC is a totally data-protective computing.
III. **Universal data access and device impedance**—All the stored records or documents are always accessible from any place you are and there is basically no compelling reason to take your reports or documents with you.

Finally, the proposed framework and developed tool enables scheming and monitoring learning contents as well as developing a stage for reconnoitering concepts and has a service-oriented architecture that streamlines the effective utilization of underlying web resources effectively.

5 Conclusions

Collaborative learning through the MCC targets to authorize the mobile user by ensuring unified functionality, without considering the constrained resources on the mobile device. There are abundant new mobile applications where mobile cloud is enriching various activities for the applications. On the other side, now Indian government is also promoting the low-cost or almost no-cost solutions for the masses to educate them in various fields.

The work has to meet the mentioned objectives by rigorously studying the operations in mobile cloud computing and related data management issues. The upcoming model has to improve the end-user's demand for collaborative learning. Before proposing the model, various research methodologies, techniques and tools have to be used critically to evaluate the issues specially related to mobile cloud computing, Web2.0, and collaborative learning.

One has to start with a detailed study and analysis of operations, data management issues, and end-user's demand of mobile cloud computing for collaborative e-learning will be done, where (i) operations will be evaluated in areas like offloading methods, managing mobility, cost-benefit analysis, and connection protocols; (ii) data management issues will be evaluated for issues related to accessing the data, mobile cloud storing user's personal data, portability of data and interoperability and scope and use of interleaved mobile databases, (iii) end-user views will be helpful to collaborate, present and use the data anywhere-anytime [29].

During research, some other cloud-based frameworks will need to be evaluated like Hyrax [30] for media streaming, Cuckoo [31] to offload mobile device application, MobiCloud [32] for ad hoc networks, etc. For the developed model, an appropriate methodology has to be used to identify individual performance issues.

Finally, existing models and newly developed enhanced model have to be compared for various performance issues.

Since, development of physical mobile cloud-based collaborative e-learning model and then assessing its performance issues for evaluation is a practical limitation, hence in the future, we will analysis and compare these results through specified simulators like Cloud Analyst, CloudSim, and Green Cloud and try to improve some of the observed quality factors by using current phenomenon in the area of collaborative learning and mobile [33] cloud computing.

References

1. Canepa G, Lee D (2010) A virtual cloud computing provider for mobile devices. In: Proceedings of the 1st ACM workshop on mobile cloud computing & services: social networks and beyond (MCS), vol 6
2. Department of Electronics and Information Technology, Government of India. http://deity.gov.in/sites/upload_files/dit/files/Digital%20India.pdf
3. http://www.google.com/apps
4. Hayes B (2008) Cloud computing. Commun ACM 51(7):9–11
5. http://www.mobilecloudcomputingforum.com
6. Laisheg X, Wang Z (2011) Cloud computing: a new business paradigm for e-learning. In: IEEE Computer Society 2011 third international conference on measuring technology and mechatronics automation, pp 716–719
7. Hoang T, Lee C, Niyato D, Wang P (2013) A survey of mobile cloud computing: architecture, applications, and approaches. Wirel Commun Mob Comput 13:1587–1611. http://onlinelibray.wiley.com/doi/10.1002/wcm.1203
8. Safran C, Helic D et al (2007) E-learning practices and Web2.0. In: The proceeding of "the ICL conference", 26–28 Sept 2007
9. Greenhow C, Robelia B, Hughes JE (2009) Web 2.0 and classroom research: what path should we take now? Educ Res 38:246–259
10. Al-zaube M (2009) E-learning on the cloud. Int Arab J e-Technol 1(2)
11. Guoli Z, Wanjun L (2010) The applied research of cloud computing platform architecture in the e-learning area. In: The 2nd international conference on computer and automation engineering (ICCAE), Singapore, pp 356–359
12. Chandran D, Kempegowda S (2010) Hybrid E-learning platform based on cloud architecture model: a proposal. In: The 2010@IEEE international conference on signal and image processing, pp 534–537
13. Ouf S, Nasr M, Helmy Y (2011) An enhanced E-learning ecosystem based on an integration between cloud computing and Web2.0. In: Proceedings of the IEEE international symposium on signal processing and information technology (ISSPIT), pp 48–55
14. Khan A, Ahirwar KK (2011) Mobile cloud computing as a future of mobile multimedia database. Int J Comput Sci Commun 2(1):219–221
15. Chun BG, Naik M, Ashwin P et al (2011) CloneCloud: elastic execution between mobile device and cloud. In: EuroSys. ACM, New York
16. Mallikharjuna Rao N, Sasidhar C, Satyendra Kumar V (2011) Cloud computing through mobile-learning. Annamacharya P.G College of Computer Studies, Rajampet, AP, India
17. Massud MAH, Huang X (2012) An e-learning system architecture based on cloud computing. World Acad Sci Eng Technol 62
18. Rajendra Prasad M, Murthi PRK et al (2012) Mobile cloud computing implications and challenges. J Inf Eng Appl 2(7):7–15. ISSN 2224-5782 (print)/ISSN 2225-0506 (online)

19. Qureshi SS et al (2011) Mobile cloud computing as future for mobile applied implication methods and challenging issues. In: Proceeding of IEEE CCISS 2011
20. Niloofar, Reza (2013) Cloud computing performance evaluation: issues and challenges. Int J Cloud Comput Serv Archit 3(5)
21. Gani A, Buyya R et al (2013) A review on distributed application processing framework in smart mobile devices for mobile cloud computing. IEEE Commun Surv Tutor 15(3):1294–1313
22. Gani A, Shiraz M et al (2015) A study on the critical analysis of computational offloading frameworks for mobile cloud computing. J Netw Comput Appl 47:47–60
23. Ahmed E, Gani A et al (2015) Application optimization in mobile cloud computing: motivation, taxonomies and open challenges. J Netw Comput Appl 52:52–68
24. Gonzalee JA et al (2015) Cloud computing on education: a state-of-the-art survey. Comput Educ 80:132–151
25. Ahemed E, Gani A, Buyya R et al (2015) Network centric performance analysis of runtime application migration in mobile cloud computing. Simul Model Pract Theory 50:42–56
26. http://www.microsoft.com/office365
27. Vyas R, Albright S, Walker D, Zachariah A, Lee MY (2010) Clinical training at remote sites using mobile technology: an India–USA partnership. Dist Edu 31(2):211–226
28. Tsai W, Sun X, Balasooriya J (2010) Service-oriented cloud computing architecture. In: Seventh international conference on information technology: new generations (ITNG), Las Vegas, USA, pp 684–689
29. Fernado N, Rahayu W et al (2013) Mobile cloud computing: a survey. Future Gener Comput Syst 29:84–106
30. Marinelli EE (2009) Hyrax: cloud computing on mobile devices using MapReduce. Master's thesis, Carnegie Mellon University
31. Kemp R, Palmer N, Kielmann T, Bal H (2010) Cuckoo: a computation offloading framework for smartphones. In: Proceedings of the second international conference on mobile computing, applications, and services, MobiCASE'10
32. Huang D, Zhang X, Kang M, Luo J (2013) Mobicloud: building secure cloud framework for mobile computing and communication. In: Proceedings of the fifth IEEE international symposium on service oriented system engineering, SOSE, pp 27–34
33. Kim S-H, Kim JK (2016) Determinants of the adoption of mobile cloud computing services. Inf Dev
34. Chun BG, Ihm S, Naik M, Patti A et al (2011) CloneCloud: execution between mobile device and cloud. EuroSys, ACM, April 2011

Chapter 4
Modern Optical Data Centers: Design Challenges and Issues

Arunendra Singh, Richa Singh, Pronaya Bhattacharya, Vinay Kumar Pathak and Amod Kumar Tiwari

1 Introduction

In today's modern era, most data center services are publicly available for free, thereby datacenter operators face dual issues of meeting the exponential rise in data traffic while minimizing delay and optimizing Quality-of-Service (QoS) for user applications [1]. Also, there is a rise in projected network bandwidth requirements in low powered energy setups without a significant rise in infrastructure and computing costs [2–4]. A typical data centers connect tens of thousands of servers forming a massively parallel supercomputing infrastructure. Since the data requirements are huge, optical switching can address the above need by multiplexing several low bandwidth users over a fiber link, as in the case of Dense Wavelength Division Multiplexing (DWDM). DWDM combines several channels onto a single fiber, realizing bandwidth up to 40 GBPs in a single sheath ofthe fiber. Users in optical networks are tuned using WDM that tune to the appropriate receiver wavelength in a Wavelength-Routing Network (WRON). This tuning is performed by TWCs that Arrayed waveguide Grating (AWGR) allows dynamic routing of optical signals in WRON. AWGR cater the demands for tuning variety, velocity, wavelength stability, and electronic control. The data delivery could be packetized, as in the case of OPS, or aggregated at the sender, as in OBS. OPS and OBS provide many advantages over their electronic counterparts, but a large bottleneck is the lack of optical RAMs. The alternative is to circulate packets through Fiber delay Lines (FDLs) to create an optical delay.

A. Singh (✉) · R. Singh
Pranveer Singh Institute of Technology, Kanpur, Uttar Pradesh 209305, India

P. Bhattacharya
Institute of Technology, Nirma University, Ahmedabad, Gujarat 382481, India

V. K. Pathak
Dr. A.P.J. Abdul Kalam Technical University, Lucknow, Uttar Pradesh 226031, India

A. K. Tiwari
Rajkiya Engineering College, Churk, Sonbhadra, Uttar Pradesh 231206, India

V. K. Giri et al. (eds.), *Computing Algorithms with Applications in Engineering*, Algorithms for Intelligent Systems, https://doi.org/10.1007/978-981-15-2369-4_4

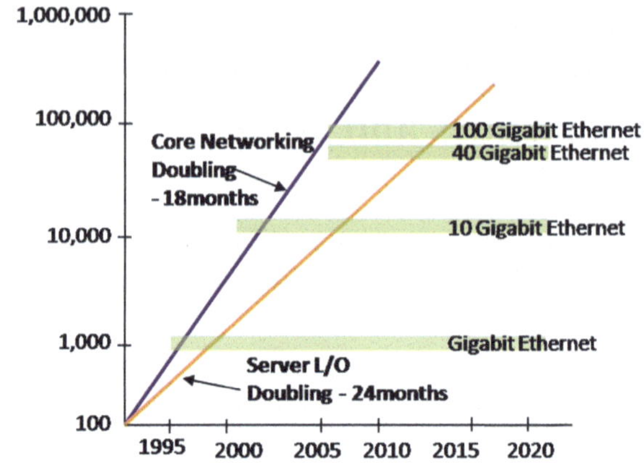

Fig. 1 Economies of scaling data center architectures

Whereas millions of packets can be stored in electronic RAMs for a longer duration, only some hundreds of packets can be stored in fiber delay lines (for temporary optical switching storage) for very short durations. Therefore, a lossless system requires an effective design of the optical switch [5].

In the near future, there is a requirement to combine OPS and OBS approaches to route data optically through end-to-end applications. These optical interconnections require a hybrid controller that utilizes both electronic and optical circuitry for effective use in data centers [6, 7]. Figure 1 shows the increasing trends for 25 years. For example, a data center with more than 50,000 servers, each fitted with a transfer speed of 40 Gb/s, would require an inner network with a complete transmission capability of 2 Petabits/s to support complete bandwidth communication between each server. While the innovation on the software [8–10] and machinery [11–13] side is obviously exceptional, the scalability is a key issue while considering the current switching and optical interconnects.

In a typical data center, multiplexed parallel servers are required in the crosswise design to support various heterogeneous applications and services. An individual rack-house of a data center contains several servers and is associated with a ToR switch through copper connections. The ToR switches further connect to core switch layers through an optical transceiver [14]. To develop the bigger scale networks, every ToR switch would associate with all accessible core switches. If a ToR has m uplinks, then it connects to m core switches. If each core switch has n number of ports, then it would support n ToR connections. In the event that every ToR utilizes u downlinks hosts then the total network scales to $n \times u$ ports. Figure 2a denotes the traditional data center design while Fig. 2b focuses on recent emerging designs.

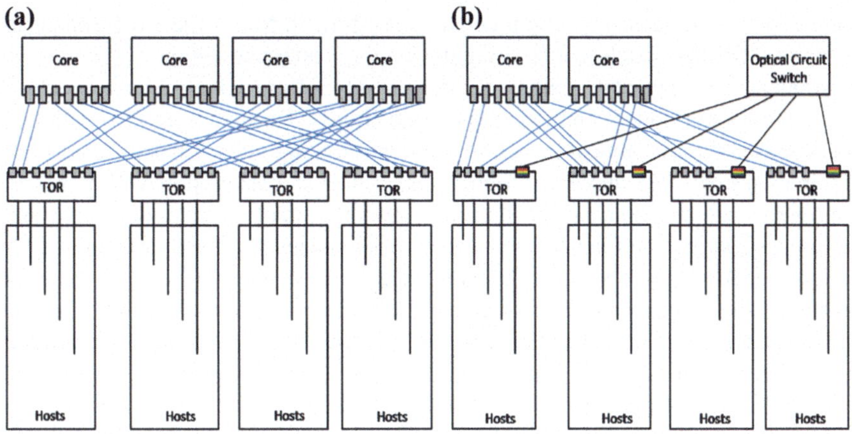

Fig. 2 **a** Traditional switch designs. **b** Emerging switch designs

1.1 Motivation

Recent technological advancements in cloud services, virtualization, and edge services augur for demands of all-optical interconnections at core switches in data centers. The server interconnections scalability is a key issue to handle the rise in the overflowing influx of traffic. Also, the interconnections need to be space-efficient, so that switch cost, powering and cooling resources are minimized. With the global roll-out of 5G commercial services, the communication latency is reduced between hops in the network. In optical, the router needs to process data very fast to meet these stringent requirements. This motivates research and innovation towards designing the switch fabric with proper scaling parameters and costing parameters. An optimal trade-off is required to maintain the scalability of users, while at the same time minimizing the computation and communication costs in switches.

1.2 Related Work: Optical Switches and Buffer Designs

In the past, Hemenway et al. [15] proposed an optical interconnect that provides high scalability at lower error rates. The drawback of the scheme is high complex architecture and less cost-effective solutions. Lira et al. [16] proposed Electro-Optical Networks (EON) for the conversion of data from electrical to optical domain and data is routed optically through the core network. Ye et al. [17] proposed *DOS* architecture to minimize optical packet delays and presented Label Extractors (LEs) to label incoming packets. The architecture is not scalable at high loads due to high power consumption. Farrington et al. [18] presented *Helios* to statically route traffic patterns based on WDM reconfiguration table. The drawback is that the time to route depends on gathering output connections through Micro-Electro-Mechanical Systems (MEMS) switches, increasing the loading pattern condition.

Wang et al. [19] improved the switch design of [18] by providing a loose synchronization between optical and electronic counterparts. Kachris et al. [20] authored WDM-based passive optical switch to minimize routing latency but suffered from requirements of ToR bandwidth requirements. Liu et al. [21] proposed *REACToR* to improve switch proposed in [20] by following a flat topology communication that minimizes communication latency between ToR connections. Rastegarfar et al. [22] designed an All-Optical Negative Acknowledgement scheme (AO-NACK) scheme at the sender site to reduce the transmission of recirculating packets through FDLs. The drawback of the scheme is that the design of AO-NACK is complicated in hardware. Bhattacharya et al. [23] simplified the AO-NACK scheme by proposing algorithms for transmission of optical packets at the sender site and at the core network. To mitigate the crosstalk components and link losses, Singh et al. [24–26] provided performance evaluations on energy considerations of an optical switch based on various buffering conditions. A proper demarcation between switching among electronic and optical buffer is not present. The problems are magnified as there is an absence of Optical RAM [22]. A considerable improvement to solve recirculation and NAK issues of bufferless designs is presented by Bhattacharya et al. [27, 28] by proposing a dual buffer-based optical switch and simplified AO-NACK mechanism for loss notification. Spectral density and power losses in case of dual buffered slots are not considered. Table 1 presents a comparative analysis of related work in Optical Switch designs.

1.3 Research Contributions

The optical interconnect needs to be designed to support hybrid operations for both OPS and OBS switches. The following are the main research contributions of the paper:

1. The paper presents a systematic survey of some notable switch designs in the past, highlighting the pros and cons of the proposed switches.
2. The survey also highlights some novel recent switch architectures to minimize the optical interconnection latency as well as reduce the physical and link-level losses.
3. Finally, the survey highlights the design challenges for future optical switches to support the high bandwidth requirements projected for data-driven applications.

Table 1 A comparative review of optical switches and buffer designs in optical data centers

Related work	Year	Advantages	Limitations
Hemenway et al. [15]	2004	Proposed an Optical Design Interconnect that provides high throughput at lower error rates	High complex switch design
Lira et al. [16]	2009	Proposed Electro-Optical Network (EON) for the conversion of E/O and O/E at end hosts	Huge losses at the physical layer and heavy impairments suffered at edge nodes
Ye et al. [17]	2010	Proposed *DOS* architecture to minimize buffer delays between ToR switches and reconfigurable TWCs patterns	High power consumption during peak traffic due to E/O conversions
Farrington et al. [18]	2010	Proposed *Helios* to statically route traffic through WDM transceivers	Reconfiguration time more for MEMS switches, higher starting configuration cost
Wang et al. [19]	2011	Proposed *C-Through* as a loose hybrid synchronization between electrical and optical switching, cost-effective	Less scalable, ToR switches need full bandwidth to connect, hence not flexible and faces bottleneck issues at peak traffic
Kachris and Tomkos [20]	2011	Proposed WDM-passive optical networks (*WDM-PON*) to minimize peak latency	Bandwidth is wasted and architecture not flexible
Rastegarfar et al. [22]	2013	AWGR-based scheme with physical layer NACK scheme	The design of the NACK scheme in hardware is complicated
Liu et al. [21]	2014	Proposed *REACToR* to support high packet switching and latency is reduced through flat topology	Synchronization between data and control channels are not scalable
Bhattacharya et al. [23]	2017	Proposed buffer-based solutions to the AO-NACK scenarios in recirculating FDLs	Energy considerations in link-level losses were not considered in the design
Singh et al. [24]	2018	Energy consumption analysis over-amplified switch	Proper demarcation of context switching from electrical to optical is missing

(continued)

Table 1 (continued)

Related work	Year	Advantages	Limitations
Singh and Tiwari [25]	2018	Performance analysis against traffic arrivals in the electronic and optical buffer	Loading conditions and PLR not discussed at hybrid scenarios of buffer storage
Bhattacharya et al. [27]	2019	Dual buffer-based optical design for storage of a large number of contending packets in the buffer	Simulation regarding power spectral density in case of dual buffer slots is not considered
Bhattacharya et al. [28]	2019	Dual buffer design with primary and secondary buffer designs for one slot duration in AWG	Link-level analysis in case of contenting packets is absent
Singh et al. [26]	2019	A hybrid-based buffer scheme for Internet-of-Things (IoT)-based environments	The issue to address crosstalk components in AWG is not addressed with optical buffer

1.4 Organization of the Paper

The remainder of the paper is organized into five sections. Section 2 of the paper presents notable switch designs in the past. Section 3 presents recently proposed all-optical switch designs. Section 4 proposes the design challenges of designing modern optical switches for future traffic projections. Finally, Sect. 5 provides the desired conclusions and future research directions.

2 Notable Switch Designs

2.1 DOS: *Datacenter Optical Switch*

Ye et al. [17] presented *DOS* switch based on packet-based optical architecture. *DOS* switch consists of AWGR which multiplexes big amount of wavelengths into a single optical fiber at the transmission end and collects individual channels at the receiving end and de-multiplexes them.

In addition to the AWGR, the switching fabric also includes a range of Label Extractors (LEs), a shared loopback Synchronous Dynamic Random Access Memory (SDRAM) buffer, TWCs, and control plane. The high-level diagram of the *DOS* architectures is shown in Fig. 3. AWGR routes optical signals from an input port to an output port based on wavelength decided by TWCs. LEs presents optical labels to each packet coming from ToR switches to get source and destination addresses. The control plane provides the required bandwidth and *DOS* operates in arbitration

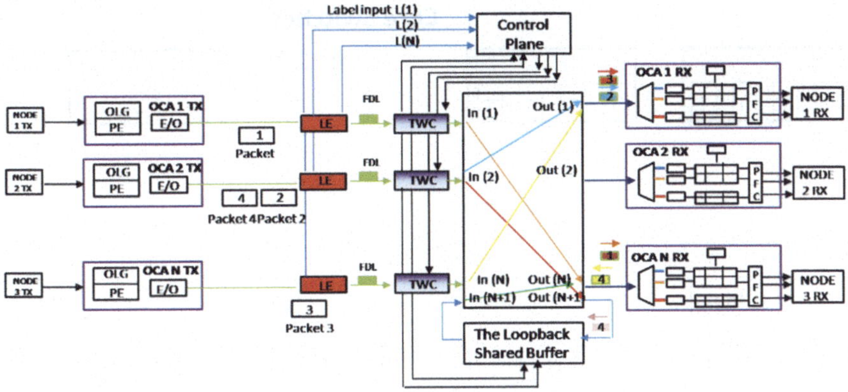

Fig. 3 Switch design of *DOS* [17]

mode for contention resolution. The advantage of *DOS* is that it has low latency and is independent of input size. No electrical switch buffer delay is needed because the ToR packets are traveling through optical switches. The reconfiguration time of TWC is a few nanoseconds. This is beneficial in case of heavy traffic. The drawback is conversions number of O/E and vice versa in case of heavy traffic. *DOS* has a large number of TWC components increasing the switching cost.

2.2 Helios: *Hybrid Electrical/Optical Switch*

Helios switch combines the hybrid advantages of both electrical and optical switches. It employs WDM links at ToR switches to distribute bursty traffic [18]. Every ToR switch employs transceivers where half of the transmitters connect to electronic core switches and the other half is used to connect optical transceivers. The software of *Helios* control scheme is based on three primary components: Pod Switch Manager, Circuit Switch Manager, and Topology Manager as shown in Fig. 4. Pod Switch Manager provides statistical data about sent traffic sent and interfaces with the Topology Manager which configures the switch appropriately based on input traffic and routing decision. The switch operates on MEMS technology, which makes it consume less power due to bandwidth independence. In MEMS, there is no optical-electronic conversion which leads to high performance and low delays. The main drawback is regarding the variable starting reconfiguration time of MEMS switches.

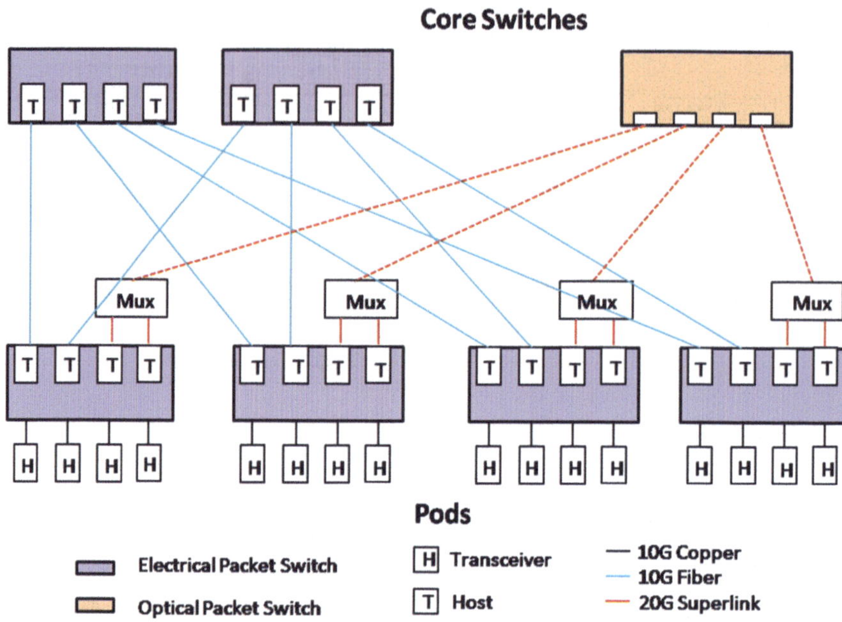

Fig. 4 Switch designs of *Helios* [18]

2.3 WDM-PON: *WDM-Passive Optical Network*

A new hybrid architecture was proposed by Kachris et al. [20] to improve the design of c-through [19] and *Helios* [18] architecture by employing several racks operating on optical WDM and a ToR switch. The interconnections rout data based on AWGR patterns. Two types of flows are suggested—intra flow based on *WDM-PON* to routing data between senders and receivers on the same ToR rack, and inter-flow to connect users on different ToR switches. Studies suggested that there is a 10% power reduction by using combined flows. The basic system design is shown in Fig. 5. The power consumption is greatly reduced as bandwidth is not wasted between ToR and intra-rack communication. Analogically, if we compare the switch architecture with a telecommunications system, the aggregate switch will be used as an optical connection terminator and the ToR switch will be used as an optical network unit. For each individual server, the WDM transceiver is set for inter-rack communication and a commodity Ethernet transceiver is set for intra-rack communication. Power consumption in *WDM-PON* network is primarily energy dissipation in aggregate switches, Ethernet transceiver, ToR, and WDM transceiver. The drawback is the switch is not flexible for adding new connections, thus wastes a lot of bandwidth.

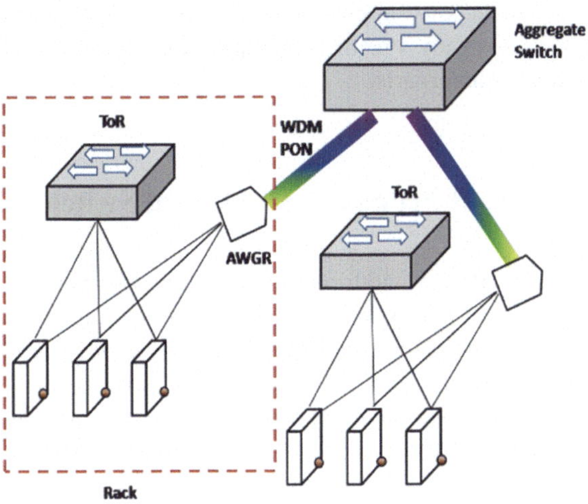

Fig. 5 A passive *WDM PON* architecture [20]

2.4 REACToR: *Hybrid Switch*

REACToR combines the benefits of packet and circuit switching and synchronizes end-to-end connections based on a packetized approach [21]. In terms of reaction design, it is much superior to other designs as indicated in Fig. 6. It is the enhanced version of Mordia switch [23]. It reduces the cost and need of optoelectronic transceivers by performing optical switching directly to connect ToRs in the data center. It provides low cost buffering and supports Optical Circuit Switching (OCS) at 100 Gbps and EPS at 10 Gbps. The access mechanism is TDMA, which reduces operational latency. Thus, *REACToR* switch has a low waiting time, but suffers from heavy

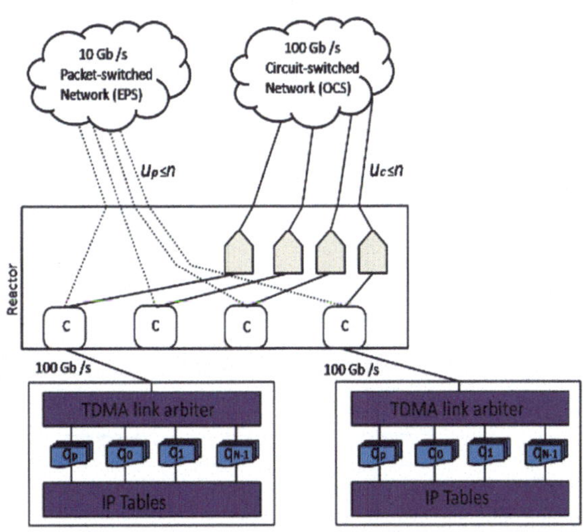

Fig. 6 Design of *REACToR* switch [21]

interconnects to support both OCS and EPS operations. Also, the switch needs to maintain steady synchronization between the flow of data between electronic and optical counterparts.

3 Recent Proposed All-Optical Switch (AOS) Designs

The rise of bandwidth requirements requires not only the backbone network to be deployed in purely optical fashion, but also the buffered data in the electro-optical switch needs to be redesigned. Thus, AOS becomes a reality in today's communication infrastructure. Although AWGR supports optical parallelism, lack of optical RAM requires contending packets to wait for optical delay slots in Fiber Delay Lines (FDLs) and then pass through output port of AWGR [23]. As data is sent in large bursts, the control information is routed earlier of actual data. If the control information does not find an output slot to pass from AWGR, it indicates a loss at the sender site.

Proietti et al. [29] presented an AO-NACK scheme to reduce retransmission at the sender site. The problem is that the hardware design was complicated, thus the switch was not scalable. Later, Bhattacharya et al. [23, 27] modified the scheme by adding FDL slots from 1 to B optical delay. The schematics of the switch is shown in Fig. 7a. The switch core is a $2N \times 2N$ AWGR for scheduling and another extra $N \times N$ at the sender is an additional input line port called the 'on-demand' line ports and used for the optical buffer creation, where N denotes the switch radix. The switch has a retransmission mechanism to inform the sender of control packet loss, through a simple edge detection mechanism. In a similar direction, Singh et al. [24, 25] presented a hybrid buffer based AOS switch, with contending packets are present in B buffer slots in FDL and presented a $2N \times 2N$ AWGR to route data optically. Instead of notifying the loss to sender, the switch has a built electronic RAM. Through analysis is performed on the design of the buffer and the queuing

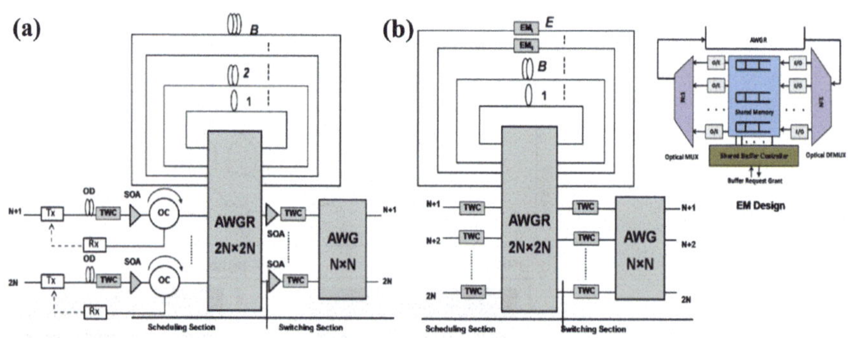

Fig. 7 **a** Proposed AOS switch with AO-NACK scheme [27]. **b** Placement of EM module in electro-optical hybrid buffer switch [24]

of packets in the shared electro-optical buffer. Figure 7b shows the placement of the electronic buffered (EM) module in the switch.

4 Challenges in Data Centers

The servers in a typical data center are at the lowest level and are organized into server racks of 40–80 blades. Each blade server connects with ToR switches with a 1 Gbps link. They further connect to aggregation switch, and the traffic through these switches are forwarded to core routers. Thus, the capability of data centers depends on optimizing the optical infrastructure. The cost and performance of data centers depend on distance as well as a hierarchy. These structures are optimized for performance at the link-layer domain. The spare capacity of data centers is provisioned for per individual service so that each service can scale out to nearby servers to respond rapidly to demand spikes or to failures. Due to this, each service is provided by nearby servers for the rapid response but it could lead the adverse effect when the demand for resources increases [30]. There are various challenges in the hierarchical data center system which are discussed below.

4.1 Scalability

Traffic management through the switch is very much difficult, as the traffic flowing in the data center continuously increases. The traditional hierarchical system is not able to scale linearly with the rapid growth of network traffic. With the rise in multicore architectures and multicore servers, the Next Generation Data Centers (NGDCs) should be optimized to support millions of microprocessor cores.

4.2 Energy Efficiency

According to Information and Communication Technology (ICT), power consumption is a critical issue at the data centers. The network elements like electronic links, switches are power-hungry. So, in this case, if the power consumption of network elements is saved then it makes a good impact on the overall consumption of data center sites. In future data centers [31, 32], energy efficiency is required for power saving.

4.3 *Space Management*

The data center communicates with millions of server cores, thus the issue of non-blocking of data is a prime concern. Thus, connectivity requires a large amount of Ethernet cables which leads to problems like—maintenance issues, implementation of server, management configurations. Due to this, other technologies are very much needed that helps to reduce the cabling requirement in future systems.

5 Conclusions and Future Work

As the hardware chip size is decreasing to nano-scales, more hardware could be integrated which makes the fabrication of optical switches a possible reality. The use of optical technology is not only limited to switching technology but also in data center applications. But due to limited technological advances, optical switching cannot be applied. Therefore, both electrical and optical technologies are used simultaneously. The paper thus discusses the needs and requirements of future optical data centers. The paper investigated the rise in data requirements and challenges in switch design to optimize network performance. The recent switch designs are discussed along with the advantages and disadvantages. From the above discussion, it is conclusive that AWGR will be an integral part of the optical switch and data center design in the future. The commercialization of TWC will be breakthrough in the field of optical data center design. As part of the future work, the authors would investigate the hybrid buffer approach for both optical and electrical packets. For high-speed bursty data, OCS will be preferred and ECS will be used only to support TWCs losses. Also, the emergence of SDN in the optical domain would increase the efficiency of future optical switch designs and high-speed transfer operations.

References

1. Al-Fares M, Loukissas A, Vahdat A (2008) A scalable, commodity data center network architecture. In: Proceedings of SIGCOMM
2. Aronson L, Lemoff B, Buckman L, Dolfi D (1998) Low-cost multimode WDM for local area networks up to 10 Gb/s. IEEE Photonics Technol Lett 10(10):1489–1491
3. Barker K et al (2005) On the feasibility of optical circuit switching for high performance computing systems. In: Proceedings of SC
4. Vishwanath KV, Greenberg A, Reed DA (2009) Modular data centers: how to design them? In: Proceedings of the 1st ACM workshop on large-scale system and application performance (LSAP)
5. Astfalk G (2009) Why optical data communications and why now? Appl Phys A 95(4):933–940
6. Davis A (2010) Photonics and future datacenter networks. In: HotChips 22 symposium (HCS). IEEE, pp 1–38
7. Kachris C, Tomkos I (2012) A survey on optical interconnects for data centers. IEEE Commun Surv Tutor 14(4):1021–1036

8. Cvijetic N, Tanaka A, Ji PN, Sethuraman K, Murakami S, Wang T (2014) SDN and OpenFlow for dynamic flex-grid optical access and aggregation networks. J Lightwave Technol 32(4):864–870
9. Cui L, Yu FR, Yan Q (2016) When big data meets software-defined networking: SDN for big data and big data for SDN. IEEE Netw 30(1):58–65
10. Channegowda M, Nejabati R, Simeonidou D (2013) Software-defined optical networks technology and infrastructure: enabling software-defined optical network operations. J Opt Commun Netw 5(10):A274–A282
11. Richardson DJ, Fini JM, Nelson LE (2013) Space-division multiplexing in optical fibres. Nat Photonics 7(5):354
12. Priolo F, Gregorkiewicz T, Galli M, Krauss TF (2014) Silicon nanostructures for photonics and photovoltaics. Nat Nanotechnol 9(1):19
13. Schares L, Kuchta DM, Benner AF (2010) Optics in future data center networks. In: 2010 IEEE 18th annual symposium on high performance interconnects (HOTI). IEEE, pp 104–108
14. Srivastava R, Singh YN (2010) Feedback fiber delay lines, and AWG based optical packet switch architecture. Opt Switch Netw 7(2):75–84
15. Hemenway R, Grzybowski R, Minkenberg C, Luijten R (2004) Optical-packet-switched interconnect for supercomputer applications. J Opt Netw 3(12):900–913
16. Lira HL, Manipatruni S, Lipson M (2009) Broadband hitless silicon electro-optic switch for on-chip optical networks. Opt Express 17(25):22271–22280
17. Ye X, Yin Y, Yoo SB, Mejia P, Proietti R, Akella V (2010) DOS: a scalable optical switch for datacenters. In: Proceedings of the 6th ACM/IEEE symposium on architectures for networking and communications systems, p 24
18. Farrington N, Porter G, Radhakrishnan S, Bazzaz HH, Subramanya V, Fainman Y, Papen G, Vahdat A (2010) Helios: a hybrid electrical/optical switch architecture for modular data centers. In: Proceedings of ACM SIGCOMM
19. Wang G, Andersen DG, Kaminsky M, Papagiannaki K, Ng TS, Kozuch M, Ryan M (2011) c-Through: part-time optics in data centers. ACM SIGCOMM Comput Commun Rev 41(4):327–338
20. Kachris C, Tomkos I (2011) Power consumption evaluation of hybrid WDM PON networks for data centers. In: 16th European conference on networks and optical communications (NOC). IEEE, pp 118–121
21. Liu H, Lu F, Forencich A, Kapoor R, Tewari M, Voelker GM, Papen G, Snoeren AC, Porter G (2014) Circuit switching under the radar with REACToR. In: NSDI, v.ol 14, pp 1–15
22. Rastegarfar H, Leon-Garcia A, LaRochelle S, Rusch LA (2013) Cross-layer performance analysis of recirculation buffers for optical data centers. J Lightwave Technol 31(3):432–445
23. Bhattacharya P, Singh A, Kumar A, Tiwari AK, Srivastava R (2017) Comparative study for proposed algorithm for all-optical network with negative acknowledgement (AO-NACK). In: Proceedings of the 7th international conference on computer and communication technology, pp 47–51
24. Singh A, Tiwari AK, Srivastava R (2018) Design and analysis of hybrid optical and electronic buffer based optical packet switch. Sādhanā 43(2):19
25. Singh A, Tiwari AK (2018) Analysis of hybrid buffer based optical data center switch. J Opt Commun. https://doi.org/10.1515/joc-2018-0121
26. Singh A, Tiwari AK, Bhattacharya P (2019) Bit error rate analysis of hybrid buffer-based switch for optical data centers. J Opt Commun. https://doi.org/10.1515/joc-2019-0008
27. Bhattacharya P, Tiwari AK, Singh A (2019) Dual-buffer-based optical datacenter switch design. J Opt Commun. https://doi.org/10.1515/joc-2018-0023(2019)
28. Bhattacharya P, Tiwari AK, Srivastava R (2019) Dual buffers optical based packet switch incorporating arrayed waveguide gratings. J Eng Res 7(1):1–15
29. Proietti R, Yin Y, Yu R, Ye X, Nitta C, Akella V, Yoo SB (2011) All-optical physical layer NACK in AWGR-based optical interconnects. IEEE Photonics Technol Lett 24(5):410–412
30. Xu L, Zhang W, Lira HL, Lipson M, Bergman K (2011) A hybrid optical packet and wavelength selective switching platform for high-performance data center networks. Opt Express 19(24):24258–24267

31. Beloglazov A, Abawajy J, Buyya R (2012) Energy-aware resource allocation heuristics for efficient management of data centers for cloud computing. Future Gener Comput Syst 28(5):755–768
32. Chen J, Gong Y, Fiorani M, Aleksic S (2015) Optical interconnects at the top of the rack for energy-efficient data centers. IEEE Commun Mag 53(8):140–148

Chapter 5
TOO-MUCH: Testing Online Over MNNIT Unified Coding Hub

Deepanshu Gupta, Dhruv Sharma, G. Sree Deepthi, Yogeshwar Patel, Pawan Subedi, Abdul Aleem and Manoj Madhava Gore

1 Introduction

Online judges are software systems designed for the reliable evaluation of algorithmic source code submitted by developers [1]. There are various web-based examination systems [2], which conduct exams of various subjects, but could not evaluate programming. On the other hand, online judges compile and test the submitted code, and could be used for practical examination purpose. Online judges are becoming popular in various applications involving today's programming paradigm. The studies show that online judge systems can be successfully applied to solve complex academic and science-driven challenges with accuracy and efficiency.

Testing Online Over MNNIT Unified Coding Hub (TOO-MUCH) is an online judge system built specifically for the evaluation of source code that is submitted by the developer. The submitted code may have been written by students in daily laboratory work or in practical coding examinations or in some other programming contest. TOO-MUCH is an asset for the academic practical examinations held in colleges for the subjects that involve coding. TOO-MUCH blends the current state-of-the-art scenario of competitive programming and the modern style of academic practical examinations in the field of computer science. Most of the computer lab examinations in the colleges/schools today are being conducted manually, despite there being a plethora of buzz about online programming, compiling, and judging platforms. This is mainly due to the fact that almost all of these platforms are paid or owned by other private bodies. Having a personalized open platform like TOO-MUCH for colleges can prove to be quite an edge in leveling up with the current world practices in the programming paradigm. TOO-MUCH addresses the following problems related to code evaluation:

- Lack of platform for hosting coding examinations in university/college.
- Complexity and delay during the code evaluation of lab examination.

D. Gupta · D. Sharma · G. Sree Deepthi · Y. Patel · P. Subedi · A. Aleem (✉) · M. M. Gore
Computer Science and Engineering Department, MNNIT Allahabad, Prayagraj 211004, India

V. K. Giri et al. (eds.), *Computing Algorithms with Applications in Engineering*, Algorithms for Intelligent Systems, https://doi.org/10.1007/978-981-15-2369-4_5

- Dependency on the error-prone manual code evaluation methods for the score obtained by the student in an examination.

Besides providing the most important features relating to code evaluation, TOO-MUCH aims at achieving goals, which some of the other similar platforms do not provide. The following features sum up the research objectives of TOO-MUCH:

- Designing an online judge for hosting any examination related to programming in academics;
- Providing an easy interface to teachers as well as students for programming and evaluation;
- Removing the factor of bias/partiality or human error on the part of the evaluator;
- Automating the examination, its evaluation and the marks awarding process;
- Providing an online editor as well as a compiler for code building along with the facility of code file submission;
- An interface for the teachers to set test questions and test-cases to be checked.

This article provides a solution for the evaluation of code instantly written by students for any contest, examination or programming practice session. The code may be written in any of the four possible languages—C, C ++, Java or Python. The evaluation is done for the correctness of source code and the number of test-cases passed by the submitted code. The proposed online judge has been implemented and successfully tested at MNNIT Allahabad, India [3] for the evaluation of various coding examinations along with few practice sessions for programming assignments.

The rest of the article is organized as follows. Section 2 discusses about the related work of online judges. It also describes how TOO-MUCH stands with respect to the missing necessities in various previously built products. Section 3 proposes the TOO-MUCH software along with the description of the functionalities, design, and implementation details. Section 4 provides the output of TOO-MUCH software and details of the testing done to verify the software. Finally, Sect. 5 concludes the article along with a listing of future works that can be done to enhance the usability of TOO-MUCH.

2 Related Work

The first programming contest was organized by Texas AM University in 1970, under the name of the ACM International Collegiate Programming Contest (ICPC). In a few decades, the event grew to be the most reputed and biggest programming contest [4]. Cheang et al. in 2003 provided an online judge which automatically graded the programming assignments submitted by the students [5]. University of Valladolid (UVa) Online Judge [6] is the first most popular code evaluation software being used across the world for code evaluation. UVa online judge has been designed for the evaluation of programming contests like ICPC.

Edwards and Perez-Quinones in 2008 have introduced an open-source tool Web-CAT [7], which automatically grades programming assignments submitted by students. Wu and Chen in 2012 presented the design and implementation of an online judge [8] specific for ACM programming contests. Recently, the most extensively used software for code evaluation is National Tsing Hua University Online Judge [1]. It is an online platform, similar to the online judge of ACM ICPC. The software supports only two programming languages—C & C++ and could be used for training as well as contesting purposes. The number of challenges published through this online judge is more than 10,000.

As the research progressed for online judges, the systems expanded to provide training to learners for programming. Kosowski et al. in 2007 have used the online judges for academic tuition in Gdan´sk University of Technology [9]. The authors introduced an improvised SPherical Online Judge (SPOJ), which is a blend of two concepts—an E-learning platform and an online judge. Luo et al. in 2008 developed a system called Programming Grid (PG) [10], which is based on the Peking University (PUK) Online Judge system, referred to as POJ in short. PG aims for computer-aided education and training in programming courses. Combéfis and Wautelet in 2014 have discussed the application of programming contests for the training on programming [11].

Alemán in 2011 has extended the functionality of online judge by automatically assessing few assignments related to the programming tools [12]. The automated assessment provided students and instructors the immediate feedback which help the students in learning. Antonucci et al. in 2015 have developed an incremental hint system [13] for solving the coding assignments. The code-revealing hints were made available to students on request. The students could solve the assignments and submit it for instant evaluation.

In summary, the utilization and extension of online judges and online practices have been addressed in a few research works. Wasik et al. in 2018 have done a survey on various online judge systems and their applications [1]. Similarly, Staubitz in 2015 has done a survey on various automated programming assessment tools that have been employed in various massive open online courses [14]. Wilcox in 2016 has summarized online judges under eight testing strategies that have been used for the evaluation of code [15].

TOO-MUCH combines the evaluation feature of online judges and training features of the programming assignments to form a novel and systematic software tool. Compared with existing researches, TOO-MUCH is the first tool that could evaluate code written in either of the four accepted programming languages. The assessment of TOO-MUCH is based on the number of test-cases passed by the code on execution, which makes it faster than others to produce results in a short time for a large number of candidates. Besides, TOO-MUCH is intended to be an open-source tool that could be easily used by examiners. The local copy of TOO-MUCH could be easily amended for assisting students in coding assignments. The coders (on TOO-MUCH) could also see the evaluation score of other coders, who are coding for the same problem. The candidates can resubmit the solution more than once in the allotted time for examination. Thus, TOO-MUCH builds a competitive environment for effective learning which could not be achieved through existing online judges. As per the knowledge and understanding of the authors, no similar systematic code evaluation software has been proposed.

3 Proposed Work

TOO-MUCH aims to provide an online judge which could be easily used in every college. The software is not only student-friendly but also teacher-friendly, so conducting computer-based examinations is easier. TOO-MUCH accepts the code written in either C, or C++, or Java, or Python. The software also provides the facility of uploading files apart from the built-in editor so that students could write code in their familiar coding environment and upload the code file to TOO-MUCH server. TOO-MUCH accepts the solution for a single coding assignment in any of the four accepted languages. Students can upload files to the server as many times as required in the allotted time. To instill competitiveness, the scoreboard of all the students is visible to all. The proposed software could be used for training students in programming lab, for helping students complete their coding assignments and also for conduction programming examinations or contest.

3.1 Software Functionalities

TOO-MUCH software can be used by students, faculty, and administrator (admin). Figure 1 shows the workflow of the web application—TOO-MUCH. It illustrates the

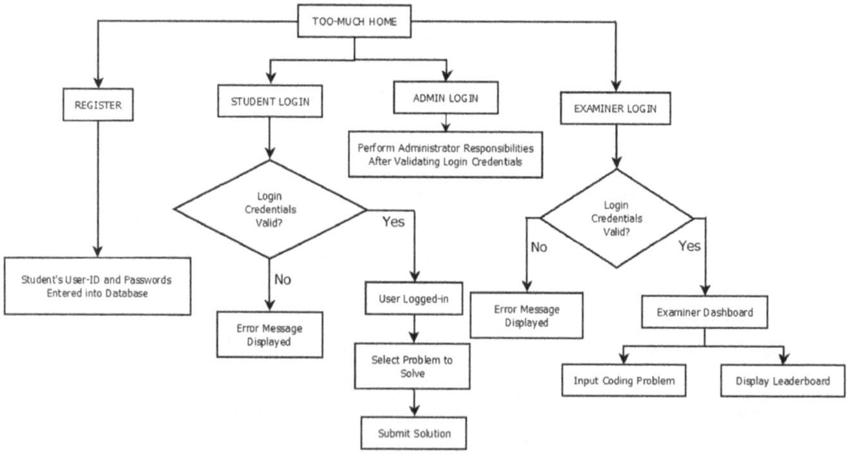

Fig. 1 Workflow diagram of TOO-MUCH

basic functionalities of the online judge along with the sequence of events through which information flows in the project. The following is an overview of the various functional requirements provided by TOO-MUCH.

1. Login: The user would be greeted with a login page in the software. Login credentials for faculty would be different. Users can use their credentials to login to their respective accounts. Successful login will take user to the home page. An error message will be displayed on an unsuccessful login.

2. Signup: There is a signup option for a new user. Signup can be done as a student only. The user must fill the details including the user-id and password. If the details entered already exists, a message will be displayed to change the details. For faculty, the login credentials would be provided by the admin.[1] The login credentials are used for logging into the system.

3. Problem Selection: After a successful login, a list of questions to be solved is displayed to the users. On clicking the solve button, the user is redirected to respective question page. Also, a leaderboard of students who had already solved the questions is displayed along with their score.

4. Question Details: On the question page, the coding problem to be solved is displayed along with three sample test-cases and their outputs. The entry for a question is done by a faculty login only. Students can choose the coding language of their choice for solving the problem. Apart from the built-in editor, there is another option available for students to upload the code file. The language selected must be the same as that of file uploaded.

5. Result: On submitting the code, the code is compiled and executed with input from hidden test-cases. The displayed result shows whether the test-case has passed or failed. If a test-case passes on the provided code, then the score obtained for that test-case is incremented.

6. Leaderboard: On the dashboard of the examiner, a leaderboard page is displayed. The usernames of all the students who are participating in the test are displayed in descending order along with the total score obtained by them. The leaderboard can be used to monitor the progress of all the participants.

7. Administrative Responsibilities: An admin could control the functionalities required at the back-end of the proposed online judge—TOO-MUCH. An admin needs to login to the system through a hidden URL variation of the proposed system. The tasks performed by admin include adding records of faculty members, removing students, and faculty records from the database (when required), and changing the password for the admin login.

3.2 Implementation Details

To make the user interface of TOO-MUCH easy and friendly along with a stable and secure server, the modern technologies which are most widely used and accepted

[1] Password can be changed by faculty.

were investigated. As the online judge is mainly focused on the college environment, prior research was made to understand the needs of the teachers as well as students. The server of TOO-MUCH is to be hosted inside the laboratory to manage a strength of around 100–150 clients, which leads to defining the specifics of the system requirements needed for running the server smoothly and feasibly. With the aforementioned preliminary investigation, our team was encouraged to include the following components for TOO-MUCH:

1. Use of Node Js server, which uses asynchronous client-server model, having scalable, fast and easily coded functioning [16].
2. Angular Js coupled with HTML for the front-end development which gave us an edge over easily modifying the interface according to the user requirements [17].
3. Using the MS Excel Sheets for database management, to make it easy and efficient for the teacher to have the records maintained feasibly.
4. Include an appealing user interface to encourage the students for making full utilization of the judge features.
5. Having a privileged login for faculty members to monitor the status and progress of the ongoing examination.

The development environment for the implementation of TOO-MUCH online judge is as follows:

- **System Configuration**: Linux (Ubuntu 16.04 LTS) installed with Node Js v10.10.0 and dependencies resolved with NPM v6.4.1. Angular JS v1.3.14 with HTML5 used for front-end rendering.
- **Front-end**: HTML, CSS, Bootstrap, and Angular JS have been used for front-end development. For the data management purpose, CSV files have been used in MS Excel format.
- **Back-end**: The online judge uses Node Js server as a back-end and the framework used in this project is express Node Js server.
- **Browser**: Any browser would do, a typical example being Google Chrome, Mozilla Firefox, etc.
- **Compiler used**: GCC compiler is used for the compilation of a typical C and C++ code. Javac compiler is used for the compilation of Java programs. Python compiler is used for compiling code written in python.

The framework—express Node Js server, links the front-end and the back-end working as follows:

1. Some event is registered on the front-end based on the action of the client.
2. After this, appropriate request is made from the client side to server side.
3. Back-end server matches the request from the set of available API and performs the appropriate functioning and returns back the response to the client.
4. Javascript file linked to the UI is responsible for rendering the changes needed over UI based on response from the server-side.

3.3 Code Evaluation Through Online Judge Engine

The core part of the TOO-MUCH lies in the execution of the uploaded code and awarding marks to the developer. Many solutions to distinct problems were found at [18] that have been applied in the proposed system. The various phases of the code evaluation have been discussed next.

1. **Code Upload**: A code file written by the user is being uploaded through the upload button. The upload steps includes the following:

   (a) After selecting programming language and file to upload, the user uploads the file by clicking the "**Upload**" button. Alternatively, code could be written in the built-in editor and submitted.

   (b) The server host receives the file, extracts the file name and extension and stores the file in a folder named **temp**.

2. **Code Compilation and Testing**: After the code file is received, the TOO-MUCH server runs a **bash** command, to execute a **bash** script, which compiles and tests the received code file. The bash script for compilation and execution of C code is as follows:

```
#!/bin/bash
If [$3 =="".c"]; then
gcc temp/$4 -o $2
a = "testcases/"
b = $a$1
c = $b"/"
d = $c"in_"
e = $d$2
f = $c"out_"
g = $f$2
h = "temp"$2
./$2 < $e > $h
diff $h $g | wc -l
rm $2 $h
fi
```

3. **Result Generation**: After the complete execution of the bash script, the server updates the score of the user as per the binary score in every test-case in an excel sheet named score.csv. For every passed test-case, the corresponding score is incremented by 1, and does not change in case the test-case fails.
4. **Sending Response**: Server fetches data from the score.csv sheet, makes a string object with ones and zeros as characters, and sends this data via **http** to the client end. Client-end updates the HTML page and shows the results accordingly.

4 Testing and Results

Real-time beta testing was performed for the validation of TOO-MUCH. Two lab examinations and two lab practice sessions were conducted for students of two courses—MCA (1st year), with approximately 100 students and M. Tech. (1st year), with approximately 50 students. The duration of the exams and practice sessions were an hour each and it included algorithmic problem statements to be solved in C language. The students were instantly evaluated on the TOO-MUCH software and were asked for feedback of the system. All the lab exams and practice sessions successfully concluded with the percentage of successful submissions being around 90%. The overall result of real-time testing was satisfactory. The key findings on testing various components of TOO-MUCH are as follows:

1. The Node Js server is able to handle 90 clients' request simultaneously and asynchronously.
2. The UI of TOO-MUCH is user-friendly from the students' perspective (known through feedback from users).
3. Students were able to login to their profile (shown in Fig. 2), select the problem (shown in Fig. 3), view the problem details and select the programming language to code (shown in Fig. 4).
4. The students coded in either local editor or the online editor (provided by TOO-MUCH) and submitted their solutions (shown in Fig. 5). Approximately 90% of students submitted code file at least once in the allotted time, which were used for evaluation.
5. Around 45% of students resubmitted their code files. Out of which 70% of students were able to improvise their score.

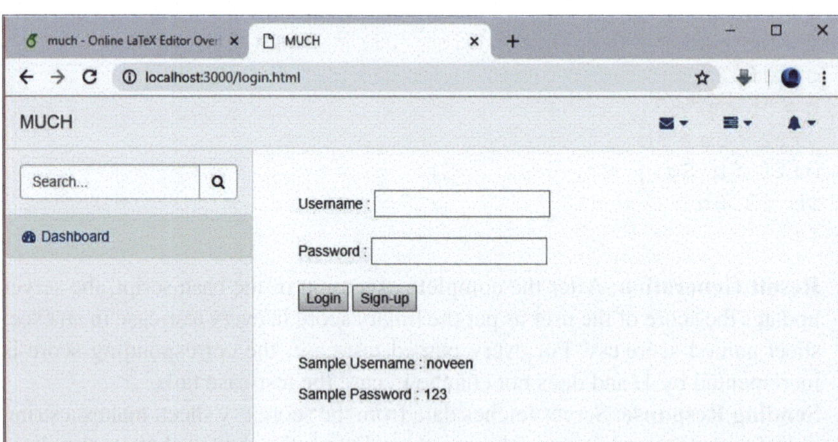

Fig. 2 Login interface

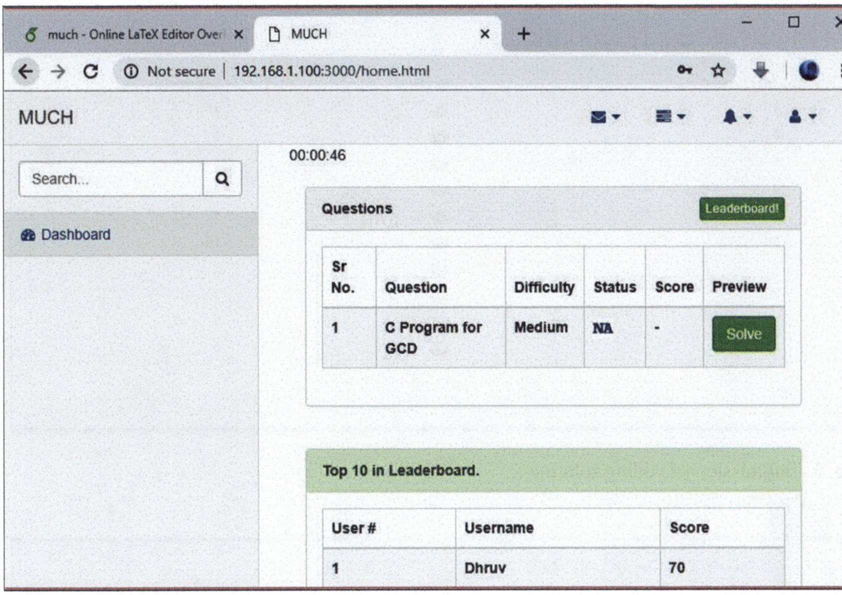

Fig. 3 Coding problem selection interface

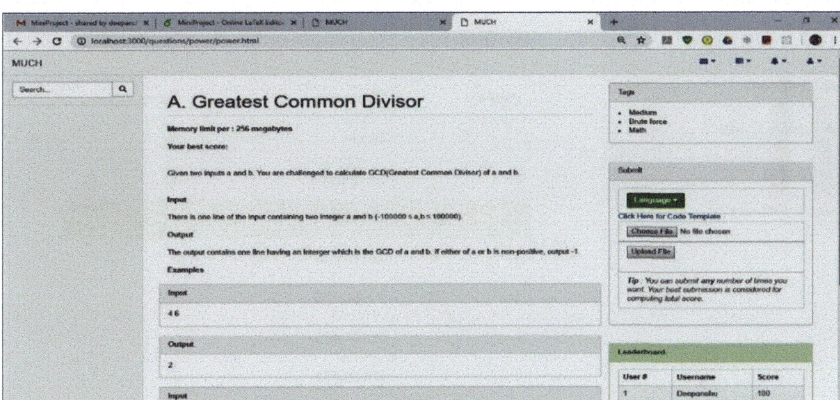

Fig. 4 Details of coding problem to be solved

6. Final results of all students were calculated and students got marks out of 100 on the basis of a number of passed test-cases. The final score of students was displayed on submission in a leader-board (as shown in Fig. 6).

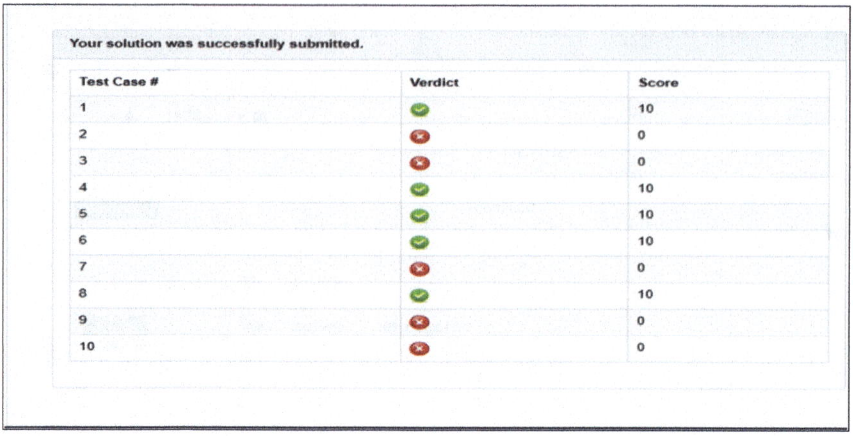

Fig. 5 Submission of coding solution

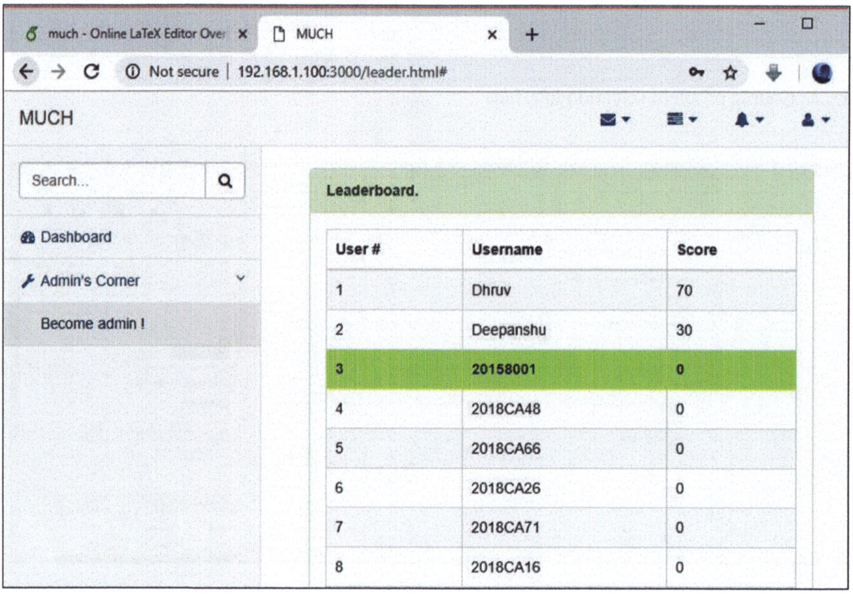

Fig. 6 Leaderboard

5 Conclusion and Future Work

The proposed online judge—TOO-MUCH facilitates the conduction of college-level computer-based examinations. It provides the teachers as well as students an easy, interactive, and interesting system. TOO-MUCH reduces the manual and cumbersome evaluation in the computer lab and brings in the requisite functionality and competitive environment to the students to make their exposure to the bigger-scale coding exams. The proposed software could also be used for lab practice sessions and in organization of programming contests. TOO-MUCH is a blend of state-of-the-art scenarios of programming contests and the code training done in lab. With the employment of TOO-MUCH in academics, there can be quite an improvement in the examination levels in terms of quality, time and resource requirements. In the future, TOO-MUCH could be enhanced for many other languages and tested for large-scale examinations, where students are more than 200.

References

1. Wasik S, Antczak M, Badura J, Laskowski A, Sternal T (2018) A survey on online judge systems and their applications. ACM Comput Surv (CSUR) 51(1):3
2. Rashad MZ, Kandil MS, Hassan AE, Zaher MA (2010) An arabic web-based exam management system. Int J Electr Comput Sci IJECS-IJENS 10(01):48–55
3. Gupta D, Sharma D, Deepthi GS, Patel Y, Subedi P (2019) B. Tech. project: Testing One Online—MNNIT Unified Coding Hub (TOO-MUCH). Tech. rep., MNNIT Allahabad, India
4. ICPC Foundation—ICPC Fact Sheet (2016, Feb) The 40th annual world finals of the ACM ICPC. https://icpc.baylor.edu/worldfinals/pdf/Factsheet.pdf. Accessed 19 Apr 2019
5. Cheang B, Kurnia A, Lim A, Oon W-C (2003) On automated grading of programming assignments in an academic institution. Comput Educ 41(2):121–131
6. Revilla MA, Manzoor S, Liu R (2008) Competitive learning in informatics: the UVa online judge experience. Olympiads Inform 2(10):131–148
7. Edwards SH, Perez-Quinones MA (2008) Web-cat: automatically grading programming assignments. In: ACM SIGCSE bulletin, vol 40. ACM, pp 328–328
8. Wu J, Chen S, Yang R (2012) Development and application of online judge system. In: 2012 international symposium on information technologies in medicine and education, vol 1. IEEE, pp 83–86
9. Kosowski A, Małafiejski M, Noiński T (2007) Application of an online judge & contester system in academic tuition. In: International conference on web-based learning. Springer, pp 343–354
10. Luo Y, Wang X, Zhang Z (2008) Programming grid: a computer-aided education system for programming courses based on online judge. In: Proceedings of the 1st ACM summit on computing education in China on first ACM summit on computing education in China. ACM, p 10
11. Combéfis S, Wautelet J (2014) Programming trainings and informatics teaching through online contests. Olympiads in informatics, vol 8
12. Alemán JLF (2011) Automated assessment in a programming tools course. IEEE Trans Educ 54(4):576–581
13. Antonucci P, Estler C, Nikolić D, Piccioni M, Meyer B (2015) An incremental hint system for automated programming assignments. In: Proceedings of the 2015 ACM conference on innovation and technology in computer science education. ACM, pp 320–325
14. Staubitz T, Klement H, Renz J, Teusner R, Meinel C (2015) Towards practical programming exercises and automated assessment in massive open online courses. In: 2015 IEEE international conference on teaching, assessment, and learning for engineering (TALE). IEEE, pp 23–30
15. Wilcox C (2016) Testing strategies for the automated grading of student programs. In: Proceedings of the 47th ACM technical symposium on computing science education. ACM, pp 437–442

16. Node Js Official Website (2019) https://nodejs.o-rg/en/. Accessed 25 Apr 2019
17. Angular JS Official Website (2019) https://angular-js.org/. Accessed 21 Apr 2019
18. Stack Overflow (2019) http://www.stackover-flow.com. Accessed 23 Apr 2019

Chapter 6
An Overview of Quality of Service with Load Balancing in Cloud Computing Environment

Tazein Azmat and Vijay Kumar Dwivedi

1 Introduction

CCN is a dynamic information technology (IT) paragon that delivers on-demand computing resources to a user over a network infrastructure. The Cloud Service Provider (CSP) offers applications that can be accessed online to users. Such applications can be shared by supplemental than one user. CSPs provide programming interfaces that allow customers to build and deploy applications on the cloud; as well as providing massive storage and computing infrastructure to users. Users usually have no curb on how data is stored in the cloud or where the underlying resources are located. With this limited curb, customers' engrossment and Quality of Service (QOS) expectations from CSPs are spelt outmaneuvering an SLA. Thus, it is imperative to have adequate QOS covenants from a CSP. This paper examines trends in the area of CCN QOS and provides a guide for future research. A review and survey of subsisting works in literature are done to identify these Cloud QOS trends. The finding is that the ultimate expectation of any QOS metrics or model is related to cost perturb for both the CSP and user.

2 Cloud Computing Overview

The web server, the application server, and the database back-end are the constituent elements routinely deployed as virtual server set, i.e., virtual machine (VM) utilized by web-based applications. The utilization of virtualization techniques encourages for the marvelous allocation and relocation of utility components inside the Cloud. It is activated every time when a consumer generates a request. The logic by which even issuance of workload covering the VM is fortified is known as a load-balancing.

T. Azmat (✉) · V. K. Dwivedi
United College of Engineering and Research, Naini, Allahabad, India

V. K. Giri et al. (eds.), *Computing Algorithms with Applications in Engineering*, Algorithms for Intelligent Systems, https://doi.org/10.1007/978-981-15-2369-4_6

The stratification of Cloud models subsists but generally the cloud provider's side of the enterprise's IT side most presented subsisting research works, not a global point of view for the Cloud services customers, suppliers, researchers, and developers. In the comprehensive view of CCN is like any robust system that is defined by these symptomatic, CCN is based on five salient idiosyncrasies: On-demand self-service, broad network access, resource pooling, rapid elasticity, and measured service.

2.1 Layers and Services of Cloud Computing

The diagram is given below, the clashing layers of CCN architecture can be seen in cloud layered architecture in Fig. 1.

The services such as application, platform, infrastructure, etc. are allowed by clashing categories of the cloud. In real-time, regardless of the device or the consumer's location over the Internet, the services are consumed and delivered.

Fig. 1 Cloud layered architecture

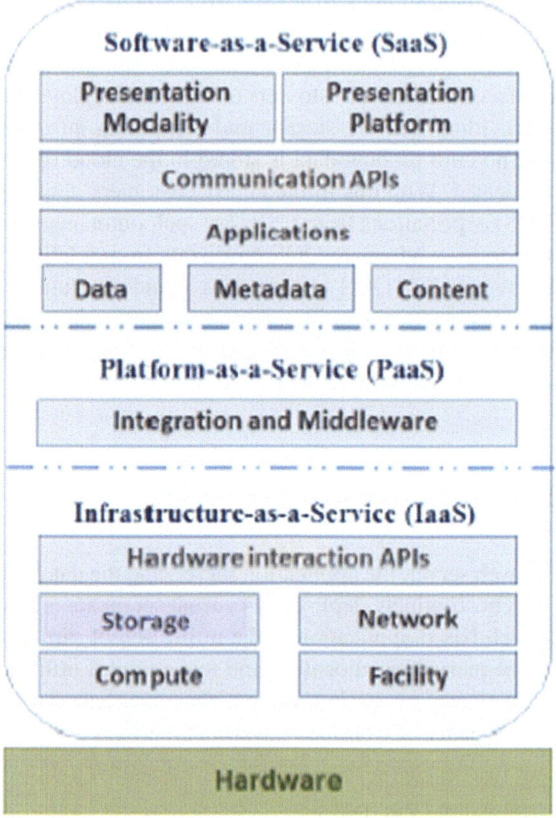

Software as a Service (SaaS), Consumers are allowed by the CCN system to use provider's applications installed and run on a CCN infrastructure.
Platform as a Service (PaaS), Consumers are allowed by the CCN system to create maneuvering programming languages, libraries, services, and tools upheld by the provider deploying into the CCN infrastructure procure applications.
Infrastructure as a Service (IaaS), Consumers are allowed by CCN system to furnish the fundamental computing resources where software is deployed and run arbitrarily, in which not only operating systems and applications are comprehended but also storage, processing, networks are comprehended.

2.2 Deployment of Cloud Computing

To choose the type of Cloud to be implemented is the most important decision. There subsist four types of Cloud that could be deployed which are private, hybrid, public, and community Clouds. The difference between these models can be seen in the following diagram in Fig. 2:

Private Cloud, for exclusionary exploitation by a single organization like business units, this CCN infrastructure is available.
Community Cloud, for the exclusionary use by a peculiar community of consumers from organizations that have shared, perturbs, this CCN infrastructure is available.

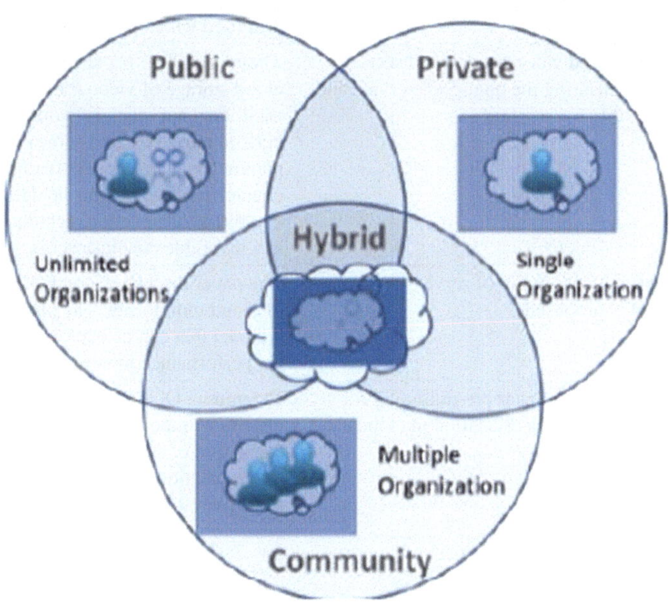

Fig. 2 Cloud deployment models

Public Cloud, for open use by the general public, this CCN infrastructure is available. ***Hybrid Cloud***, it is a composition of two or supplemental distinct public, community, or private CCN infrastructures, which remain a unique entity.

The main benefactions of paper, in a nutshell, are intended as follows: In Sect. 2, it describes the challenges related to exploitation of LB; In Sect. 3, it provides a review of the algorithms focused on LB, QOS superintend; In Sect. 4, it presents an inclusive skull session about the QOS work in CCN. Finally, in Sect. 5, it wraps up the paper and contours the key research acclimatization where future exertions can optimize the virtue of LB tack.

3 Related Work

A coherent delineating study is accomplished to find the related literature reviews by espousing some papers for studies that are disputed. In contemporary years, QOS approaches in CCN have become a paramount talking point in the CCN area and there endure unbolted summonses and cavity which demand future research exploration.

Paper	Proposed scheme	Findings
In [1]	A QOS-driven approach for CCN superscribing attributes of performance and security is proposed	To examine cloud security-based QOS is the main focus of this paper. A model with clashing security environment is proposed which provide QOS to the CCN
In [2]	Cloud-based video-on-demand service model ensuring the quality of service and scalability is proposed	The focus of the paper is the QOS for cloud storage of video services. The traditional approach is to optimize performance, cost and some other parameters. The paper performs a characterization of start-up delays and exploited. A modeling technique to arrive at a favorable conclusion
In [3]	Environment quality of service in the cloud is proposed	The paper discussed cloud QOS in terms of cloud monitoring. The paper proposed a model that can be exploited as a guide for performance monitoring on the cloud
In [4]	Mixed-Integer linear programming for quality of service escalation in clouds is proposed	Examining QOS maneuvering certain criteria is important to cloud providers for escalation of service. The paper exploited two escalation algorithms on some QOS objectives to obtain contrary trade-offs

(continued)

(continued)

Paper	Proposed scheme	Findings
In [5]	A quality of service-based cloud resource provisioning called Q-aware is proposed	In the work, QOS was considered an important factor in the provisioning of resources by cloud providers and then proposed a QOS metric-based technique for analysis of workloads
In [6]	Quality of service approaches in CCN: A systematic mapping study is proposed	QOS is an affair that must be superscribed properly to enhance trust in the cloud. The paper analyzed several QOS approaches to determine the area of supplemental focus and suggested the way forward
In [7]	QOS in Software Defined Networking (SDN): A survey is proposed	Several solutions were suggested in terms of QOS. Relevant surveys were carried out in diverse areas for QOS, highlighting challenges and lessons
In [8]	Secure and quality-of-service-upheld service-oriented architecture for mobile cloud handoff process is proposed	The paper focuses on QOS in terms of mobile CCN, energy and handoff affairs. The paper proposes a four-layer model for energy efficiency and QOS in mobile CCN

4 Quality of Services

The capability of clinching the maximum bandwidth and superintending other network elements like latency, error rate and uptime refers to QOS of CCN. QOS includes the superintending of other network resources by dispensing priorities to a peculiar type of data (audio, video, and file). Salient implementation of QOS needs three paramount components:

- QOS within single CCN element.
- QOS policy and superintend functions to curb end-to-end traffic across CCN.
- Recognition techniques for dovetailing QOS from end-to-end between network elements.

QOS is the ability to provide clashing priority to clashing applications, users, or dataflow, or to covenant a certain level of performance. QOS is the entirety of symptomatic of a service that bears on its potential to appease the stated and tacit needs of the user of the service (service quality assurance). QOS is determined by the fulfilment of each functional and non-functional compelling. Meeting the user's requirement with regards to functionality will depend on the description of the services. The amount of nonfunctional services that have to be considered in CCN service providers is very high.

Five key QOS attributes were identified and these are: reliability, flexibility, performance, security, and usability:

- Reliability superscribe system availability, fault tolerance, user experience levels, privacy, and safety.
- Flexibility entails system scalability, portability, interoperability among others.
- The performance deals with system efficiency, response time, throughput and compliance to pre-agreed conditions of service. It often provides metrics for drawing SLAs and measuring QOS compliance. This attribute is of high importance to both the users and the CSPs, as users are mainly interested in the response time, processing time or throughput of the applications running on top of CSPs, whilst maneuvering these performance metrics to rate the CSPs.
- Security includes accountability, confidentiality, integrity, audit trails, etc.
- Usability focuses partly on user experience and value for money. Beyond these attributes, a paramount perturbs for CCN users and CSPs alike is that of resource dissipation and techniques like monitoring exploitation to identify over-provisioning or underperformance. Dissipation pattern identification is a vital step in ascertaining and maintaining certain levels of QOS.

5 Challenges of CCN in Achieving QOS

The consequences of amalgamating cloud computing and services paragon can help service providers to attain QOS by amalgamating these paragons. The following are the challenges in a CCN framework.

- The challenges such as QOS management and security demands therewithal researches for CCN to superscribe.
- Challenges and affairs in CCN development using service-oriented methods, such as QOS monitoring, availability, confidentiality, integrity and service reliability.
- For QOS monitoring, as the contrary workflow obligatory to host services dynamically and multitudinous providers with contrary approaches need to handle the services, it is arduous to manage contrary QOS requirements.
- The CCN QOS challenges, embracing the provision of QOS and the lack of application management and the lack of approaches to cloud deployment escalation services with contrary QOS metrics like cost and performance.

In CCN multitudinous factors are there which reveals the necessity of further enhancement so as to achieve QOS. Paramount factors are as follows:

Parameter	Challenge	Function	Necessity
Security and privacy	As the network and data applications are mobile and it can lose its curb over its data due to contrary types of resources and security policies	Security and Privacy parameter makes the data and resources secure and confidential over CNN	An effective model should be introduced which can overcome the CCN issue of security and privacy
Performance	Performance of CNN is very crucial for customer satisfaction	It deliberates the capability of the cloud organization	The outcome may be poor due to not have appropriate assets viz. limited bandwidth, memory, diminutive CPU speed, etc
Efficient LB	Efficient LB is mandatory as to have even distribution of traffic load over CCN	It ensures the optimized distribution of resources so that no node undergoes overloading condition	It mainly focuses on the optimized distribution of resources over the virtual machines which are very essential for user contentment
Resource management and scheduling	Proper resource management and scheduling are very crucial to achieve the QOS and customer satisfaction	it can be considered at several levels viz. software, hardware, virtualization level with performance, privacy, security, and other attributes being dependent on the resources and superintend	It includes superintend of disk space, memory, CPU's, cores, VM images, threads, I/O devices, etc.
Require a constant and fast internet speed	To ease the resource allocation and data transfer, it is mandatory to have a constant and fast internet speed	With the help of the cloud system, the business gets the capability to save money on software and hardware but still requires spending additional on the bandwidth	This is not possible to fully exploit the services of the cloud without high-speed communication channels

(continued)

(continued)

Parameter	Challenge	Function	Necessity
Quality of Service management	Quality of Service management is mandatory for customer satisfaction and standard performance. It enables the consumer to throw data into the cloud	Although cloud computing has significantly eased the competency-based method, a lot of challenges of quality of service superintend. Quality of service means the levels of concert, availability, and reliability on hand by the platform and use of infrastructure that hosts it	Quality of service is elementary for cloud consumers and to be expecting from the providers to provide the declared idiosyncrasies

5.1 QOS Parameters

When QOS is considered simultaneously, we need to have knowledge about its essential parameters. To determine the properties of QOS services it is mandatory to define the QOS parameters. There are many qualitative and quantitative QOS parameters are discussed here, which are given as follows:

(1) Throughput (TP) can be understood as the countable maximum number of client requests which is handled efficiently by the system expressed in defined unit time, generally per minute is considered for calculation purpose.
(2) Latency (L) can be understood as the time taken by the system expressed in unit time to store/extract data packets; generally, a millisecond is considered for calculation purpose.
(3) Jitter (J) can be understood as the dissimilitude in the retardation of received packets, generally, a millisecond is considered for calculation purposes.
(4) Response time (RT) can be understood as consumption of time the processor while conducting the received request and capitulating the response, generally, a millisecond is considered for calculation purposes.
(5) Availability (A) can be understood as the calculation of the probability of availability of web service, which is manifested as counts in which the server is available per unit time.
(6) Packet Loss (PL) can be understood as the ratio of the total number of sent packets that failed to hit the destination to the total number of packets that have been undefeated delivered to the destination without fail or loss over peculiar time interim.

(7) Reliability (RL) can be understood as the probability of undefeated satisfactory accomplishment of the task.
(8) Reputation (RP) can be understood as an ethical measurement of stature services
(9) Cost (C) can be understood as payment that a service requester has to do for the operation/services utilization generally cents per request is considered for calculation purpose.

5.2 QOS Methods Analysis

To evaluate the CSPs performance in terms of a large set of QOS parameters, we examined many research works which are explained below. When we talk about the covenant of QOS, we should define the QOS parameters to determine the properties of the service. There are many parameters that can be used for analysis but here Scheduling, Admission control, Monitoring, Resource Allocation, Queuing system is considered.

5.3 Scheduling

In CCN for appurtenant resource allocation, effective scheduling is mandatory. To satisfy the cloud exploiters, it is crucial to opt for the best scheduling approach so that tasks would be able to accomplish optimally and attain QOS necessity. The designation of genuine services and hypothetical time for the establishment of the task is involved in scheduling. It profusions profit to the cloud providers. CCN service scheduling is systemized into two designations, i.e., user level and system level. The dilemma triggered due to dispensing services between both service provider and customer is referred to user-level scheduling whereas the resource management in datacenter is managed by the system-level scheduling.

Work	Proposed model/framework	Strengths	Weakness
[9]	A framework of the Optimal decision rule of data center scheduling for quality of Service	Maneuvering the parameters for the optimal decision rules in the real-time CCN. All applications of sensor-cloud were considered	Limited to the temporal parameters of QOS performance

(continued)

(continued)

Work	Proposed model/framework	Strengths	Weakness
[10]	A novel QOS demand customizable cloud workflow, scheduling model	Highlighting the cloud user's QOS demand preference and their real needs by dividing the model scheduling into two levels	Lacked variation of QOS attributes only temporal parameters
[11]	A set-based particle swarm escalation approach scheduling the problem in CCN	Provides a flexible way for users to specify a QOS escalation preference and define contrary QOS constraints	No security mechanism is implemented after the update of velocity and position

5.4 Admission Control

An overload condition is protected by a mechanism called admission control. When an under peak workload condition is rejected by QOS admission control mechanism to maintain QOS standards and protecting it from its abasement. This is the first type of mechanism out of two which is known as the infrastructure-provider admission control. The other one is known as infrastructure-user admission control. It is utilized as an ultimate admission control mechanism. It is most beneficial when there is some remarkable detain in acquiring resources. The foremost impetus of admission control is to provide robust performance.

Work	Proposed model/framework	Strengths	Weakness
[11]	An admission curb protocol to prevent over-exploitation of system resources for queues and schedules applications based on resource quality engrossment	Maneuvering an open multi-class queueing network to support a QOS-aware admission curb on heterogeneous resources, increasing system throughput	Maneuvering FIFO-ordered scheduling that can provoke problems such as ignoring some available resources or in demand in the queue
[12]	An admission curb test for the availability of resources based on the horizontal elasticity requirement	Considering a mix of trust, risk, eco-efficiency and cost factors for an optimum allocation	One direction to collaborate: from the customer to the provider limited to the customer's application elasticity policy

(continued)

(continued)

Work	Proposed model/framework	Strengths	Weakness
[13]	An admission curb framework based on SLAs to run on a cloud-based database service provider	Embracing transactional and non-transactional application domains. optimizing the inclusive provider's revenue	They did not superscribe the fundamental affair of acquiring distributions without running the queries

5.5 *Monitoring*

A dynamic stalking technique of QOS parameter is known as CCN monitoring i.e. running or hosted on the VM kindred to CCN services. The developer, administrators can facilitate themselves with this technique as to.

- Maintain the Cloud applications and services running at the extortionate efficiency,
- Maintain track of the SLA defiance,
- Perceive the performance of specific QOS parameters,
- Recon the percentage of leave and join operation of services, and
- Discern the phenomenon source of these operations.

Work	Proposed model/framework	Strengths	Weakness
[14]	A novel cloud monitoring techniques and services enabling automated application QOS to superintend under uncertainties	A comprehensive study to develop this monitoring application, describing the idea: motivation, question, and approach and methodology related to the proposed application	No numerical evaluation is presented to validate the QOS model
[15]	A lightweight monitoring framework for public cloud	Resource optimized, covenanting better performance in the public cloud	No monitoring of QOS temporal parameters (response time, processing time)
[16]	Cloud computing solution for cross-layer QOS: A monitoring application for QOS parameters	Providing to users to monitor the performance of service providers	Limited only to the available transfer rate and one-way delay as QOS parameters

5.6 *Resource Allocation*

Resource allocation requires a strategy to be obeyed through which resources are provisioned dynamically. Every time when a strategy is poor or not obeyed genuinely then it may cause a service agonize. The CCN application is designated by the best obtainable resources via resource allocation strategy which demands request and allocation time. It allows integration of CSP, services, and activities to designate slender resources to the tremendous need for CCN.

Work	Proposed model/framework	Strengths	Weakness
[9]	An algorithm for network-aware resource scheduling in clouds and grid	Able to handle multitudinous resource engrossments	In the execution phase, the conditions are assumed ideal. The Failures after resource allocation are not considered
[17]	A complete model to highlight the key affair of resource prediction, allocation, pricing and refunding for cloud brokerage	Up to ten clashing types of services with clashing customers were considered	No varied parameters under supplemental heterogeneous environment
[10]	Profit-based analysis of iaas cloud performance of resource allocation on QOS	A new method proposed to analyze the impact of resources provisioning	The resource allocation strategy is not mentioned and not discussed
[18]	A method for the efficient mapping of resource requests in CCN networks	Appropriates reconfiguration and evaluation strategy has been adopted to deal with a highly dynamic networked cloud environment	No taking into account dynamic heterogeneous environments, infrastructures and the stochastic nature of the corresponding resources

6 Conclusions

- CCN provides scalable, on-demand, elastic, and metered services to cloud users over the Internet.
- There are contrary service types and deployment models to ensure that appropriate services are delivered by the CSP.
- In this paper, QOS, as it relates to CCN, was discussed. Contrary performance metrics were highlighted.

- Contrary Cloud QOS models and applications were also discussed.
- Finally, it can be concluded that QOS could mean completely clashing things when viewed from the CSPs and the users' perspectives, there should be a balance such that the CSP can minimize cost (maximize profit) by efficiently utilizing.
- QOS of a CCN application can be ameliorated by manoeuvring multitudinous techniques such as scheduling by superintending the supply and demand of CSPs. Admission curb technique taking guardianship about the achievement of the CSPs.

References

1. Batista BG, Ferreira CHG, Segura DCM, Leite Filho DM, Peixoto MLM (2017) A QoS-driven approach for cloud computing addressing attributes of performance and security. Future Gener Comput Syst 68:260–267
2. Barba-Jimenez C, Ramirez-Velarde R, Tchernykh A, Rodríguez-Dagnino R, Nolazco-Flores J, Perez-Cazares R (2016) Cloud based video-on-demand service model ensuring quality of service and scalability. J Netw Comput Appl 70(10)
3. Kalekuri M, Rao KR (2016) Environment quality of service in cloud. In: 7th international conference on communication, computing and virtualization 2016, procedia computer science, vol 79, pp 118–126
4. Guérout T, Gaoua Y, Artigues C, Da Costa G, Lopez P, Monteil T (2017) Mixed integer linear programming for quality of service optimization in Clouds. Future Gener Comput Syst 71(1)
5. Singh S, Chana I (2015) Q-aware: quality of service-based cloud resource provisioning. Comput Electr Eng 47:138–160
6. Abdelmaboud A, Jawawi DNA, Ghani I, Elsafi A, Kitchenham B (2015) Quality of service approaches in cloud computing: a systematic mapping study. J Syst Softw 101:159–179
7. Karakus M, Durresi A (2017) Quality of Service (QoS) in software defined networking (SDN): a survey. J Netw Comput Appl 80:200–218
8. Razaque A, Rizvi SS, Khan MJ, Hani QB, Dichter JP, Parizi RM (2017) Secure and quality-of-service supported service-oriented architecture for mobile cloud handoff process. Comput Secur 66:169–184
9. Adami D (2012) A hybrid multidimensional algorithm for network-aware resource scheduling in cloud and grid. In: IEEE ICC communication QoS, reliability and modelling symposium
10. Li J (2012) Profit-based experimental analysis of iaas cloud performance: impact of software resource allocation. In: IEEE ninth international conference on services computing
11. Delimitrou C (2013) QoS-aware admission control in heterogeneous datacenters. In: International conference on autonomic computing
12. Konstantel K (2012) Admission control for elastic cloud services. In: IEEE fifth international conference on cloud computing
13. Xiong P (2011) ActiveSLA: a profit-oriented admission control framework for database-as-a-service providers. In: Proceedings of the 2nd ACM symposium on cloud computing
14. Alhamazani K (2012) Cloud monitoring for optimizing the QoS of hosted applications. In: IEEE 4th international conference on cloud computing technology and science
15. Ma K (2012) Toward a lightweight framework for monitoring public clouds. In: Fourth international conference on computational aspects of social networks (Cason)
16. An OpenNetInf-based cloud computing solution for cross-layer QoS: monitoring part using iOS terminals, IEEE, 2012

17. Aazam M (2014) Advance resource reservation and QoS based refunding in cloud federation. In: Workshop—cloud computing systems, networks, and applications
18. On the optimal allocation of virtual resources in cloud computing networks. IEEE transactions on computers, vol. 62, 2013

Chapter 7
An Overview of Neuro-Fuzzy-Based DTC for Matrix Converter-Fed PMSM Drives

Kannan Selvam and Subhanarayan Sahoo

1 Introduction

The recent scenario in drives suggests the use of PMSM due to various advantages such as higher torque density and power density [1], higher torque for starting, at low speed and wider speed range and higher efficiency over wide speed and torque ranges. So, the PMSM drives are mostly found useful in electric vehicles (EV) [2], hybrid electric vehicles (HEV) [3] and aerospace [4] applications.

When high-performance and high-efficiency are the main concerns, it is usually characterized by means of smooth rotation over the wide speed range, effective torque control while speed is zero, and quick acceleration and deceleration. In PMSM drives to obtain the above said control, vector control techniques are employed to the digitally controlled power converter such that the stator supply of PMSM will be maintained according to load variations. The DTC method or field-oriented control (FOC) method is generally used along with the inverter circuit in order to achieve the vector control.

But main drawback of PMSM drive is pulsating torque which is produced due to high-frequency component and commutation transient between every 60° angle. These causes are mainly induced while we appoint inverter-based control which must require the pulse width modulation (PWM) strategy. So, MC topology has been considered and discussed here. Also, DTC has been chosen among FOC [5], as it is having fast dynamic response than FOC, and it required less parameters, so it is easy to implement whereas FOC needs more measured components to further calculations.

K. Selvam (✉) · S. Sahoo
Gujarat Technological University, Ahmedabad, India

S. Sahoo
Department of EE, Adani Institute of Infrastructure Engineering, Ahmedabad, India

V. K. Giri et al. (eds.), *Computing Algorithms with Applications in Engineering*, Algorithms for Intelligent Systems, https://doi.org/10.1007/978-981-15-2369-4_7

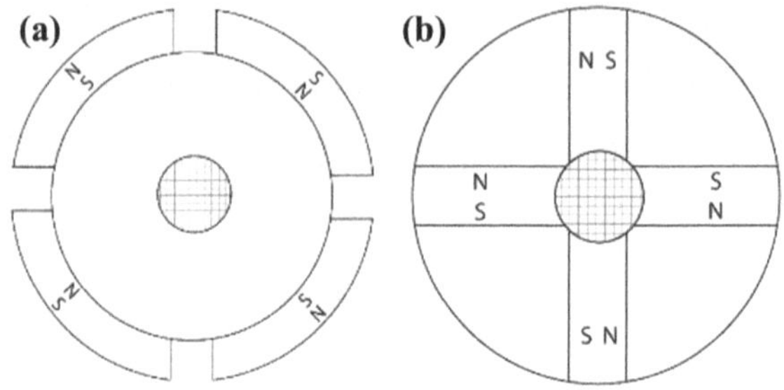

Fig. 1 PMSM rotor configurations: **a** surface-mount; **b** buried [6]

1.1 Permanent Magnet Synchronous Motor (PMSM)

In PMSM, the rotor is excited from magnets placed over it, instead of DC excitation circuit. So that the core loss and iron losses are reduced and the absence of the field losses helps to improve the thermal characteristics as well [6]. Usually, the PM motors are classified based on back-EMF. If the machine produces sinusoidal waveform, then it is called as PMSM and the back-EMF is trapezoidal for PMDC motor. Further, PMSM can be categorized as surface-mounted PMSM, in which the magnets are fixed on the surface of the rotor and interior magnet type PMSM, in which magnets are fixed inside the rotor like the salient pole type (Fig. 1).

1.2 Matrix Converter

The regular power converter such as inverter, cyclo-converter cannot produce the sinusoidal output and also need power capacitors for energy storages, i.e., DC-link capacitor in case of rectifier-inverter combination, whereas the (MC) directly works.

The main advantage of MC is, input and output waveforms mostly sinusoidal, due to which the higher-order harmonics are reduced and sub-harmonics are absent. Also, it has bidirectional energy flow capability so that control over power factor at input side is easiest [7]. The major advantages are [8], essential four-quadrant operation, pure sine-in and pure sine-out waveforms and displacement factor is always unity.

MC uses $m \times n$ switches, which are bidirectional and usually composed with IGBTs connected in anti-parallel [9] as shown in Fig. 2. The input phases should be chosen at least three, whereas the number of output phase n can vary. Using nine switches, the MC can make 512 (2^9) different switching states, but only few of them are useful. The relation between the input and output voltage and current of MC is,

$$\begin{pmatrix} v_a \\ v_b \\ v_c \end{pmatrix} = \begin{pmatrix} S_{Aa}(t) & S_{Ab}(t) & S_{Ac}(t) \\ S_{Ba}(t) & S_{Bb}(t) & S_{Bc}(t) \\ S_{Ca}(t) & S_{Cb}(t) & S_{Cc}(t) \end{pmatrix} \begin{pmatrix} v_A \\ v_B \\ v_C \end{pmatrix} = M \,.\, v_i$$

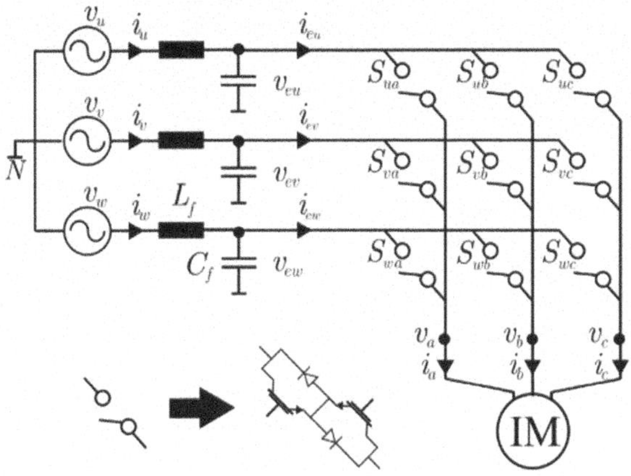

Fig. 2 Matrix converter topology [9]

and

$$\begin{pmatrix} i_a \\ i_b \\ i_c \end{pmatrix} = \begin{pmatrix} S_{Aa}(t) & S_{Ba}(t) & S_{Ca}(t) \\ S_{Ab}(t) & S_{Bb}(t) & S_{Cb}(t) \\ S_{Ac}(t) & S_{Bc}(t) & S_{Cc}(t) \end{pmatrix} \begin{pmatrix} i_A \\ i_B \\ i_C \end{pmatrix} = M^{\mathrm{T}} . i_o$$

where $s_{kj}(t)$ is the state of switch S_{kj}, $k \in \{A, B, C\}$, $j \in \{a, b, c\}$, and $\boldsymbol{M}^{\mathrm{T}}$ is the transpose of transfer matrix $\boldsymbol{M}$. The single switch follows the pattern as mentioned below:

$$S_{kj} = \begin{cases} 1; \text{ Closed} \\ 0; \text{ Open} \end{cases} \quad S_{Aj} + S_{Bj} + S_{Cj} = 1$$

Basically, the MC switching states combinations are must follow the following rules for the safest operation: (i) the input phases should not be short-circuited, since the converter is supplied by a voltage source and usually feeds an inductive load, and (ii) the output currents should never be interrupted. So that only one switch per output phase can be operated at any instant. With respect to this constraint, only 27 states are possible switching combinations.

2 Mathematical Modeling of PMSM & MC

2.1 PMSM Model

In the arbitrary reference frame, the dynamic equations of PMSM (Surface mounted) can be derived as [10]

$$\frac{\mathrm{d}\omega_r(t)}{\mathrm{d}t} = -\frac{B_m}{J_m}\omega_r(t) + \frac{1}{J_m}(T_e(t) - T_m)$$

$$\frac{\mathrm{d}i_d(t)}{\mathrm{d}t} = -\frac{R_s}{L}i_d(t) + \omega_e(t)i_d(t) + \frac{1}{L}v_d(t)$$

$$\frac{\mathrm{d}i_q(t)}{\mathrm{d}t} = -\frac{R_s}{L}i_q(t) - \omega_e(t)i_d(t) - \frac{1}{L}\omega_e(t)\lambda_m + \frac{1}{L}v_q(t)$$

where $\omega_r(t)$ = rotor speed, $\omega_e(t)$ = electrical rotational rotor speed, $i_d(t)$ and $i_q(t)$ = stator current in *d-q* frame. The electromagnetic torque of PMSM is derived as [11]

$$T_e = \frac{3p|\psi_s|}{4L_dL_q}\left[2\psi_f L_q \sin\delta - |\psi_s|(L_q - L_d)\sin 2\delta\right]$$

where δ = displacement angle between the flux linkage of stator and PM ψ_f = flux of PM, L_d and L_q = stator inductances in *d-q* frame, p = No. of pole pairs.

3 Conventional Approach of DTC for MC

The representation of the conventional MC-fed PMSM is depicted in Fig. 3. The measured components are transformed in terms of synchronous reference frame from the so-called arbitrary reference frame of PMSM. The torque and flux are calculated with the help of Eq. (6).

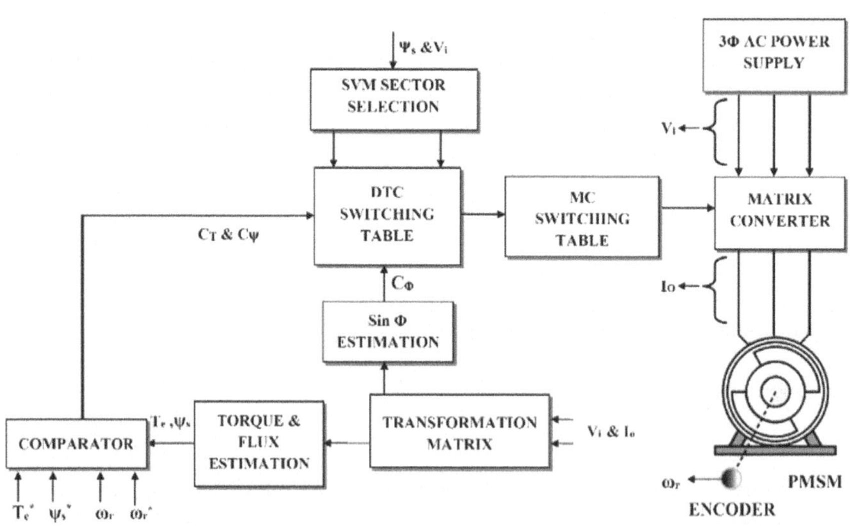

Fig. 3 Conventional approach of DTC for MC-fed PMSM

Table 1 DTC switching table

C_ψ	C_T	Stator flux sectors h_θ					
		①	②	③	④	⑤	⑥
+1	+1	V_2	V_3	V_4	V_5	V_6	V_1
	0	V_7	V_0	V_7	V_0	V_7	V_0
	−1	V_6	V_1	V_2	V_3	V_4	V_5
−1	+1	V_3	V_4	V_5	V_6	V_1	V_2
	0	V_0	V_7	V_0	V_7	V_0	V_7
	−1	V_5	V_6	V_1	V_2	V_3	V_4

These values are compared with the reference values such that C_T and C_ψ are determined. The desired space vector modulation (SVM) sector has been further chosen from Table 1. Then, required MC switching state is chosen based on the virtual SVM sector from offline-calculated table as mentioned in [12]. Normally the MC-DTC scheme uses the hysteresis and SVM to minimize the torque ripple and speed variations in PMSM. In which the main drawback is, virtual voltage vectors are obtained using discrete SVM and also the switching table is completely prepared offline with more accurate calculation. In order to achieve low torque ripples, fixed switching frequency and improved power factor with ease of control, fuzzy sliding mode control (FSMC) [13] and adaptive network-based fuzzy inference system (ANFIS) [14] are discussed here.

4 Role of Fuzzy Logic Control

In order to rectify the burden of offline calculation and to achieve accuracy, Fuzzy Logic Control (FLC) is introduced as shown in Fig. 4.

4.1 Fuzzy Sliding Mode Control (FSMC)

The notable feature of a FSMC is less sensitive to load variation once the switching operation has started [13]. Also, a small derived value of defined ambiguities will be enough to maintain switching operation smooth.

i. Structure of FNN

Input layer: For each *i*th node, the net input and the net output are expressed as,

$$\text{net}_i^1 = x_i^1,\, y_i^1 = f_i^1(\text{net}_i^1), \quad i = 1, 2$$

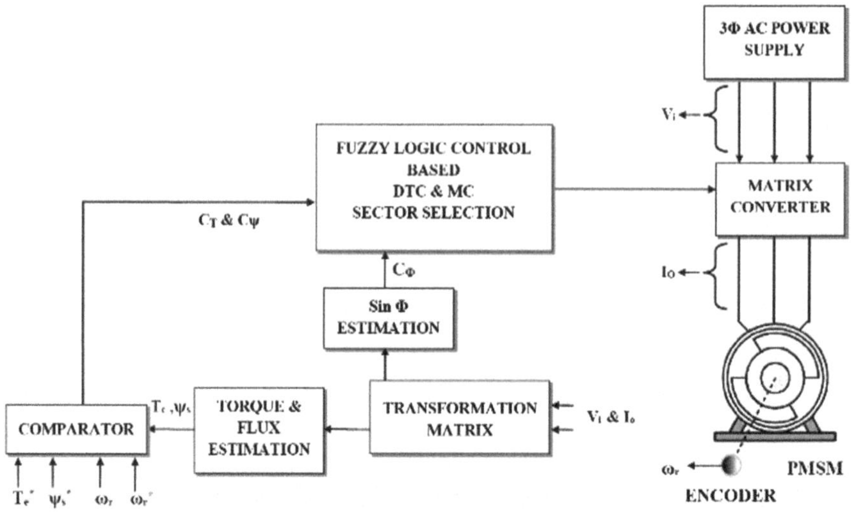

Fig. 4 Fuzzy logic-based approach of DTC for MC-fed PMSM

Membership layer: Since each node performs a membership function, the Gaussian approach is adopted to the membership function. So jth node can be expressed as,

$$\text{net}_j^2 = -\frac{(x_i^2 - m_{ij})^2}{(\sigma_{ij})^2},\ y_j^2 = f_j^2(\text{net}_j^2) = \exp(\text{net}_j^2),\ \ j = 1, 2, \ldots, n$$

where m_{ij} = mean deviation, σ_{ij} = standard deviation

Rule layer: In this, every node multiplies input signals and provides the result of the product as output. Every kth rule node,

$$\text{net}_k^3 = \prod_j \omega_{jk}^3 x_j^3,\ y_k^3 = f_k^3(\text{net}_k^2) = \text{net}_k^3,\ \ k = 1, 2, \ldots, l$$

External layer: Here every node labeled as Σ and it calculates the complete output as an real value of the input signals, so

$$\text{net}_o^4 = \sum_k \omega_{ko}^4 x_k^4,\ y_o^4 = f_o^4(\text{net}_o^4) = \text{net}_o^4,\ \ o = 1$$

ii. Live algorithm

The main motive behind the live algorithm is just to understand online, how apparently the gradient vector is achieved. Whereas each element is derivative part of such energy function with respect to a pre-mentioned parameter of the machines model by using the chain rule. The proposed method minimizes the torque ripple finely

[13] and also helps to maintain the fast response of conventional DTC method. Also, frequency of switching is controlled and can be maintained.

4.2 Adaptive Network-Based Fuzzy Inference System (ANFIS)

The ANFIS is derived from combining FLC and neural network together. It mainly helps to reduce the difficulties found in the MC-DTC switching issues. In this approach [14], the speed of the motor has been determined by using *d-q* components of current and voltage [15, 16]. The generated data is used in FLC block and according to the rules defined by user, the membership function parameters have been produced. This system is developed [17] as follows:

i. Training Step: The derived arbitrary components of voltage and current of the drive have been sent to the neural network to convert into vector form. With the help of the least square method and backward method, the data have been made compatible with training algorithm as per the reference parameters. By using the trained data, the FIS generates the control fuzzy rules and parameters of membership function [18].
ii. Testing Step: The appropriate speed estimation of PMSM has been derived from fuzzy control rules and parameters of membership function [19–21]. The five-layer approach of ANFIS algorithm has been shown in Fig. 5 [22], where square and circle symbols are called as adaptive and fixed nodes, respectively.

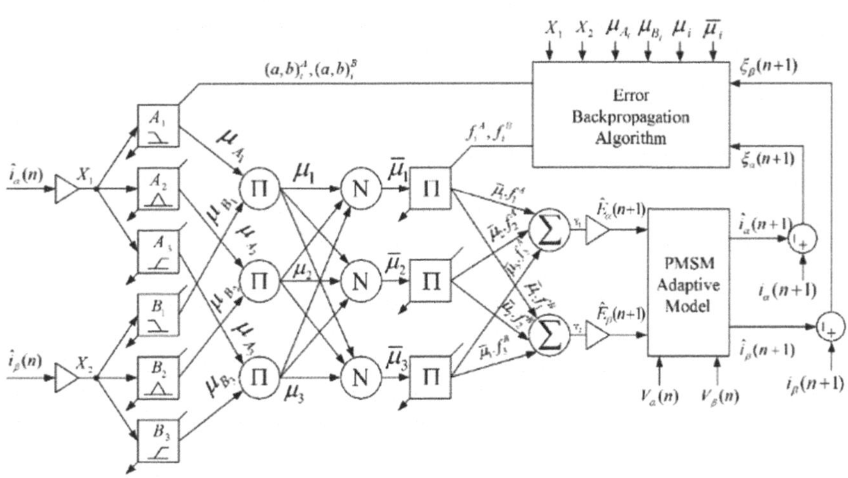

Fig. 5 Schematic of ANFIS architecture [22]

5 Comparison

The main advantage of sliding mode controller for PMSM [13] is minimized torque ripple, very low flux variation, rigid control over variations and constant switching frequency. The results show that the sliding mode control of FLC can deliver higher dynamic performance and rigid with respect to parameter changes in the drive. Also, the chattering phenomenon has much reduced, so that switching frequency maintained constant and it helped to improve the power factor. Whereas ANFIS [14] shows the effectiveness in the speed and torque control and it completely relies on a stator electrical components. Since the ANFIS having the self-learning mechanism, it is more convenient and economically could be applied in PMSM drives.

6 Conclusions

In this paper, latest trends of neuro-fuzzy controlling techniques are discussed for the DTC-MC-fed PMSM drives. In order to narrow down the comparison, FSMC [13] and ANFIS [14] are taken into account, and their performance evaluation is compared. Both the algorithms may estimate the DTC-MC switching sectors position effectively. But compared to FSMC, the ANFIS find better [22] in performance and immunity against the variation in electromagnetic torque ripples.

References

1. Thomas M, Seok-Hee H, Ayman M, Jei-Hoon B, Metin A, Mustafa K, Wen L (2006) Design and experimental verification of a 50 kW interior permanent magnet synchronous machine. In: Conference record of the IEEE industry applications conference 41st IAS annual meeting, Tampa, FL, pp 1941–1948
2. Abir H, Yemna B, Seifeddine B, Mohamed N (2019) Sliding mode observer based sensorless control of five phase PMSM in electric vehicle. In: 19th international conference on sciences and techniques of automatic control and computer engineering (STA), Tunisia
3. Husain I (2003) Electric and hybrid vehicles design fundamentals.CRC Press. ISBN 0-8493-1466-6
4. Xinyun L, Andrew J (2018) Active stabilisation of a PMSM drive system for aerospace applications. In: IEEE power electronics specialists conference, pp 283–289
5. Fatih K, Ismail T, Faruk M, Riza G (2013) Comparative performance evaluation of FOC and DTC controlled PMSM Drives. In: IEEE Proceedings of 4th international conference on power engineering, energy and electrical drives, Istanbul, Turkey, pp 705–708
6. Pragasen P, Krishnan R (1991) Application characteristics of permanent magnet synchronous and brushless dc motors for servo drives. IEEE Trans Ind Appl 27(5)
7. Alesina A, Venturini M (1989) Analysis and design of optimum-amplitude nine-switch direct Ac-Ac converters. IEEE Trans Power Electr 4(1):101–112
8. Casadei D, Grandi G, Serra G, Tani A (1993) Space vector control of matrix converters with unity input power factor and sinusoidal input/output waveforms. In: Proceedings of IEEEPE'93, vol 7, pp 170–175

9. Mehdi F (2015) Power electronic converters and systems—frontiers and applications. The Institution of Engineering and Technology. ISBN 978-1-84919-826-4
10. Zhoua J, Wang Y (2005) Real-time nonlinear adaptive back stepping speed control for a PM synchronous motor. Control Eng Pract 13:1259–1269
11. Zhong L, Rahman M, Hu W, Lim K (1997) Analysis of direct torque control in permanent magnet synchronous motor drives. IEEE Trans Power Electron 12(3):528–536
12. Xia C, Zhao J, Yan Y, Shi T (2014) A novel direct torque control of matrix converter-fed PMSM drives using duty cycle control for torque ripple reduction. IEEE Trans Ind Electron 61(6):2700–2713
13. Hongkui L, Qinlin W (2010) Sliding mode controller based on fuzzy neural network optimization for direct torque controlled PMSM. In: Proceedings of the 8th world congress on intelligent control and automation, pp 2434–2438
14. Waleed A, Gomaa F, Shawki F (2018) Adaptive neuro-fuzzy inference system based field oriented control of PMSM & speed estimation. In: Twentieth international middle east power systems conference (MEPCON), pp 626–631
15. Ushanandhini R (2016) Adaptive neuro fuzzy inference system with self turning for permanent magnet synchronous motor. Int J Emerg Technol Eng Res (IJETER) 4(3)
16. Goswami Y, Deshmukh S (2015) adaptive neuro fuzzy inference based direct torque control strategy for robust speed control of induction motor under highly variable load conditions. Int J Sci Res (IJSR) 4(12)
17. Ashok K, Kodad S, Sankar R (2010) Modeling, design & simulation of an adaptive neuro-fuzzy inference system (ANFIS) for speed control of induction motor. Int J Comput Appl (0975–8887) 6(12):29–44
18. Jang J (1993) ANFIS: adaptive-network-based fuzzy inference systems. IEEE Trans Syst Man Cybern 23(3):665–684
19. Md M, Islam M (2009) Development and Implementation of a new adaptive intelligent speed controller for IPMSM drive. IEEE Trans Ind Appl 45(3)
20. Giribabu D, Kumar K, Chandra S (2015) ANFIS based modified voltage model RFMRAS speed observer for induction motor drive. In: International conference on energy, power and environment, IEEE
21. Mohsen S, Davood A, Marco R, Jose R (2017) A computationally efficient lookup table based FCS-MPC for PMSM drives fed by matrix converters. IEEE Trans Ind Electron 64(10):7645–7654
22. Mukhtiar S, Ambrish C (2010) Comparative study of sliding mode and ANFIS based observers for speed & position sensorless control of variable speed PMSG. In: Proceedings of CCECE' 10, pp 1–4

Chapter 8
Switching of Solar PV Fed Cascaded H-Bridge Multilevel Inverter from Grid Connected to Islanding Mode and Its Control

Alok Kumar Singh

1 Introduction

Development in the field of renewable energy is the present requirement of the world. SPV system which is a clean and green source of renewable energy connected to utility network has to maintain some standards with respect to safe running performance, power quality along with islanding protection. In addition, grid side converter and their control structure [1]. When the SPV system operated independently of the grid, then system is known as islanded. For provide energy within the islanded system the H-bridge inverters control (a phase-locked loop algorithm) have to became enough capable to provide voltage and frequency regulation and maintain synchronization in the absence of the grid voltage [2, 3].

Before islanding operation makes sure about the changes in the amplitude of grid voltage, phase, and frequency and also about the harmonics. Due to this issue which also mentioned in IEEE standards by a few national codes, utility does not like to perform islanding operation. So many anti-islanding algorithms are proposed in different literature [4–7]. In case of an islanding utility grid is not available. The SPV system units should be able to fulfill the local load power demand. This can be done by adjusting reference which is the output voltage of H-bridge inverter in terms of delivered power [8].

This SPV system work in current control mode for grid-connected system and in voltage control mode for islanding system. In SPV system, H-bridge inverter must

A. K. Singh (✉)
Electrical Engineering Department, Adani Institute of Infrastructure Engineering, Ahmedabad, Gujarat 382421, India
e-mail: alok.singh@aiim.ac.in

V. K. Giri et al. (eds.), *Computing Algorithms with Applications in Engineering*, Algorithms for Intelligent Systems, https://doi.org/10.1007/978-981-15-2369-4_8

be able to detect the islanding or time of separation or cut off from the main grid and start operating in voltage control mode. At this time, the main goal is to provide constant voltage to existing local load [9]. In voltage control mode at a time they control the harmonics of line current along with LCL filter capacitor voltage in output side of H-bridge inverter and by regulating the capacitor voltage output power of the existing generating system also controlled. This also gives hybrid control method, with the help of this method they do not require any type of extraction process to compensate local harmonic loads [10]. In micro-grid along with SPV generating system some load also connected. At the time of islanding micro-grid disconnected from main grid [11]. Some emergency load connected with micro-grid. Then to save that load make sure other than emergency load no other type of load connect from inverter output otherwise inverter maybe a collapse.

In grid-connected mode of operation current tracking and for it reference generation is one issue. To address these problems, a simple control strategy for current control is used. In this, also see the performance for normal solar radiation condition and partial shading condition. Voltage fluctuation is the major problem in the islanding mode of SPV generation systems. To address these problems, a simple control strategy for voltage control is used then sees the performance for switching from one mode to other like grid-connected mode to islanding mode with voltage control.

2 Control Technique Used

Figure 1 shows the complete control and circuit diagram of the system. In the power circuit, first part from right in Fig. 1 shows the SPV system which is a series and parallel combination of SPV module to achieve the desired voltage and current output. The second part shows the boost DC/DC converter which ensures maximum power point tracking (MPPT) from SPV system even under partial shading conditions (PSC). The third part shows the IPM (intelligent power module) first leg name as W use as a part of DC/DC converter and remaining two legs V and U are H-bridge or a part of DC/AC converter. Here we proposed to use cascaded H-bridge multilevel inverter (CHBMI) for transfer of large power generated from the SPV system. With one H-bridge three-level of output is generated. With cascading of two H-bridges five-level of output is obtained and with cascading of three H-bridges seven-level of output is obtained and so on. The number of output levels is equal to $(2n + 1)$ where n is the number of cascaded H-bridge. Fourth part from right shows the emergency load which may be resistive, inductive or nonlinear and the last part in Fig. 1 is the grid [12].

The first part from the right shows the control diagram in Fig. 1. In the control circuit, upper part of the circuit is for grid-connected system and lower part of the control circuit is for the islanded condition. At a time, any will work according to control signal at control terminal. If control signal is one (1) then it operates in current control grid-connected mode and if zero (0) then operates as voltage control or islanding mode. In control circuit, upper part shows a series and parallel

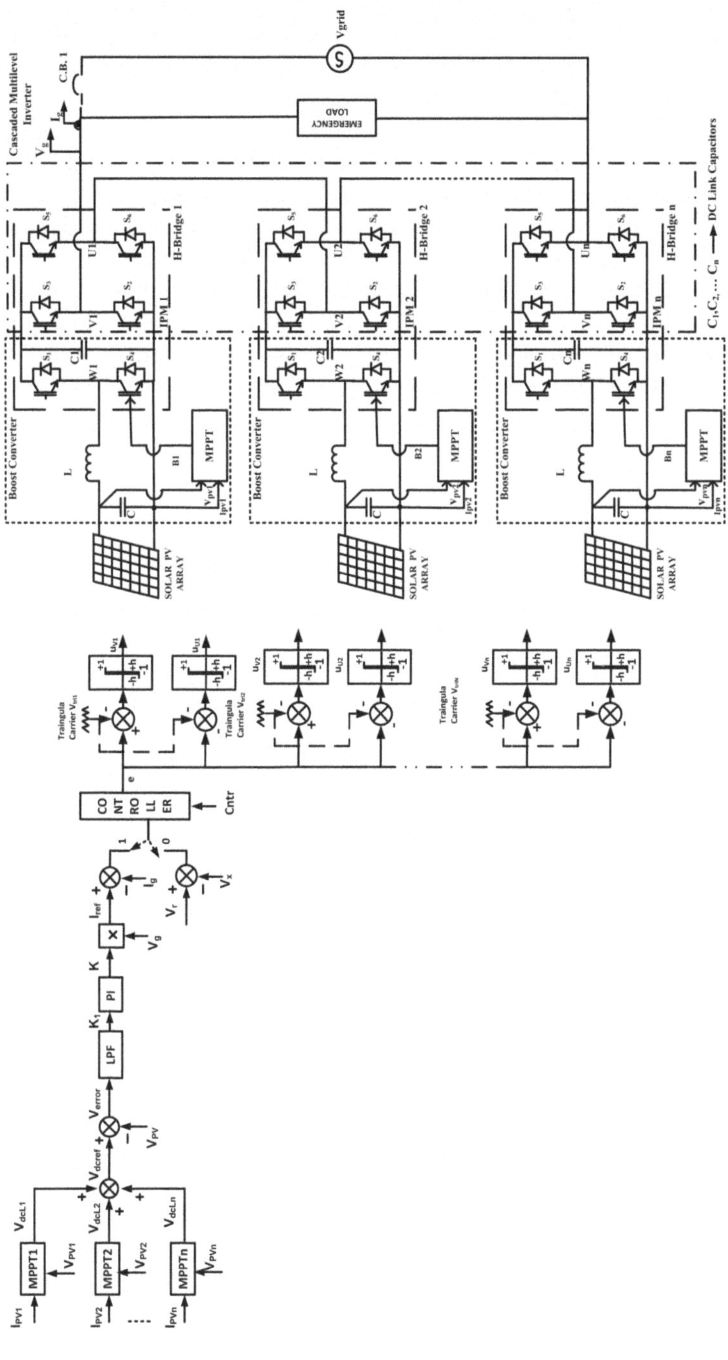

Fig. 1 Solar PV array, boost converter, cascaded multilevel inverter, emergency load, and grid

combination of solar photovoltaic module voltage and current is used as an input for achieving MPPT and accordingly set dc link voltage. Addition of all dc link voltage gives reference for V_{dc}. Reference dc voltage V_{dcref} compares with actual PV voltage V_{PV} then voltage error V_{error} generates V_{error} signal pass through low pass filter to remove noise signal. Filter signal K_1 then passes through PI controller that generates gain constant K. Gain constant K multiply with grid voltage V_g to generate current reference I_{ref}. V_g also used phase lock inverter current to grid voltage V_g. I_{ref} compares with grid current I_g gives error signal [13]. When the controller allows operating in islanding mode then lower control circuit will start work. In this compare existing reference value of voltage controller to output voltage of cascaded H-bridge multilevel inverter and generate an error signal. According to controller error signal, hysteresis controller generates a pulse for CHBMI switch. For this error, signal compare with triangular carrier signal which is generally high-frequency signal here we use 3000 Hz. Then error signal passes through hysteresis band to generate pulse. That pulse controls the switching of CHBMI switch and their output [14].

3 Simulation Result

Perform simulation for grid-connected mode of operation. In this also perform simulation for normal solar radiation condition and partial shading condition. Then perform simulation for switching from one mode to other like grid-connected mode to islanding mode without voltage control and with voltage control. The performance shows the importance of voltage control operation in islanding mode and the effectiveness of the control technique used in both grid-connected and islanding mode of operation.

3.1 Grid-Connected Mode

The grid-connected mode of operation for current tracking and for it reference generation PI controller with phase lock loop is used. This is a simple control strategy for current control. This also performs for normal solar radiation condition and partial shading condition.

Under Normal Solar Radiation Condition. In Fig. 2a shows Solar PV array MPPT performance during normal solar radiation condition and from result output voltage of SPV tracking V_{mpp} with the help of P&O MPPT algorithm. In Fig. 2b shows during normal solar radiation conditions after stable operation constant grid voltage and Inverter current and Fig. 2c zoom view of Fig. 2b shows clearly the effectiveness of the control technique used. Inverter current phase lock with grid voltage. Inverter current and grid voltage both have constant value so transfer the power to the grid is also constant and smooth.

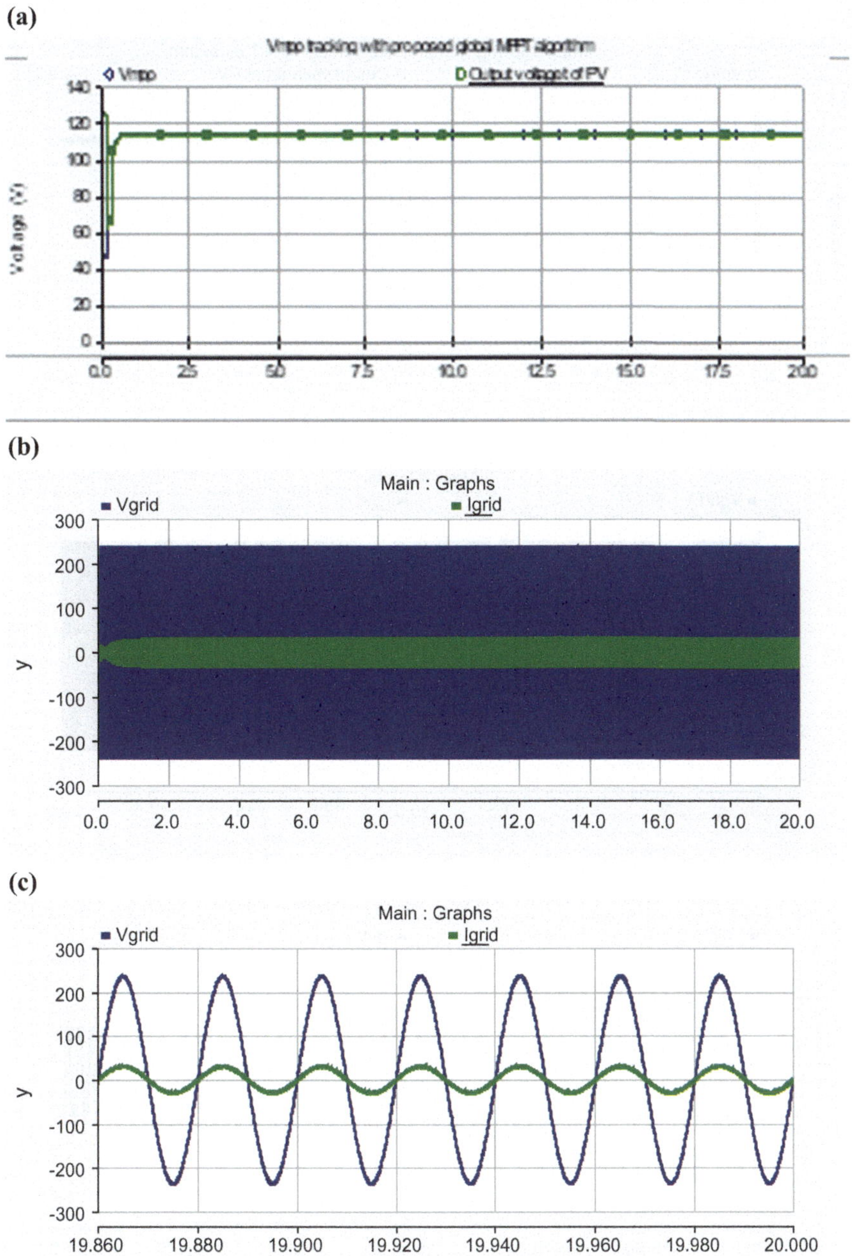

Fig. 2 Solar PV array during normal solar radiation condition **a** V_{mpp} tracking with P&O MPPT algorithm, **b** After stable operation constant grid voltage and inverter current and **c** Zoom view of **b** shows phase lock of inverter current with grid voltage

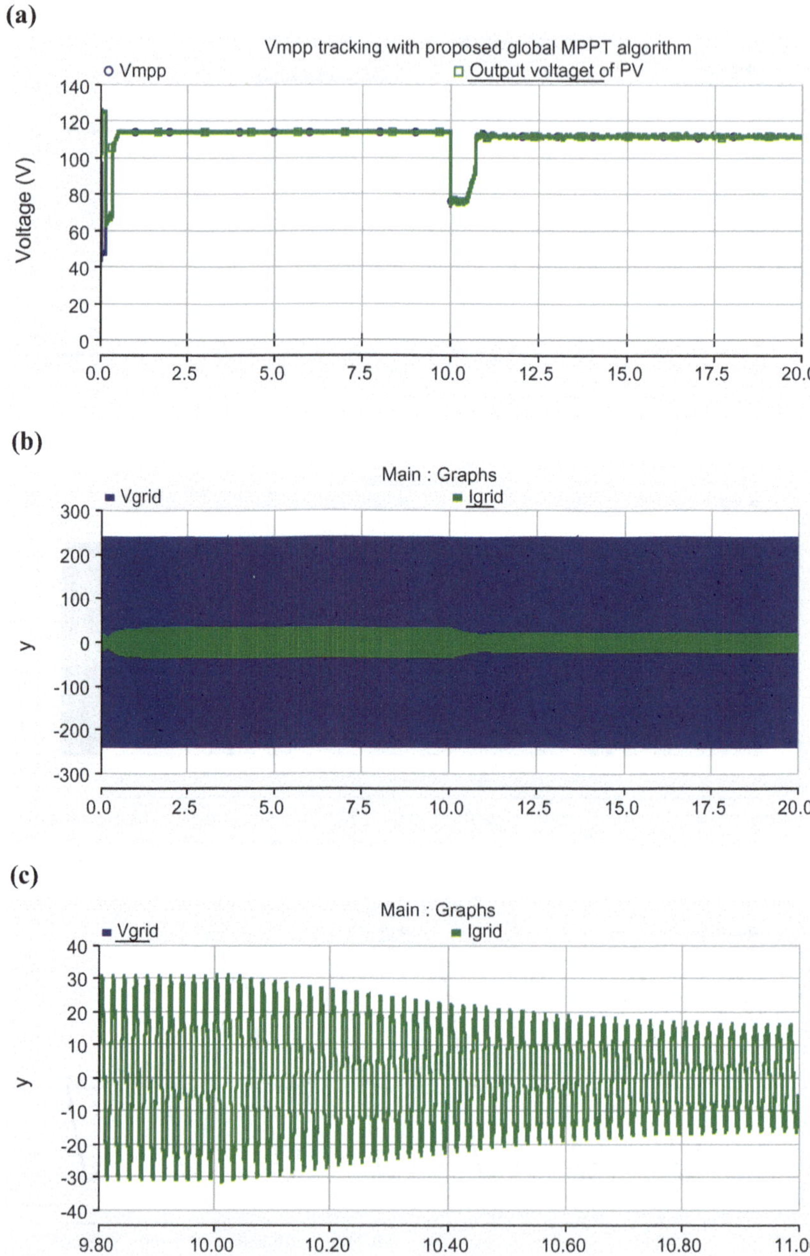

Fig. 3 Solar PV array during partial shading conditions (PSC) opération **a** V_{mpp} tracking with P&O MPPT algorithm, **b** After stable operation constant grid voltage and inverter current and **c** Zoom view of **b** without grid voltage shows changing current from one stable region to another stable region and **d** Zoom view of **b** shows phase lock of inverter current with grid voltage

(d)

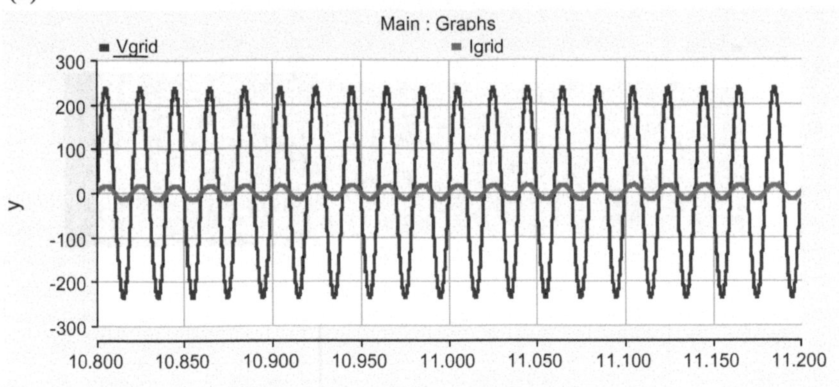

Fig. 3 (continued)

Under Partial Shading Conditions (PSC). Figure 3a shows solar PV array MPPT performance during partial shading condition and from result output voltage of SPV tracking V_{mpp} with the help of P&O MPPT algorithm. In Fig. 3b shows during partial shading condition after stable operation constant Inverter current and grid voltage again become constant and stable within second. Figure 3c zoom view of Fig. 3b without grid voltage shows changing current from one stable region to another stable region. Figure 3d zoom view of Fig. 3b shows clearly the effectiveness of the control technique used. Inverter current phase lock with grid voltage. Grid voltage and inverter current both have constant value so transfer the power to the grid is also constant and smooth.

3.2 Switching from Islanding Mode to Grid-Connected Mode with Voltage Control

In this, first run program for 12.5 s as islanding mode of operation. Then, switch to grid-connected mode. So, in first-half reference generate for voltage tracking and then in second half in grid-connected mode generate a reference for current tracking. In Fig. 4a, results clearly show smooth and stabile operation in output voltage and current tracking of CHBMI.

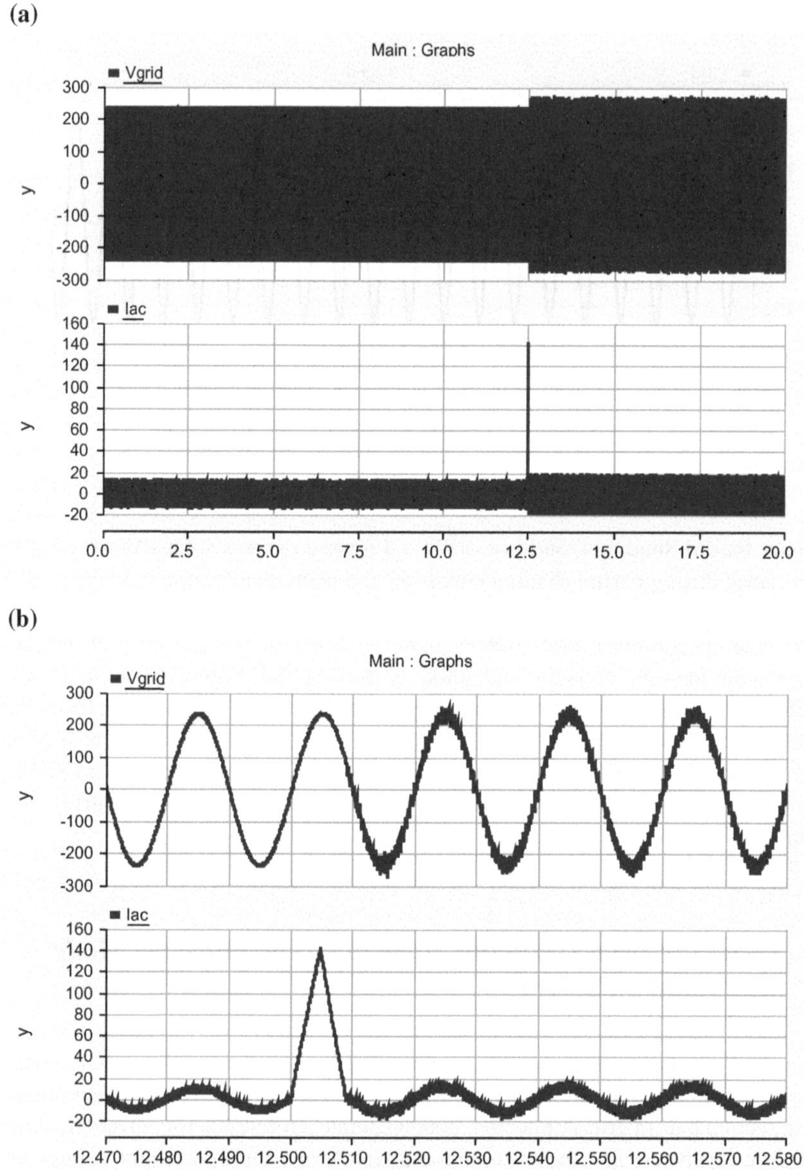

Fig. 4 Current and voltage tracking, **a** Up to 12.5 s generate current reference and track after that generate voltage reference and track, **b** Zoom view of **a**

Table 1 System parameters

Parameter	Rating
SPV array	4.5 kW
Inductor (L)	10 mH
DC link capacitor (C_1)	3300 μF
Switching frequency of DC-DC (boost) converter	20 kHz
Switching carrier frequency of H-bridge inverter, f_c	3 kHz
Reference voltage state vector, Z_{m2}	$167\sqrt{2}\sin(2 \cdot \pi \cdot 50 \cdot t)$
Filter inductance, L_o	20 mH

The Magnitude of output current is seven times from initial value that damages equipment and load associated with this type of system. Figure 4b zoom view of Fig. 4a up to 12.5 s generate voltage reference and track after that second generate current reference and track. Results show clearly the effectiveness of the control technique used (Table 1).

4 Conclusions

In this paper, the performance of CHBMI with hysteresis current and voltage control has been demonstrated and found to be effective solution for grid and islanding mode of operation. The proposed topology produces the source current become sinusoidal so as a result the harmonics also reduced. The performance for switching from one mode to other like grid-connected mode to islanding mode with voltage control shows the importance of voltage control operation.

References

1. Blaabjerg F, Teodorescu R, Liserre M, Timbus AV (2006) Overview of control and grid synchronization for distributed power generation systems. IEEE Trans Ind Electron 53(5):1398–1409
2. Dong D, Wen B, Boroyevich D, Mattavelli P, Xue Y (2014) Analysis of phase-locked loop low-frequency stability in three-phase grid-connected power converters considering impedance interactions. IEEE Trans Ind Electron 62(1):310–321
3. Pozzebon GG, Goncalves AFQ, Pena GG, Mocambique NEM, Machado RQ (2013) Operation of a three-phase power converter connected to a distribution system. IEEE Trans Ind Electron 60(5):1810–1818
4. Liserre M, Pigazo A, Dell'Aquila A, Moreno VM (2006) An anti-Islanding method for single-phase inverters based on a grid voltage sensorless control. IEEE Trans Ind Electron 53(5):1418–1426
5. Kim SK, Jeon SH, Ahn JB, Lee B, Kwon SH (2010) Frequency-shift acceleration control for anti-islanding of a distributed-generation inverter. IEEE Trans Ind Electron 57(2):494–504

6. Koizumi H, MizunoT, Kaito T, Noda Y, Goshima N, Kawasaki M, Nagasaka K, Kurokawa K (2006) A novel microcontroller for grid-connected photovoltaic systems. IEEE Trans Ind Electron 53(6):1889–1897
7. Estébanez EJ, Moreno VM, Pigazo A, Liserre M, Dell'Aquila A (2011) Performance evaluation of active islanding-detection algorithms in distributed-generation photovoltaic systems: two inverters case. IEEE Trans Ind Electron 58(4):1185–1193
8. Guerrero JM, Matas J, de Vicuna LG, Castilla M, Miret J (2007) Decentralized control for parallel operation of distributed generation inverters using resistive output impedance. IEEE Trans Ind Electron 54(2):994–1004
9. Balaguer IJ, Lei Q, Yang S, Supatti U, Peng FZ (2011) Control for grid-connected and intentional islanding operations of distributed power generation. IEEE Trans Ind Electron 58(1):147–157
10. He J, Li YW (2013) Hybrid voltage and current control approach for DG-grid interfacing converters with LCL filters. IEEE Trans Ind Electron 60(5):1797–1809
11. Sunkara JS, Gudey S (2018) Seamless power transfer to a critical load in a microgrid system using SMC. In: 2018 international conference on computing, power and communication technologies (GUCON), Greater Noida, Uttar Pradesh, India, pp 814–819
12. Singh AK, Gupta R (2014) Transformer-less inverter design for solar PV system application. In: Solar energy society of india (SESI) sponsored international congress on renewable energy (ICORE-2014), Manekshaw Centre, New Delhi, India
13. Rajasekar S (2014) Solar photovoltaic power conversion system using DC-DC and DC-AC converters control. Ph.D. Thesis, under the guidance of Dr. Rajesh Gupta, Department of Electrical Engineering, Motilal Nehru National Institute of Technology Allahabad
14. Gupta R, Ghosh A, Joshi A (2008) Switching characterization of cascaded multilevel-inverter-controlled systems. IEEE Trans Ind Electron 55(3):1047–1058

Chapter 9
Realization of New Third-Order Sinusoidal Oscillator Based on OTRA

Gurumurthy Komanapalli and Akash Tomar

1 Introduction

Current mode (CM) sinusoidal oscillators (SOs) with features of electronic tunability, independent control between the condition of oscillation (CO), and frequency of oscillation (FO) have gained special interest in the technical literature [1, 2]. The SOs are widely used in a variety of electronic system applications pertaining to control, instrumentation, and measurement communications [2]. Further, to achieve lower harmonic distortion, better frequency response TOSOs are preferred over second-order oscillators [2] for these applications. A number of TOSO topologies have been reported in the literature using various CM analog building blocks (ABBs) [3, 4], and references cited therein]. However, in this study, we exclusively focus on TOSOs based on one of the fascinating active blocks called OTRA [5, 6].

The OTRA is a current-controlled voltage source analog block whose input nodes are virtually grounded, thereby directing to circuits insensitive to parasitic capacitances. It also has interesting features like the independent gain-bandwidth product, and not slew limited. Due to these attributed advantages, researchers and analog circuit designers have exhibited considerable interest in developing various SOs using OTRA [4–6].

The comparative study of previously known OTRA-based TOSOs [7–13] is listed in Table 1. From Table 1, it is observed that the existing topologies

- lack independent tuning between FO and CO through resistors [7–10, 12, 13];
- do not support electronic tunability [8, 13];
- provide quadrature (Q) phase [7, 8, 10, 11], single (S) phase [9, 12, 13];
- are not MOS-C realizable [8, 12, 13].

G. Komanapalli (✉)
Audisankara College of Engineering & Technology (Autonomous), Nellore, India
e-mail: gurumurthyk.ece@audisankara.ac.in

A. Tomar
Delhi Technological University, Delhi, India

V. K. Giri et al. (eds.), *Computing Algorithms with Applications in Engineering*, Algorithms for Intelligent Systems, https://doi.org/10.1007/978-981-15-2369-4_9

Table 1 Comparison with existing OTRA-based TOSOs

Reference	No. of OTRAs utilized	Passive component count (Floating (F)/Grounded (G))	Independent tuning between CO and FO through resistors	Electronic tuning	Complete MOS-C implementation	Output phase	Highest frequency measured
[7]	2 OTRA	4R(F) + 3C(F)	No	Yes	Possible	Q	159 kHz
[8]	2 OTRA	3R(F) + 3C(F)	No	No	Not possible	Q	29.07 kHz
[9]	3 OTRA	6R(F) + 3C(F)	No	Yes	Possible	S	275 kHz
[10]	2 OTRA	3R(F) + 3C(F)	No	Yes	Possible	Q	18.85 MHz
[11]	3 OTRA	5R(F) + 3C(F)	Yes	Yes	Possible	Q	159 kHz
[12]	1 OTRA	3R(F) + 3C(1F,2G)	No	Yes	Not possible	S	1.08 MHz
[13]	1 OTRA	3R(F) + 3C(2F, 1G)	No	No	Not possible	S	15.92 MHz
Proposed	2 OTRA	5R(F) + 3C(F)	Yes	Yes	Possible	S	10.53 MHz

From the comparison table, it is clear that no OTRA-based TOSO is existing in technical literature which has features of uncoupled control of FO, and CO through resistors, electronic tuning, and complete MOS-C implementation. Therefore, to fill this gap, this work presents a new OTRA-based third-order SO topology, which fulfills all the above-quoted features.

In the following Sect. 2, the proposed TOSO circuit and its MOS-C implementation are presented. The non-ideality and sensitivity calculations are carried out in Sects. 3 and 4, respectively. Non-ideal calculations are done by taking a single-pole model of OTRA into account. In Sect. 5, the functionality of the proposed design is demonstrated through PSPICE simulations. The paper is concluded in Sect. 6.

2 Proposed Configuration

This section describes the overview of the OTRA and a detailed description of the proposed TOSO structure. The OTRA is a high gain (R_m) current input (I_p, I_n) voltage output (V_o) analog block [6, 11] whose input terminals (V_p, V_n) are internally grounded. The output response limitations due to stray capacitances are reduced because of grounded input terminals. This interesting feature makes OTRA feasible for high- frequency applications. The schematic symbol is depicted in Fig. 1. The relation between input ports and output ports is given in (1) using matrix representation. The notation R_m signifies the transimpedance gain of OTRA, which approaches

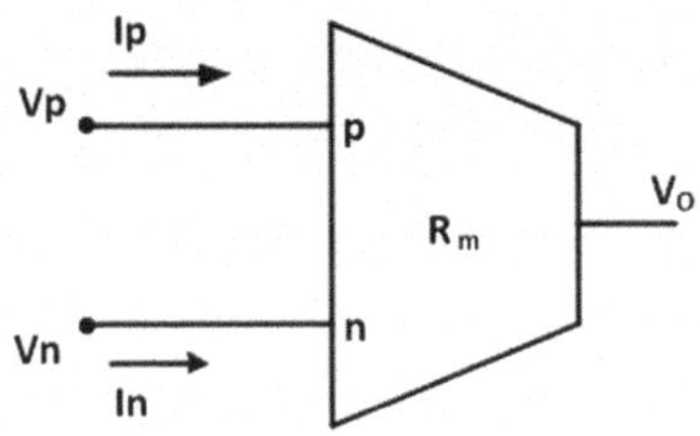

Fig. 1 Symbolic notation of OTRA

infinity under ideal conditions. Thus, OTRA must be operated in a negative feedback configuration.

$$\begin{pmatrix} V_n \\ V_p \\ V_o \end{pmatrix} = \begin{pmatrix} 0 & 0 & 0 \\ 0 & 0 & 0 \\ -R_m & R_m & 0 \end{pmatrix} \begin{pmatrix} I_n \\ I_p \\ I_o \end{pmatrix} \quad (1)$$

The proposed TOSO is given in Fig. 2. It entails two OTRAs and eight passive components. Considering ideal conditions for OTRA and by performing routine analysis, the characteristic equation (CE) for Fig. 2 is given by

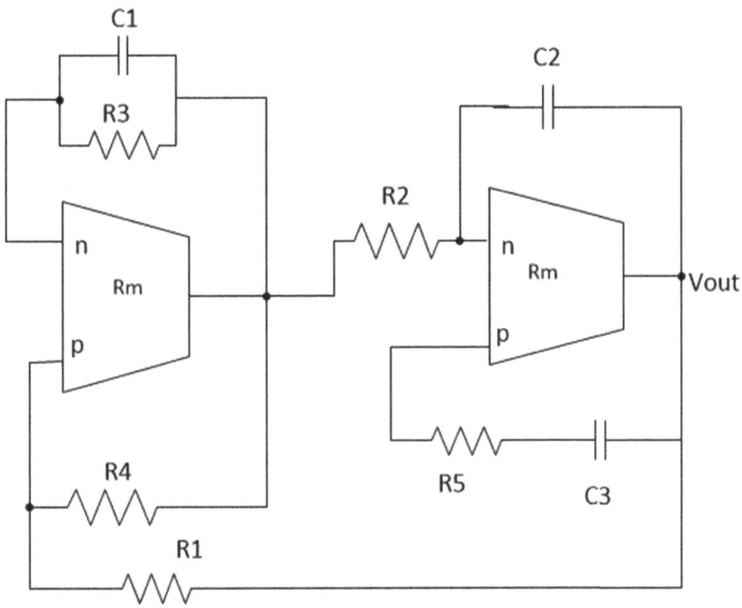

Fig. 2 Proposed TOSO topology

$$s^3C_1C_2C_3R_1R_2R_3R_4R_5 + s^2[C_2C_3R_1R_2R_5(R_4 - R_3) - C_1(C_3 - C_2)R_1R_2R_3R_4] + s[R_3R_4R_5C_3 - (C_3 - C_2)R_1R_2(R_4 - R_3)] + R_3R_4 = 0 \quad (2)$$

Assuming $C_1 = C_2 = C_3 = C$, the CE reduces to (2)

$$s^3C^3R_3R_4R_5 + s^2[C^2R_5(R_4 - R_3)] + s\left[\frac{R_3R_4}{R_1R_2}R_5C\right] + \frac{R_3R_4}{R_1R_2} = 0 \quad (3)$$

Therefore, the FO and CO are given by

$$\text{FO: } f_o = \frac{1}{2\pi C\sqrt{R_1R_2}} \quad (4)$$

$$\text{CO: } R_3 = \frac{R_5R_4}{R_4 + R_5} \quad (5)$$

From (4) and (5), it is interesting to note that oscillation frequency can be tuned using R_1, R_2 without disturbing oscillation condition, similarly, oscillation condition can be adjusted using R_3, thereby achieving complete independent tuning between CO and FO.

2.1 Complete MOS-C Implementation

By utilizing the current differencing feature of the OTRA, the resistors associated with the input nodes of OTRA can be easily designed using MOS transistors [14]. The proposed design fully ensures the complete MOS-C implementation, which is very beneficial from the IC integration viewpoint. Each resistor which is connected at the input node of the OTRA requires two matched n-MOSFETs, as depicted in Fig. 3.

The value of the resistor can be specified as

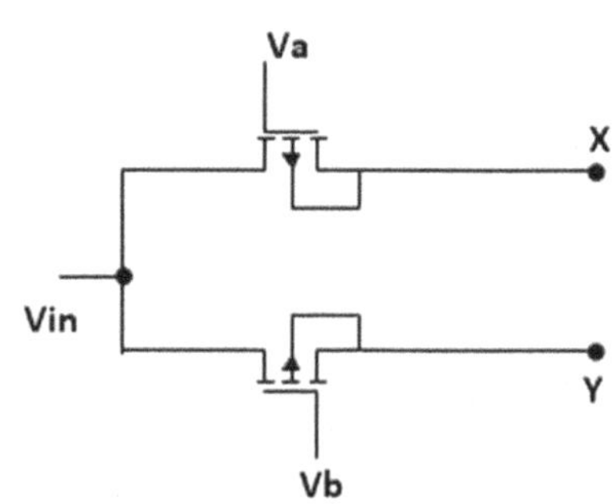

Fig. 3 Implementation MOS-based resistance [14]

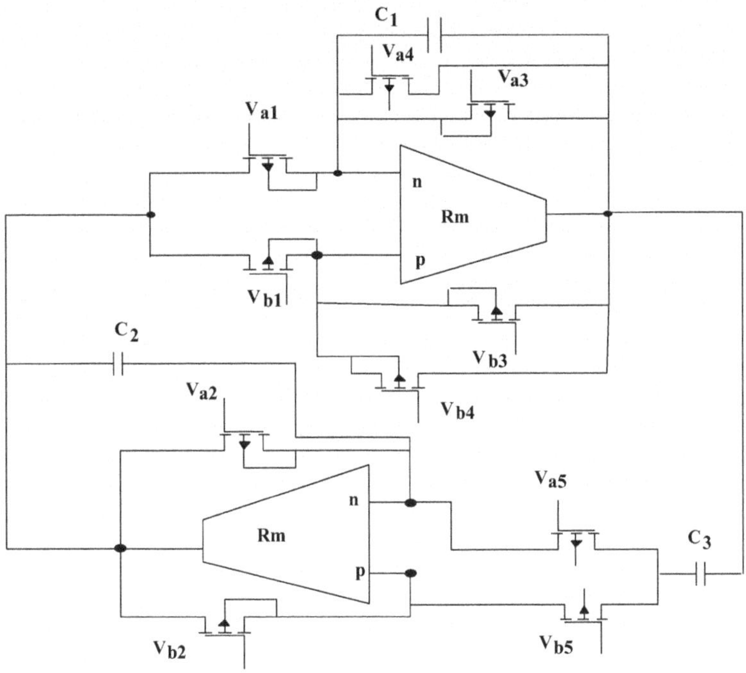

Fig. 4 Complete MOS-C schematic of proposed TOSO topology

$$R = \frac{1}{\mu_n C_{\text{ox}} \frac{W}{L} (V_a - V_b)} \tag{6}$$

Equation (6) has well-known parameters such as C_{ox} = Oxide capacitance per unit area; μ_n = Electron mobility; W/L represents effective (channel width/channel length), and V_a, V_b are the gate controlling voltages.

The complete MOS-C implementation of proposed OTRA-based TOSO is illustrated in Fig. 4.

3 Non-ideal Analysis

The presence of OTRA non-idealities will affect the performance of the proposed oscillator. The internally grounded input terminals of the OTRA minimize the effect of stray/parasitic capacitances at the input nodes and paving the way for high-frequency operation. Ideally, R_m tends to infinity; however, in practical cases, its value is frequency-dependent.

By taking singe-pole model into account, the circuit is analyzed and R_m can be defined as

$$R_m(s) = \frac{R_0}{1 + \frac{s}{\omega_0}} \tag{7}$$

where R_o is the dc trans-resistance gain. For applications pertaining to high frequencies, the R_m in s-domain reduces to

$$R_m(s) = \frac{1}{sC_p}; \text{ where } C_p = \frac{1}{R_o\omega_o}$$

Considering the aforementioned single-pole model for $R_{m,}$ the CE (3) modifies to:

$$\begin{aligned} & s^3(C + C_p)^2 C R_1 R_2 R_3 R_4 R_5 + s^2[(C + C_p))C R_1 R_2 R_5 (R_4 - R_3) \\ & - (C + C_p)(C - (C + C_p))R_1 R_2 R_3 R_4] \\ & + s[R_3 R_4 R_5 C - (C - (C + C_p))R_1 R_2 (R_4 - R_3)] + R_3 R_4 = 0 \end{aligned} \tag{8}$$

The FO and CO turns out to be

$$\text{FO: } \widehat{f_o} = \frac{1}{2\pi(C + C_p)\sqrt{R_1 R_2}} \tag{9}$$

$$\text{CO: } R_3 = \frac{R_5 R_4}{R_4 + R_5} \tag{10}$$

From (8), it can be seen that the effect of *Cp* can be removed by pre-adjusting the component values of the capacitor C, i.e., (C_1, C_2, and C_3). Therefore, the effect of stray capacitances can be nullified without using any extra elements.

4 Sensitivity Calculations

The sensitivity [1] calculations are key performance measure of any network. The component sensitivity of FO (f_o) say Y is given by

$$S_Y^{f_o} = \frac{\partial f_o}{\partial Y} \cdot \frac{Y}{f_o} \tag{11}$$

Sensitivity analysis of the proposed TOSO with respect to the passive components yields

$$\left|S_C^{f_0}\right| = 1; \left|S_R^{f_0}\right| = \left|S_{R_3}^{f_0}\right| = 0; \left|S_{R_1}^{f_0}\right| = \left|S_{R_2}^{f_0}\right| = \frac{1}{2} \tag{12}$$

From (11), it is perceived that all passive sensitivities for proposed TOSO in Fig 2 are 1 or 1/2 in magnitude. Therefore, it validates that the sensitivity performance is good.

5 PSPICE Simulation Results

The proposed topology functionality is tested through simulations using the CMOS implementation of the OTRA [6] depicted in Fig. 5. The 0.5 μm CMOS process technology node parameters by MOSIS-AGILENT are utilized for SPICE simulations. The supply voltages (V_{DD}, V_{SS}) set as ±1.5 V and bias voltage (V_B) as −0.5 V.

The proposed TOSO was simulated on PSPICE, and the simulated oscillation frequency for the passive component values as ($R_3 = 5\,\text{k}\Omega$, $R_i = 10\,\text{k}\Omega$ for $i = 1$, 2, 4 and 5, $C_i = 100\,\text{pF}$ for $i = 1$ to 3) was noted to be 160 kHz against the targeted theoretical value of 159 kHz. The corresponding output steady-state waveform and the frequency spectrum are depicted in Figs 6a and b, respectively. The typical graph of variation of FO with respect to tuning resistor R_1 is depicted in Fig. 7. For Fig. 7, the resistor R_1 is varied from 1 to 20 kΩ while maintaining C at 100 pF.

From simulation results, the percentage THD is observed to be 1.1.

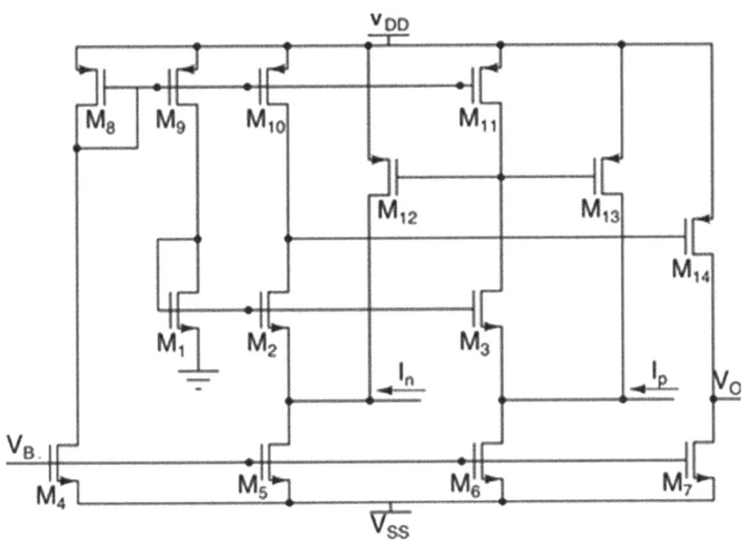

Fig. 5 OTRA CMOS implementation [6]

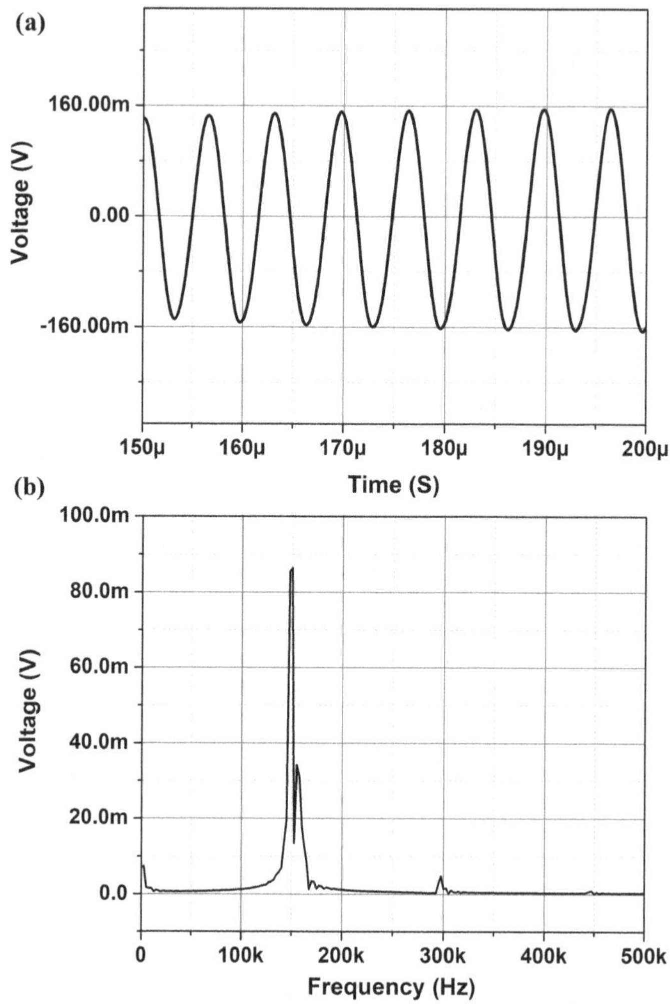

Fig. 6 **a** Output waveform. **b** Output frequency spectrum of the oscillation signal

6 Conclusions

A new third-order sinusoidal oscillator based on OTRA is presented in this paper. It bares independent control of oscillation frequency without altering the oscillation condition through resistors. Sensitivity calculations are carried out, and it is observed that the proposed oscillator topology has low passive sensitivities with respect to the frequency of oscillation. The output results from the PSPICE simulations are included to check the functionality of the proposed configuration. The % THD is observed to be 1.1.

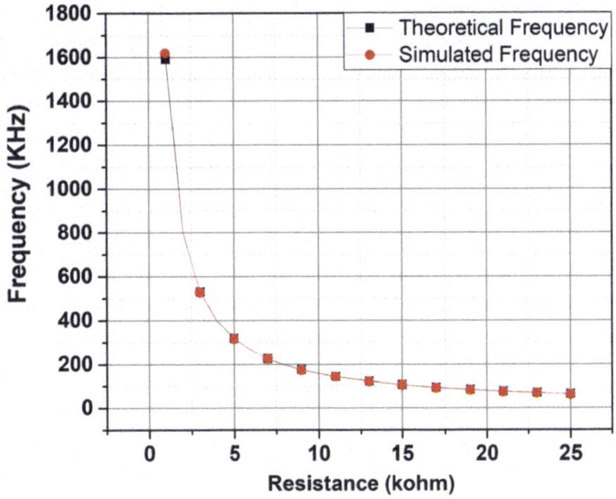

Fig. 7 Frequency tuning with resistor (R_1)

References

1. Sedra AS, Smith KC (2004) Microelectronic circuits, 2nd edn. Oxford University Press, New York
2. Senani R, Bhaskar DR, Singh VK, Sharma VK (2016) Sinusoidal oscillators and waveform generators using modern electronic circuit building blocks, 2nd edn. Springer, Switzerland
3. Toumazou C, Makris A, Lidgey FJ, Haigh DG (1990) Towards a new generation of analogue IC design architectures. In: EE colloquium on Analogue IC design: obstacles and opportunities, vol 9. IET, London, pp 1–916
4. Bashir SA, Shah NA (2012) Active device usage in filter design—An overview. Int J Sci Res Publ 2(6)
5. Salama KN, Soliman AM (1999) CMOS operational transresistance amplifier for analog signal processing. Microelectron J 30(3):235–245
6. Mostafa H, Soliman AM (2006) A modified CMOS realization of the operational transresistance amplifier (OTRA). Frequenz 60(3–4):70–76
7. Pandey R, Pandey N, Paul SK (2012) MOS-C third order quadrature oscillator using OTRA. In: 2012 Third international conference on computer and communication technology, vol 24. IEEE, Allahabad, pp 77–80
8. Nagar BC, Paul SK (2016) Voltage mode third order quadrature oscillators using OTRAs. Analog Integr Circuits Signal Process 88(3):517–530
9. Gurumurthy K, Pandey N, Pandey R (2017) OTRA based second and third order sinusoidal oscillators and their phase noise performance. In: AIP conference proceedings, vol 1859. AIP Publishing, India
10. Nagar BC, Paul SK (2018) Realization of OTRA-based quadrature oscillator using third order topology. In: Konkani A, Bera R, Paul S (eds) Advances in systems, control and automation. Lecture Notes in Electrical Engineering. Springer, Singapore, pp 86–375
11. Pandey R, Pandey N, Gurumurthy K, Anurag R (2014) OTRA based voltage mode third order quadrature oscillator. ISRN Electron Hindawi Publ Corp
12. Chien H (2016) Third-order sinusoidal oscillator using a single CMOS operational transresistance amplifier. J Appl Sci Eng 19(2):187–196
13. Gurumurthy K, Pandey N, Pandey R (2018) New realization of third order sinusoidal oscillator. Int J Electron Commun 93:182–190
14. Gurumurthy K, Pandey R, Pandey N (2019) New sinusoidal oscillator configurations using operational transresistance amplifier. Int J Circ Theor Appl 47(5):666–685

15. Lahiri A, Jaikla W, Siripruchyanun M (2012) First CFOA-based explicit-current-output quadrature sinusoidal oscillators using grounded capacitors. Int J Electron 100(2):259–273
16. Gupta S, Sharma R, Bhaskar DR, Senani R (2008) Sinusoidal with explicit current output employing current-feedback op-amps. Int J Circuit Theory Appl 38(2):131–147
17. Maiti S, Pal R (2015) Voltage mode quadrature oscillator employing single differential voltage current controlled conveyor transconductance amplifier. IJEEE 3(5):344–348
18. Chaturvedi B, Maheshwari S (2012) Second order mixed mode quadrature oscillator using DVCCs and grounded components. Int J Comput Appl 58(2):42–45
19. Celma S, Martinéz P, Carlosena A (1992) Reply: minimal realisation for single resistor controlled sinusoidal oscillator using a single CCII. Electron Lett 28(13):1265
20. Senani R, Singh VK (1996) Comment: synthesis of canonic single-resistance-controlled-oscillators using a single current-feedback-amplifier. IEEE Proc -Circuits, Devices Syst 143(1):71
21. Senani R, Bhaskar DR (1991) Single op-amp sinusoidal oscillators suitable for generation of very low frequencies. IEEE Trans Instrum Meas 40(4):777–779
22. Srivastava DK, Singh VK, Senan R (2015) New very low frequency oscillator using only a single CFOA. Am J Electr Electron Eng 1(3):1–3
23. Elwakil AS (1998) Systematic realization of low-frequency oscillators using composite passive-active resistors. IEEE Trans Instrum Meas 47(2):584–586

Chapter 10
Comparative Analysis of Wavelet and OFDM-Based Systems

Mrinalini Srivastava, Rafik Ahmad and Kamlesh Kumar Singh

1 Introduction

Multicarrier modulation method such as orthogonal frequency division multiplexing (OFDM) technique is a wideband method for digital wireless communication [1]. In this modulation technique, input data is segmented into different frequency bands in which modulation is carried out and further multiplexed at different carrier frequencies so the information is transmitted on each subcarrier [2]. Due to the multiplexing of N number of data modulated subcarriers, OFDM signal experiences high fluctuation of amplitude which results in a large peak-to-average power ratio (PAPR) [3]. Various methods are developed such as clipping, filtering, peak windowing, and peak cancelation to solve PAPR problem in OFDM [4]. Researchers had given several modified OFDM formats by converting and modulating OFDM waveform. These waveforms are changed into frequency or the phase of a single-carrier signal. Because of applications of phase or frequency modulation, these formats result in constant envelope signals which give a unity PAPR (0dB) [5]. The signal transformation method that converts variations in the peak power as well as average power into a constant envelope signal is known as constant envelope orthogonal frequency division multiplexing (CE-OFDM) system [6]. In CE-OFDM system, the processed OFDM signal is converted by phase modulation and this, in turn, gives a signal constructed for effective power amplification. At the receiving end, phase demodulation is applied prior to conventional OFDM demodulator. The main difference between OFDM and CE-OFDM is that the transformation of signal is carried by phase modulation and phase demodulation [7].

M. Srivastava · K. K. Singh
Amity School of Engineering Technology, Amity University, Lucknow, India
e-mail: kksingh@lko.amity.edu

R. Ahmad (✉)
EE Department, BBDNITM, Lucknow, India

V. K. Giri et al. (eds.), *Computing Algorithms with Applications in Engineering*,
Algorithms for Intelligent Systems, https://doi.org/10.1007/978-981-15-2369-4_10

In FFT-based OFDM and CE-OFDM system, the addition of cyclic prefix is done before transmitting the signal through the channel in order to minimize ISI and ICI [8]. Fourier transform (FT) is reversible transform that allows to going back and forwarding between the raw and processed signals [9]. But only one is present at any instant of time. It can only give frequency information of signal, but there is lack of information of the time when frequency component exits. FT is used for non-stationary signals but study of stationary signal is of paramount importance in which time localization of spectral components is needed. In OFDM, cyclic prefix takes the 25% bandwidth and as fast Fourier transform is applied in OFDM so there is not complete information of the signal transmitted. To overcome all problems mentioned above, discrete wavelet transform (DWT) system was proposed [10]. The wavelet-based system has a stronger ability to mitigate ISI as well as ICI than conventional OFDM and CE-OFDM system [11]. Recently, various experiments were performed on Haar orthonormal wavelets by researchers. It was found that DWT is a suitable method to reduce the ISI and ICI, which are created by loss in orthogonality between carriers. Spectrum efficiency of DWT is better than discrete Fourier-based OFDM. The DFT is exchanged by DWT so that further reductions in the level of interference and increase spectral efficiency can be observed. This paper comprises of Sect. 2 with description of CE-OFDM, STFT, wavelet-based system, DWT, Sect. 3 deals with simulation work, and Sect. 4 focuses on conclusion.

2 CE-OFDM System

OFDM is a transmission scheme that contains a high-rate data present in serial stream. This is classified into a group of low rate sub streams, individually modulated on separate single-carrier (SC) frequency division multiplexing [12]. In OFDM system, cyclic prefix (CP) is introduced which is a repetition of the end of symbol at the beginning. The main aim of CP is to permit multipath and behave as a buffer to shield OFDM signals from ISI [13]. The receiver is designed in such a way that decoding of signal is done simultaneously with the settlement of signal due to orthogonality of frequencies [14]. Constant envelope is a modulation format for digital wireless transmission. In this format, the electrical part of the carrier phase is modulated by OFDM waveform which provides 0 dB PAPR. In this system, we have to give bit sequence as input and by using modulation mapping symbols further process is generated. These symbols are then applied to the input sequence of the IDFT block [15]. Sum of orthogonal signal is generated by inverse discrete Fourier transform (IDFT) at each T second. Finally, signal $x(n)$ is produced and OFDM sequence is given to a phase modulator to produce 0 dB PAPR sequence $x(n)$. In the constant envelope signal, information message signal is considered to be a real-valued OFDM waveform [16]. Signal modulates the phase of carrier and the resulting output of the phase modulator.

The phase signal during nth interval is:

$$\Phi(t) = \theta_n + 2\Pi h C_n x(t) \tag{1}$$

where h is modulation index and C_n are a constant (Amplitude) which is used to normalized the variance. θ_n is the initial phase to make the modulation phase continuous. The modulation index h plays very important role for bandwidth description and performance of constant envelope OFDM [11]. In the case of CE-OFDM, the CP is added, and at the receiving end, CP is discarded and the reverse operation is performed.

2.1 Wavelet-Based System

Wavelet transform is a tool that gives time–frequency representation of a signal. Multi-resolution analysis (MRA) one of the most important characteristics of wavelet-based system which checks signal having different frequencies and different resolution; therefore, this results in complete information of signal [17]. WT uses MRA which is designed to give good time resolution as well as poor frequency resolution at high frequency and good frequency resolution as well as poor time resolution at low frequency [18]. WT is classified into two types:

2.2 Continuous Wavelet Transform

It is an alternate method to short-time Fourier transform (STFT) that is used for analyzing small part of signal at instant of time. STFT has resolution problem due to windowing which was overcome by CWT. This wavelet transform gives signal with overcomplete representation means in time and frequency domain along with translation as well as scale parameters. These parameters play a significant role in WT. CWT is given by the following expression:

$$\psi(f)(a, b) = \frac{1}{\sqrt{a}} \int_{\infty}^{\infty} f(t)\psi\left(\frac{t-b}{a}\right)\mathrm{dt} \tag{2}$$

where a is considered as scaling factor and b is translation factor.

Mother wavelet is used to generate other window functions. Translation is correlated to location of window that corresponds to time information [19]. Scaling is done to dilate or compress a signal which is related frequency information.

2.3 Discrete Wavelet Transform

The discrete wavelet transform (DWT) is defined as linear transformation method which performs operation on data vector with the length of an integer power of two finally giving it into vectors that differ each other numerically but have the same length [20]. It is a technique that categorizes data in several frequency components by thoroughly studying each and every component with resolution completely matching to its scale.

It is a method of decomposition of discrete-time signal with the help of analysis scheme known as sub-band coding.

$$x[n] * h[n] = \sum_{k=-\infty}^{\infty} x[k] \,.\, h[n-k] \tag{3}$$

$x[n]$ = Original Signal
$h[n]$ = Impulse response of signal through half band digital low-pass filter.

Resolution of signal is a quantity of detailed information in the signal which can be achieved by sub-band coding. It consists of sub-sampling which means the elimination of some of the samples of signals (Fig. 1).

The IDWT signal can be generated by

$$x[t] = \sum_{m=-\infty}^{\infty} \sum_{k=-\infty}^{\infty} x(k) 2^{m/2} \psi\left(2^m t - K\right) \tag{4}$$

where $\psi(t)$ is wavelet function that consists of the compressed factor m times as well as shifted k times each subcarrier and $x(k)$ is the data that is modulated onto the wavelets at different scales [21]. To recover the data at the rate 2^m Hz, the

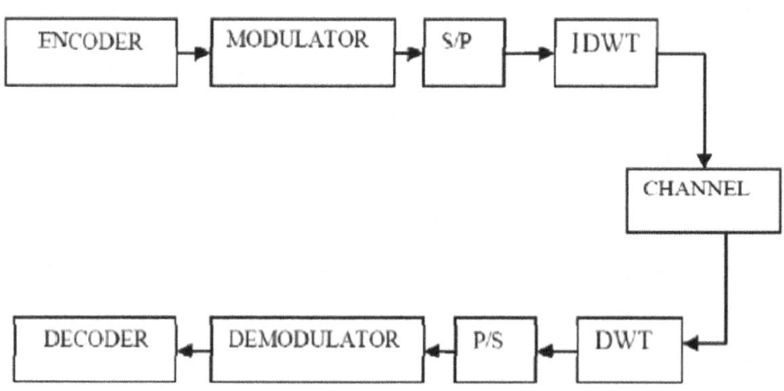

Fig. 1 Block diagram of encoded DWT-OFDM

smallest baseband bandwidth should be $2^m + 1$ Hz. In wavelet-based modulation, the subcarrier waveforms are formed with the application of wavelet transform.

The $x(t)$ signal is passed through phase modulator, and the output of phase modulator is

$$S[t] = A\exp[j2\pi fct + \varphi(t)] \tag{5}$$

$$\varphi(t) = 2\pi hx(t) \tag{6}$$

A is amplitude of signal and an arbitrary phase offset is denoted by $\varphi(t)$ that is used as a parameter to design in order to get phase continuous modulation, h is modulation index [22].

The method of signal in which inverse wavelet transform takes place is a process of synthesizing into wavelet coefficients, whereas reverse operation is known as analyzing of wavelet coefficients [23]. The waveforms determined as a result of wavelet transform (WT) are those which are longer than the transform size.

3 Simulation Results

We have given BER performance for OFDM, CE-OFDM, and wavelet-based OFDM system considering number of carriers 64 by 16 QAM modulation technique under AWGN channel. Convolutional encoding and decoding methods are used which involves multiple XOR operations so that system can give better signal transmission and security of signal [24]. The BER of conventional OFDM system and DWT-based OFDM system is analyzed and examined. MATLAB simulation is done for the comparison of both the systems and results are presented as follows:

Figure 2 describes the graphical representation of BER and SNR performance for wavelet-based system. The BER equals to 0.403 and is obtained at SNR of 2 dB; when the system is observed, value continues to decrease up to 12 dB which is equal to 0.03302.

Figure 3 describes the graphical representation of BER and SNR performance for wavelet-based system. The BER equals to 0.03751 and is obtained at SNR of 2 dB; when the system is observed, it is seen that there is continuous decrease in BER with an increase in SNR from 0.03751 to 0.0001904 for 2–8 dB of SNR.

In Fig. 4, it is clearly visible the blue-colored plot of wavelet is showing much better fall in error rate of signal with an increase in SNR value. Unlikely, in conventional OFDM the bit error rate remains constant for major portion of the plot which is not efficient signal output. From the graph, it can be easily concluded that for wavelet system bit error rate is decreasing with an increase in SNR but same case is not applicable for conventional OFDM [25, 26]. In wavelet as compared to conventional OFDM, the error in signal decreases much quicker and logarithmically and is very close to the theoretical value.

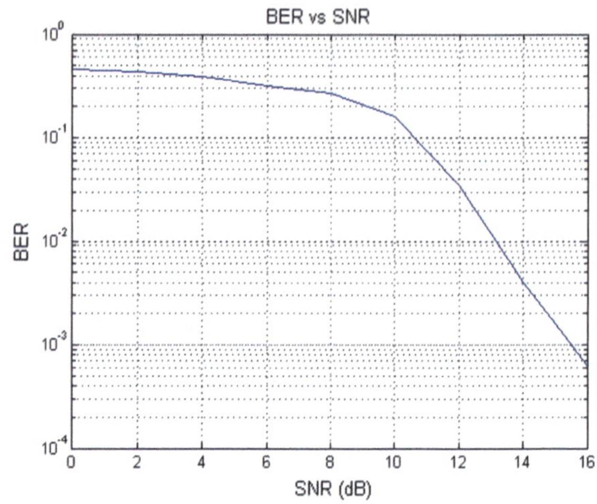

Fig. 2 Performance of conventional OFDM system

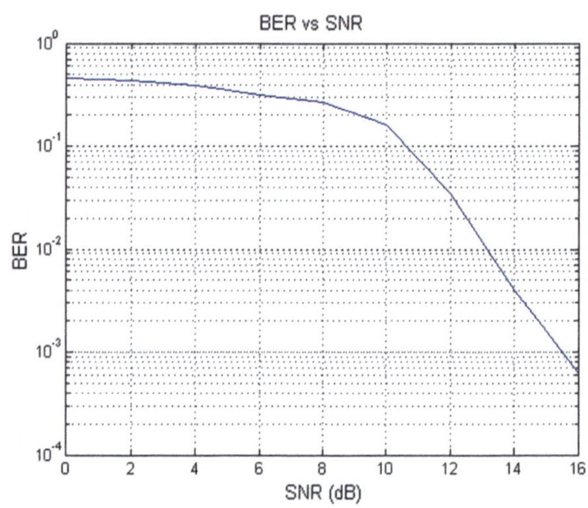

Fig. 3 Performance of wavelet-based system

The BER versus SNR graph clearly shows the value of BER at initial point for conventional OFDM which is much more than wavelet transform OFDM.

The drop or decrease in BER value with respect to the SNR is clearly visible in graphs. From 0 to 5 dB, for conventional OFDM decrease in BER is 0.6–0.4, elsewhere in case of wavelet transform OFDM fall in value of BER is from 0.1 to 0.01. In comparison, the decrease in BER with an increase in SNR value for conventional OFDM is lesser than wavelet OFDM (Fig. 5).

From 5 to 10 dB, for conventional OFDM decrease in BER is 0.4–0.1, elsewhere in case of wavelet transform OFDM fall in value of BER is from 0.01 to 0.0001. In comparison, the decrease in BER with an increase in SNR value for conventional OFDM is drastically lesser than wavelet OFDM.

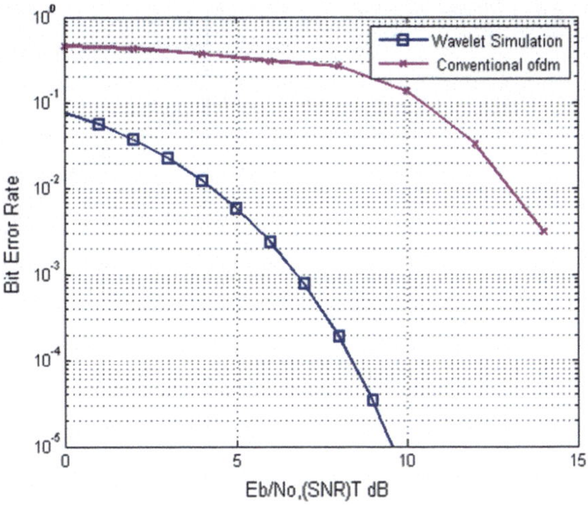

Fig. 4 Comparison of conventional OFDM and wavelet-based OFDM

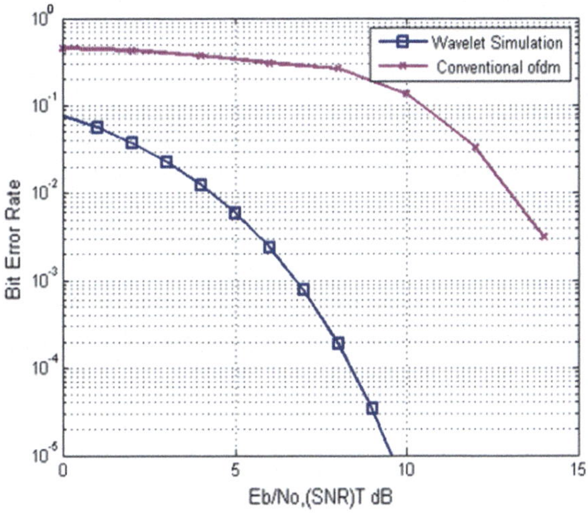

Fig. 5 BER decrease for 0–5 dB of SNR

The area covered under wavelet OFDM is smaller as compared to the conventional OFDM which clearly indicates lesser value of BER in case of wavelet OFDM which is clearly visible in graph so wavelet-based systems are better as it increases performance with moderate enhancement in computational complexity as well as delay (Fig. 6).

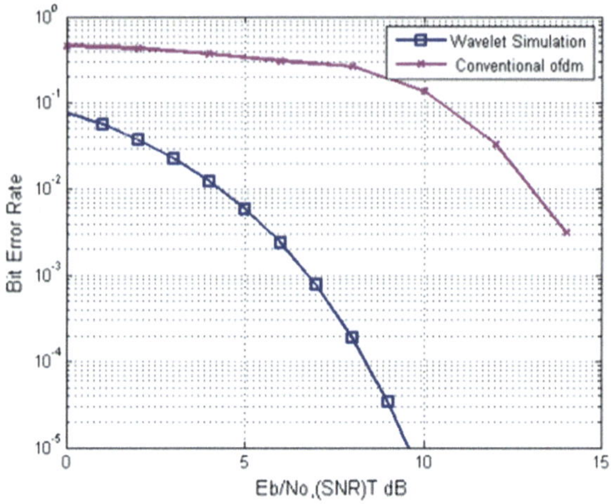

Fig. 6 BER decrease for 5–10 dB of SNR

4 Conclusion

It can be easily concluded from observations that wavelet-based systems are better in giving complete information of signal. In wavelet as compared to conventional OFDM, the error that presents in the signal degrades much quicker in logarithmically response and is very close to the theoretical value that can be evaluated from graphical study. From various graph results after simulation, we can say that bit error rate is less for wavelet system so it is far much better than conventional OFDM. With an increase in SNR, the decreasing value of BER is more in case of wavelet-based system as compared to conventional OFDM. Wavelet transform is widely used in acoustics, nuclear engineering, and sub-band coding because of its special property that is perfect reconstruction. Wavelet transform is though complex in nature, but its overall bandwidth usage makes it a better option for upcoming new cellular technologies.

References

1. Zhang H, Zhang J (2013) Performance comparison of wavelet packet transform based & conventional coherent optical OFDM transmission system. Elsevier
2. Gupta MK (2012) Performance evaluation of FFT and wavelet based OFDM system. Int J Electr Commun
3. Kumbasar V, Kucur O (2012) Performance comparison of wavelet based and conventional OFDM systems in multipath rayleigh fading channels. Elsevier
4. Ochiai H, Imai H (2001) On the distribution of the peak-to-average power ratio in OFDM signals. IEEE Trans Commun 49:282–289
5. Muquet B, Wang Z, Giannakis GB, De Courville M, Duhamel P (2002) Cyclic prefixing or zero padding for wireless multicarrier transmissions. IEEE Trans Commun 50(12):2136–2148

6. Tan J, Stuber GL (2002) Constant envelope multi-carrier modulation. In: Proc. IEEE MILCOM, vol. 1. Anaheim, pp 606–611
7. Sandberg SD, Tzannes MA (1995) Overlapped discrete multitone modulation for high speed copper wire communications. IEEE J Select Areas Comm 13(9):1571–1585
8. Dilmirghani R, Ghavami M (2007) Wavelet versus fourier based UWB systems. In: 18th IEEE international symposium on personal indoor and mobile radio communications, pp 1–5
9. Bingham JAC (1990) Multicarrier modulation for data transmission. IEEE Commun Mag 28:5–8
10. Jamin A, Mahonen P (2005) Wavelet packet modulation for wireless communications. Wirel Commun Mobile Comput J 5(2):123–137
11. Liu H, Li G (2005) OFDM-based broadband wireless networks: design and optimization. Wiley
12. Tan J, Stüber GL (2002) Constant envelope multi-carrier modulation .In: Proc. IEEE MILCOM, vol. 1, pp. 607–611
13. Ahmed N (2000) Joint detection strategies for orthogonal frequency division multiplexing, Dissertation for master of science, Rice University, Houston, Texas, pp 1–51
14. Casas EF, Leung C (1991) OFDM for data communication over mobile radio FM channels–part I: analysis and experimental results. IEEE Trans Commun 39:783–793
15. Sathananthan K, Tellambura C (2001) Probability of error calculation of OFDM systems with frequency offset. IEEE Trans Commun 49(11):1884–1888
16. Prasad VGS, Hari KVS (2004) Interleaved orthogonal frequency division multiplexing (IOFDM) system. IEEE Trans Signal Process 52(6)
17. Tan P, Beaulieu NC (2006) A comparison of DCT-based OFDM and DFT-based OFDM in frequency offset and fading channels. IEEE Trans Commun 54(11):2113–2125
18. Negash BG, Nikookar H (2001) Wavelet based OFDM for wireless channels. In: Proceedings of the IEEE vehicular technology conference (VTC), Springer, pp 688–691
19. Rao KR, Yip P (1990) Discrete cosine transform. Academic, New York
20. Chen WH, Smith CH, Fralick SC (1977) A fast computational algorithm for the discrete cosine transform. IEEE Trans Commun 25(9):1004–1009
21. Wang ZD (1984) Fast algorithms for the discrete W transform and for the discrete fourier transform. IEEE Trans Acoust, Speech, Signal Process 32(4):803–816
22. Oltean M, Isar A (2009) On the time-frequency localization of the wavelet signals, with application to orthogonal modulations In: Proceedings of ISSCS'09, Iasi, pp 173–177
23. Tsai Y, Zhang G, Pan JL (2005) Orthogonal frequency division multiplexing with phase modulation and constant envelope design. In: Proc. IEEE MILCOM, vol. 4, pp 2658–2664
24. Mirghani R, Ghavami M (2008) Comparison between wavelet-based and fourier-based multicarrier UWB systems. IET Commun 2(2):353–358
25. Oppenheim AV, Schafer RW, Buck JR (1998) Discrete-time signal processing, 2nd edn. Prentice-Hall, Englewood Cliffs, NJ
26. Smulders P (2002) Exploiting the 60 GHz band for local wireless multimedia access: prospects and future directions. IEEE Commun Mag

Chapter 11
Review in Recent Trends on Energy Delivery System and Its Issues in Smart Grid System

Kitty Tripathi, Sarika Shrivastava and Somendra Banarjee

1 Introduction

A power grid system is a platform used for realizing energy translation and communication of power over a wide range by means of optimum distribution of resources. This smart energy delivery system thus remains a gigantic system that has operative association within the numerous efficient areas of a power distribution. It is an automated system that has interoperation features. To realize efficient and reliable demand–supply operations of power system [1], there is a proposal of real-time regulation system that is effective for small-independent grid organization with unstable renewable energy generations. The smart grid structure pay emphasizes on a smart energy delivery which relies on two-way flow of electricity in addition to the information which is essential to enhance consistency, security, and efficiency [2–5]. The foremost characteristics of such system comprises of self-monitoring, self-healing, and adaptive response to a faulty condition in the distributed network. Hau et al. in [3] conversed about the real-time monitoring of the system to improve the consistency of the power delivery system along with the failure protection mechanism. They essentially focus on precise estimating of fault which can be done using several concepts like neural network. In [6–9], there has been a debate on efficacy and control with distributed generation entrenched into the system and key outlines are constructed on the incorporation of microgrid service-oriented construction intended for monitoring and control. The upgradation in this structure includes the convention of the renewable energy source that presents at distribution network and guided control with the real-time status monitoring of the system. This system comprises

K. Tripathi (✉)
Babu Banarsi Das Northern India Institute of Technology, Dr. A.P.J. Abdul Kalam Technical University, Lucknow, Uttar Pradesh, India

S. Shrivastava · S. Banarjee
Ashoka Institute of Technology and Management, Dr. A.P.J. Abdul Kalam Technical University, Lucknow, Uttar Pradesh, India

V. K. Giri et al. (eds.), *Computing Algorithms with Applications in Engineering*, Algorithms for Intelligent Systems, https://doi.org/10.1007/978-981-15-2369-4_11

information of subsystem by using phase measurement unit (PMU), smart sensors, and smart metering architecture. The notion of smart energy delivery system offers a massive improvement in the prevailing passive distributed network of power system by means of active network which has twofold communications. It also possesses the capability of self-healing through the decentralized control system [10–13]. It practices the perception of demand response and demand-side management along with the non-hierarchical dispersal of electric power and usage in the distribution generation determined by the consumer. In the future, smart grid system one of the main challenges is the handling of stochastic energy supply which is localized and thus requires highly accurate method for energy prediction that will need the assistance of decision-making tools deep belief network, reinforced learning, etc. [2]. Electric power system control procedures have always progressed in order to accomplish the necessities of the electric power industry throughout its development and thus [14] have proposed the usage of reinforced learning in the grid monitoring system to solve the problems related to control decision making in the power system. Online dispensation of large amounts of data endlessly produced by the smart grid can carry appropriate and detailed power load forecasts which are an important input for communication on the market where the energy can be slender even minutes ahead of its consumption to curtail the grid inequalities. Authors in [15] demonstrated the correctness of online support vector regression (SVR) method for short-term power load forecasting and methodically discovered its pros and cons.

2 Smart Grid Technologies

The effort in emerging the future smart transmission system embraces the three foremost components, viz. substation, control center, and transmission network. Lu et al. in [15] conferred the inter-reliant network which affects the original power fault proliferated in a wide area and upsurge the threat of cascading failure. The current perception in monitoring and control of an energy delivery system entails conception of the computerized one-line diagram, state estimation, and contingency analysis [3]. The output from this state estimation has a significant delay inside it which persists for a limited time period, and this delay is built on the local information of the control area. The investigation of voltage stability in these classifications system is simulation-based, and the precision is subjected to the model of simulation and the performance of the state estimator. Transmission line fortification uses the notion of wavelet multi-resolution examination as discussed in [7, 16]. It is an exact fault location technique where the investigation of transient wave traveling from fault to substation is studied. There are numerous distinct protection structures which are similarly used with the modifiable control strategies as discussed in [6, 17]; the simulation with PSCAD is giving good results for the protection of LV micro-grid. The monitoring role in the future smart grid system might be human-centered that would offer the operator with useful data. The protection system at the substation can record the commotion event and data at the explicit type of fault which can be of added use for the dynamic calculation. The smart sensors are positioned at several substations and transmission lines which can be analyzed for resolving the working on system equipment. The phase measurement units (PMUs) are used for monitoring the disturbance in the measurement-based analysis in [18–20]; the conversation about steadiness margin calculation is completed using the measurement-based method which is constructed on the preventive control strategies which in turn is specified by the simulation-based approach. A novel concept practices intelligent distributed autonomous power system wherever demand is supply-driven but the measure of the grid that is limited to few interrelated micro-grids [10]. In [21–24], a profligate analysis system was anticipated to be used which was a hybrid system and uses a fuzzy logic controller for exposure and diagnosis of fault in power system and the fault clearance power process prototypical was created. A few additional methods to exist in fault detection system in smart grid that embraces the technique to distinguish the fault in distributed network as well as communication network.

2.1 Self-healing Control Strategies

In the traditional control structure, the procedure and monitoring are established based on a specific problem. It is dependent on the simulation and the information remains static, and at that time, there exists a deficiency in organization among protection and control structures [25]. In [26–29] the conversation is around each disturbance in the power system that might cause innumerable protections and control structure to reply, and optimal control strategies are therefore essential to be established which can convey stable operating conditions with the smallest control efforts. The self-healing control arrangements effectually re-establish the scheme and take it back to regular circumstances of the procedure and thus advance the control proficiencies [30, 31]. A smart way of employed fault is anticipated by Glavic et al. in [14] with fault location algorithm that is used to match the power system ailment and accessibility of data. To reinstate service from permanent fault, there is an obligation of precise location of fault to overhaul the faulty line as earliest as possible. Even though the distance relay is been used, it cannot encounter the precise location lower than all the situations. Various smart arrangements have been projected to locate the transmission line fault and are anticipated to deal with the selection of optimum fault location approaches using several algorithms which are contingent upon accessibility and location of record data as well as network topology settings edging the fault [32, 33].

2.2 Data Integration for Smart Fault Location

A smart-integrated substation is furnished with innumerable types of intelligent electronic device (IED) used for monitoring and control [34–37]. The substation analog signal is restrained at high power level and is being administered using these intelligent electronic devices. The record comprises of measurement acknowledged from remote terminal unit and IEDs, and the static system data encompasses the explanation of workings and associates. This data is then associated with static system model. The smart sensors are mounted all over the power system from substation to the client location. The illustrations of synchronized phasor are provided, and then the communication is been made [38]. The accessibility of supplementary feeder data supports in cultivating the efficiency of fault location methods. The central feeder automation necessitates the data from field devices in SCADA system and then subjects a supervisory control command; however, in IED the data is administered locally, and at that time, no central station is essential. In [39, 40], the author has conferred about a huge system with manifold search running in parallel and can be accomplished using population-based optimization method like genetic algorithm in which foremost the initial population was designated by varying fault location and fault resistance randomly. Short-circuit studies were then carried out, and fitness is estimated for each possible fault location. Then genetic algorithm operator fault is pretentiousness for next iteration which once obtained is then assessed for optimum solution in [41–45].

3 Communication Infrastructure in Smart Grid

Huang et al. in [27] specified that there exists a wide assortment of wireless communication protocol for two-way communication in smart grid system. It covers from the traditional central generator through transmission network and distributed system to industrial consumer [46]. In [47–49], the author discoursed about time critical communication that is coupled with the power system protection and has miscellaneous system controlling feature by means of unified communication platform. The delay module of smart grid communication is scrutinized and analyzed, and then local area network target has complete network through enhancement of multi-flow delay and the simulation-based study augments the wireless mesh network routine [50]. This wireless admittance and relay scheme are the solution for transmitting time with two-tier architecture of mesh router and mesh client. In [51–54], there is conversation on cognitive radio network for communication in smart grid system to expand the performance of data transmission with dynamic and adaptive spectrum distribution ability. It deals with wide area monitoring, control and protection with distributed generation management, advance metering infrastructure than real-time pricing. The data routing is at 2.5 GHz (ISM) band; besides, it also improves the complete performance of data communication with dynamic and adaptive spectrum supervision proficiencies and acquires mean square proficiencies.

4 Challenges in Establishing Smart Fault Control System

The main challenges for smart energy delivery system are about installation and commissioning of smart sensor used for two-way communication [55, 56]. It also has to deal with the difficulties concerning incorporation and synchronization of advance metering infrastructure with added software application [57]. The challenges also exist in providing with the enterprise resource planning (ERP) system and message service to the user. Another challenge consists of the incorporation of low voltage control [58] and monitoring system with real-time high voltage system.

5 Integration of Renewable Energy Generation in Smart Energy Delivery System

In [59, 60], there has been discussion on the reliability of energy delivery system when it is integrated with renewable energy source. There has a problem regarding the identification of fault and failure in data transmission system when integrated with renewable energy like solar energy. The problem is resolved using Petri net model [61, 62]. In [63], there is discussion on next-generation monitoring and analysis and control in the future transmission grid. In [64, 65], a unique network architecture is discussed which has two coupled interdependent network with complex communication framework and uses iterative fault propagation. There has been trend in the development of solar micro-grid as discussed in [66–68] which uses protection strategies like adaptive, differential, distance, voltage base, and over current with high-speed communication system. In [18, 69, 70], renewable energy is studied with distributed generation regulatory framework which provides utility and power reliability.

6 Conclusion

This paper has conversed about the prevailing and future trend in fault location method in smart grid system for transmission as well as distribution system. The technologies discussed and source of data could be utilized to improve the fault location method by inter-relating the data from measurement and simulation which is obtained using power system model. It also discusses wireless communication prevailing in smart grid system with respect to the delay bounds for time critical communication in wireless access network. The vital issues and exposed trials in the SG are deficient of cognizance, consumer reception, cyber terrorism, data collection administration, dynamism metering, energetic optimization, and energy control. In view of this paper, a wide-ranging assessment is discovering data of development, technologies, and techniques in the SG. The main objective is to examine and disclose the crucial empowering technologies and to obtain a better picture of the current status of SG development.

References

1. Zoka Y, Sasaki Y, Yorino N (2016) An on-demand generation regulation for small independent power grids with a selective control. Elsevier J Int Fed Autom Control 49(27):158–163
2. Mocanu E, Nguyen PH, Kling WL, Gibescu M (2016) Unsupervised energy prediction in a Smart Grid context using reinforcement cross-building transfer learning. Elsevier J Energy Build 116:646–655
3. Hau L-C, Lee J-V, Chuah Y-D, Lai A-C (2013) Smart grid-the present and future of smart physical protection—a review. Int J Energy Inf Commun 4(4):43–54
4. Anonymous, EPRI Smart Grid Resource Centre, June 2012. Available at: http://smartgrid.epri.com/
5. Sun Q, Ge X, Liu L, Xu X, Zhang Y, Niu R, Zeng Y (2011) Review of smart grid comprehensive assessment systems. In: ICSGCE proceedings 2011, 219–229, Elsevier, Chengdu, China
6. Afsar S, Nadendla G, Sudha R (2015) A review of recent development in smart grid systems. Proc Int J Adv Res Electr Electron Instrum Eng 4(1):97–103
7. Saint B (2009) Rural distribution system planning using Smart Grid Technologies. In: IEEE rural electric power conference, REPC '2009, B3-B8. IEEE, USA
8. Sailalitha P, Kiranbabu B (2013) Smart grid. Int J Eng Trends Technol (IJETT) 4(4):883–887
9. Shafiullah G, Amanullah M (2013) Smart grid for a sustainable future. In: Smart grid and renewable energy, scientific research, 23–34, UK
10. Lu X, Wang W, Ma J, Sun L (2013) Domino of the smart grid: an empirical study of behaviour in independent network architecture. In: IEEE Smart Grid Comm 2013 Symposium—Smart Grid Standards, Co-simulation, Test-beds, and Field Trials, 612–617. IEEE, China
11. Gautam KK, Bhuria V (2011) Fuzzy logic application in power system fault diagnosis. Indian J Comput Sci Eng (IJCSE) 2554–558
12. Lu X, Wang W, Ma J, Sun L (2013) Domino of smart grid: an imperial study of system behavior in interdependent network architecture. IEEE J Smart Grid Comm, 1–6
13. Fang X, Misra S, Xue G, Yang D (2011) Smart grid—the new and improved power grid: a survey. In: Proceedings of IEEE system, 12–50, IEEE, Mexico
14. Glavic M, Fonteneau R, Ernst D (2017) Reinforcement learning for electrical power system decision and control: past consideration and perspectives. Elsevier J Int Fed Autom Control 50(1):6918–6927
15. Vrablecováa P, Ezzeddinea AB, Rozinajováa V, Šárik S, Sangaiah AK (2018) Smart grid load forecasting using online support vector regression. Elsevier J Comput Electr Eng 65:102–117
16. Kezunovic M (2011) Smart fault location on smart grids. In: IEEE Trans Smart Grid 2(1):11–22. IEEE, USA
17. Hau L-C, Lee J-V, Chuah Y-D, Lai A-C (2013) Smart grid—the present and future of smart physical protection: a review. Int J Energy Inf Commun 4(4):43–53
18. Shaukat N, Ali SM, Mehmood CA, Khan B, Jawad M, Farid U, Ullah Z, Anwar SM, Majid M (2018) A survey on consumers empowerment, communication technologies, and renewable generation penetration within Smart Grid. Elsevier J Renew Sustain Energy Rev 81(1):1453–1475
19. Zhang P, Li F, Bhatt N (2010) Next generation monitoring, analysis and control for smart grid control centre. IEEE Trans Smart Grid 1(2):186–192. IEEE, USA
20. Sahu N, Jha M, Qureshi F (2014) Fault diagnosis of power distribution system using interval type-2 fuzzy data mining. IJISET Int J Innov Sci Eng Technol 1(5):71–83
21. Albasrawi MN, Jarus N, Joshi KA, Sarvestani SS (2014) Analysis of reliability and resilience for smart grid. In: Proceedings of IEEE conference on smart grid communications, 516–521. IEEE, USA

22. Laaksonen HJ (2010) Protection principle of future micro grid. IEEE Trans Power Electron 25(12):2910–2918
23. Xu S, Qian Y, Hu RQ (2015) On reliability of smart grid neighborhood area networks. Proc IEEE, 2352–2365. IEEE, USA
24. Lee S-H, Lee S-J, Moon K-I (2013) Adaptive neural fuzzy reasoning model for smart grid. In: Proc Adv Sci Technol Lett (42):141–144
25. Noorian Z (2009) An autonomous agent-based framework for self-healing power grid In: IEEE conference on systems, man and cybernetics proceeding, 18–22, IEEE, San Antonio, USA
26. Nouri H, Wang C, Davies T (2001) An accurate fault location technique for distribution lines with tapped loads using wavelet transform. In: Proceedings of IEEE power tech vol 3, pp 1–4, Porto
27. Huang J, Wang H, Qian Y, Wang C (2013) Priority-based traffic scheduling and utility optimization for cognitive radio communication, infrastructure-based smart grid. IEEE Trans Smart Grid 4(1):78–86. IEEE, USA
28. Deng R, Chen J, Cao X, Zhang Y, Maharjan S, Gjessing S (2013) Sensing-performance trade-off in cognitive radio enabled smart grid. IEEE Trans Smart Grid 4(1):302–310
29. Nirmala M, Karthikeyan G (2012) Impact of GCIs performance on power quality of smart grid using fuzzy logic. Int J Modern Eng Res IJMER 2(2):310–314
30. Roche R, Blunier B, Miraoui A, Hilaire V, Koukam A (2010) Multi agent system for grid energy management—a short review. In: Proceedings of IEEE system and transport, 3341–3346. IEEE, France
31. Ahmadzadeh M, Mohammad Zadeh S (2014) The effect of fuzzy logic controller on power system stability; a comparison between fuzzy logic gain scheduling PID and conventional PID controller. In: international conference on computer, systems and electronics engineering ICSCEE'2014, 83–87, ICSCEE, Johannesburg, South Africa
32. Strasser T, Andrén F, Kathan J, Cecati C, Buccella C, Siano P, Leitão P, Zhabelova G, Vyatkin V, Vrba P, Mařík V (2015) A review of architectures and concepts for intelligence in future electrical energy system. IEEE Trans Ind Electron 62(4):2424–2438. IEEE, Prague
33. Gigris AA, Hart DG, Peterson WL (1992) A new fault location technique for two- and three terminal lines. IEEE Trans Power Deliv 7:98–107
34. Gharvi H, Ghafurian R (2011) Smart grid—the electrical energy system of future. Proc IEEE 99(6):917–921. IEEE, UK
35. Marken P, Marczewski J, D'Aquila R, Hassink P, Roedel J (2009) VFT—a smart transmission technology that is compatible with the existing and future grid. In: IEEE power systems conference and exposition, IEEE, USA
36. Calderaro V, Hadjicostis CN, Piccolo A, Siano P (2011) Failure identification in smart grids based on petri net modeling. EEE Trans Ind Electron 58(10):4613–4622
37. Gudzius S, Markevicius LA, Morkvenas A (2011) Characteristics of fault detection system for smart grid distribution network. Electron Electr Eng Proc 6(112):123–126
38. Kezunovic M (2011) Smart Fault location for smart grids. IEEE Trans Smart Grid 2(1):11–22
39. Chin H-C (2003) Fault section diagnosis of power system using fuzzy logic. IEEE Trans Power Syst 18(1):245–250. IEEE
40. Goldberg DE (1989) Genetic algorithms in search, optimization and machine learning. Addison-Wesley, Reading, MA
41. Lu X, Lu Z, Wang W, Ma JF (2011) On network performance evaluation toward the smart grid: a case study of DNP3 over TCP/IP. In: Proceedings of IEEE GLOBECOM, 9268–9273. IEEE, USA
42. Roger JJ (1993) ANFIS: adaptive-network-based fuzzy inference system. IEEE Trans Syst Man Cybern 23(3):665–685. IEEE, USA
43. Komorowski J, Pawlak Z, Polkowski L, Skewon A (1999) Rough sets: a tutorial. In: Pal SK, Skewon A (eds) Rough fuzzy hybridization: a new trend in decision making, 3–98. Springer, Singapore
44. Ahn MC, Ko TK (2011) Smart fault current controller with superconducting coil for smart Grid. IEEE Trans Appl Supercond 21(3):2201–2204. IEEE, Korea

45. Dua D, Dambhare S, Gajbhiye RK, Soman SA (2008) Optimal multistage scheduling of power measurement unit placement: an integer linear programming approach. IEEE Trans Power Deliv 23(4):1812–1820. IEEE, India
46. http://www.gpo.gov/fdsys/pkg/FR-2010–05-11/html/2010-11129.htm, last accessed 2019/07/22
47. El-hawary ME (2014) The smart grid—state-of-the-art and future trends. In: Electric power components and systems, 239–250. Taylor and Francis, Canada
48. Suna Q, Ge X, Liu L, Xu X, Zhang Y, Niu R, Zeng Y (2011) Review of smart grid comprehensive assessment system. In: SciVerse ScienceDirect, 219–229. Elsevier, China
49. Xu Y, Wang W (2013) Wireless mesh network in smart grid: modeling and analysis for time critical communications. IEEE Trans Wirel Commun 12(7):360–371
50. Lu X, Wang W, Ma J (2013) An empirical study of communication infrastructure toward smart grid: design implementation and evaluation. IEEE Trans Smart Grid 4(1):170–183. IEEE, North Carolina, USA
51. Wang H, Qian Y, Sharif H (2013) Multimedia communication over cognitive radio network for smart grid application. In: Proceedings of IEEE conference on wireless communication, 125–132, IEEE, USA
52. NIST (2009) Framework and roadmap for smart grid inter-operability standards, Release 1.0. U.S. Department of Commerce, Gary Locke, Secretary, NIST 1108, 1–145, NIST, USA
53. Yu R, Zhong W, Xie S, Zhang Y, Zhang Y (2016) QoS differential scheduling in cognitive-radio-based smart grid networks: an adaptive dynamic programming approach. IEEE Trans Neural Netw Learn Syst 27(2):435–443. IEEE, China
54. Liu S, Liu PX, El Saddik AM (2015) Modeling and stability analysis of automatic generation control over cognitive radio networks in smart grids. IEEE Trans Syst Man Cybern Syst 45(2):223–234
55. Kulkarni SN, Shingare P (2016) A review on smart grid architecture and implementation challenges. In: International conference on electrical, electronics, and optimization techniques, 3285–3290. IEEE, India
56. Gharavi H, Ghafurian R (2011) Smart GRID: THE electrical energy system of future. Proc IEEE 99(6):917–921
57. Sun M, Aronson JE, Mckeown PG, Drinka D (1998) A tabu search heuristic procedure for the fixed charge transportation problem. Eur J Oper Res 106:441–456
58. Laaksonen H, Kauhaniemi K, Vomia S (2011) Protection system for future LV microgrid. In: 21st international conference on Electricity distribution, 0431–0434, Germany
59. Short TA, Sabin DD, McGranaghan MF (2007) Using PQ monitoring and substation relays for fault location on distribution Systems. In: Proceedings of IEEE rural electric power conference, B3-B7. IEEE
60. Afsar S, Nadendla G, Sudha R (2015) A review of recent development in smart grid systems. Int J Adv Res Electr Electron Instrum Eng 4(1):96–103
61. U.S.-Canada Power System Outage Task Force (2012) Final report on August 14, 2003 blackout in the United States and Canada: causes and recommendations, Tech. Report
62. Calderaro V, Hadjicostis CN, Piccolo A, Siano P (2011) Failure identification in smart grids based on petri net modelling. IEEE Trans Ind Electron 58(10):4613–4624
63. Mirsaeidi S, Said DM, Mustafa MW, Habibuddin MH, Ghaffar K (2014) Review and analysis of existing protection strategies for micro-grids. J Electr Syst 10(1):1–10
64. Jensen R, Shen Q (2004) Fuzzy-rough attribute reduction with application to web categorization. Fuzzy sets and systems. Trans Rouge Set XVII 141(3):469–485
65. Strasser T, Andrén F, Kathan J, Cecati C, Buccella C, Siano P, Leitão P, Zhabelova G, Vyatkin V, Vrba P, Marík V (2015) A review of architectures and concepts for intelligence in future electric energy systems. IEEE Trans Ind Electron 62(4):2424–2438
66. Corcau JI, Stoenescu E (2007) Fuzzy logic controller as a power system stabilizer. Int J Circuits Syst Signals Process 3(1):266–273
67. Ranganathan P, Nygard K (2010) An optimal resource assignment problem in smart grid future computing. In: International conference on future computational technologies and applications, 28–34. IEEE, Lisbon

68. Hannu Jaakko Laaksonen (2010) Protection principles for future microgrids. IEEE Trans Power Electron 25(12):2910–2918
69. Zhang P, Li F, Bhatt N (2010) Next-generation monitoring, analysis, and control for the future smart control center. IEEE Trans Smart Grid 1(2):186–192
70. Ferber J (1999) Multi-agent systems: an introduction to artificial intelligence, 3rd edn. Addison-Wesley, USA
71. Kulkarni SN, Shingare P (2016) A review on smart grid architecture and implementation challenges. In: International conference on electrical, electronics, and optimization techniques (ICEEOT), 978–993, India

Chapter 12
A Performance Comparison of Segmentation Techniques for the Urdu Text

Atif Mahmood, Amod Kumar Tiwari and Sanjay Kumar Singh

1 Introduction

All text segmentation method has particular sub-procedures: scanning, analysis, preprocessing and segmentation, for the best possible division of content into its subparts [1]. In the primary phase, printed or handwritten text image is converted into jpg image by using flatbed scanner at 300 dpi. Further, in the second phase scanned text is examined for the existence of inclination or slant and they are revised eventually. As printed contents have various forms of depictions such as graph, picture, table, etc., therefore it is necessary that the pure textual regions to be distinguished independently from different delineations and could be limited and removed. In the third stage, procedures, for example, commotion and obscure expulsion, binarization, skeletonization, edge discovery, and some morphological procedure are applied to the text image, to prepare a segmentation ready picture which is free from clamor and obscure. In the final phase, image consisting of only text contents as a whole is first segmented into separate lines of text and further into words or individual letters [2, 3]. It involves the decision of the segmentation method to get this division into line word or individual characters. The consequence of the segmentation method used for text picture is displayed by the system, as appeared in Fig. 1.

In this paper, three existing text-line segmentation methods are adopted and implemented for the Urdu script. Performance comparison of all three algorithms over the gathered Urdu data samples is presented, and the best segmentation algorithm is recommended. We have used 115 text samples from different newspapers (20 samples), poetry books (35 samples), and magazines (60 samples) to test all the three text segmentation methods. Our data set contains good as well as some bad samples

A. Mahmood (✉) · A. K. Tiwari
Department of Computer Science & Engineering, Rajkiya Engineering College, Sonbhadra, Uttar Pradesh, India

S. K. Singh
Department of Computer Science, R.S.M.T., U.P College, Varanasi, Uttar Pradesh, India

V. K. Giri et al. (eds.), *Computing Algorithms with Applications in Engineering*, Algorithms for Intelligent Systems, https://doi.org/10.1007/978-981-15-2369-4_12

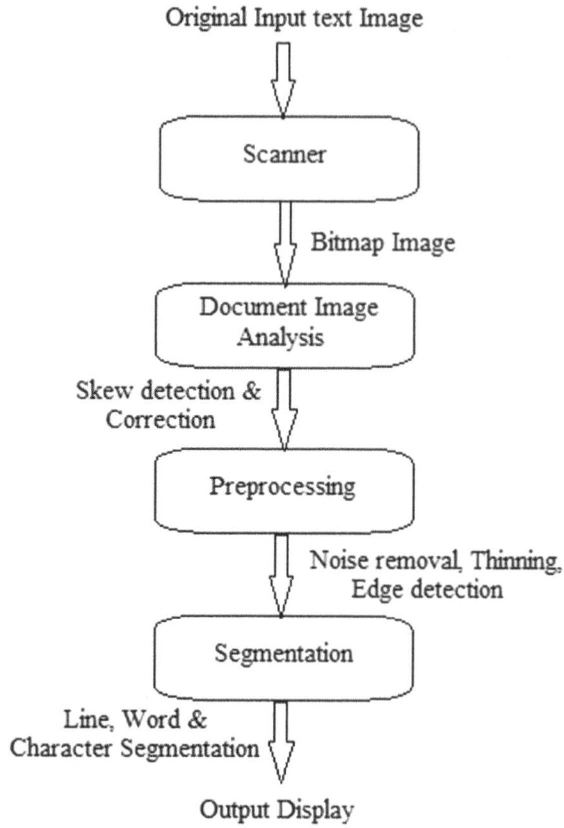

Fig. 1 Segmentation process [3]

(i.e., skew-angled text, the text having boundary noise, less interline space between text lines) which makes the data set a real data set.

The structure of the present work is as follows: In Sect. 2, detail analysis of Urdu language along with basic characteristics and complexities is described. Section 3 presents the overview of different segmentation algorithms reported in the literature for the text-line division, and also the work of implemented approach is briefly explained. Experimental results of three segmentation method over gathered data set are presented in Sect. 4. And finally, Sect. 5 gives the conclusive idea of the present work.

2 Urdu Language Explication

Urdu is a common language of South East Asia and generally spoken by individuals living in India and Pakistan. In India, around 51 million individuals generally communicate in Urdu in numerous states as a provincial lingo, while in Pakistan around 16 million individuals communicates in Urdu as a local vernacular. Base of the Urdu language is Indo-Persian, and it is created and affected by a wide range of dialects for example Hindi, Sanskrit, Arabic, Persian, Farsi, Pashto, Turkish, Malay, and English [4].

Urdu language is bidirectional in nature, where characters are created from right and numerals from left-hand side in Nastaleeq or Naskh contents ordinarily. Nastaleeq content is composed slantingly from upper appropriate to base left corner having multiple baselines, while Naskh content is composed laterally having single baseline, and both of these contents are cursive and delicate in nature. Urdu utilizes all-encompassing Arabic adjusted content; it has 39 characters as against Arabic 28 [5]. These characters are combined to make expressions of the language, one of the intriguing marvels that Urdu content display is change in essential states of characters. This difference in shape is subjected to the situation of the character in ligatures. There are four conceivable positional classes in which states of a character can be partitioned: starting, average, last places of a character in a ligature, and the separated one. One character can have a few shapes relying on going before and succeeding characters [6].

2.1 Urdu Characters and Assortment

Urdu language has ordinarily 39 essential characters, yet in the review it is discovered that characters fluctuates in the count from 38 to 58. We have assumed 38 characters as a standard in our paper compatibility to the Nastaleeq contents [5, 7]. Characters in Urdu are classified into 20 classes on the basis of their identical shapes, shown in Fig. 2. These indistinguishable shape characters differ from one another just because of various dots (diacritics) plans.

2.2 Diacritics

In the Urdu language, characters are advanced with some uncommon imprints (e.g., Nuqta (specks), Paish, Zabr, Zair, Juzm, Shud, Khari Zabr, and Chota To), called diacritics. Two of the diacritics 'Nuqta' and 'Chota To' are ordinary piece of certain characters. Numbers and position of Nuqta related with a character are imperative in recognizing a character. Rests of the diacritics are called 'Aerab'; these are vowel marks and supportive in way to express Urdu word. In Fig. 3, diacritics are recorded, and the circles speak to a character to explain the situation of diacritic. 'Aerab' are written in text when there is chaos in the articulation; else they are discretionary and typically not written in normal content [7].

ک گ	11	ا	1
ل	12	ب پ ت ٹ ث	2
م	13	ج چ ح خ	3
ن	14	د ڈ ذ	4
و	15	ر ڑ ز ژ	5
ہ	16	س ش	6
ھ	17	ص ض	7
ء	18	ط ظ	8
ی	19	ع غ	9
ے	20	ف ق	10

Fig. 2 Assortment of Urdu characters

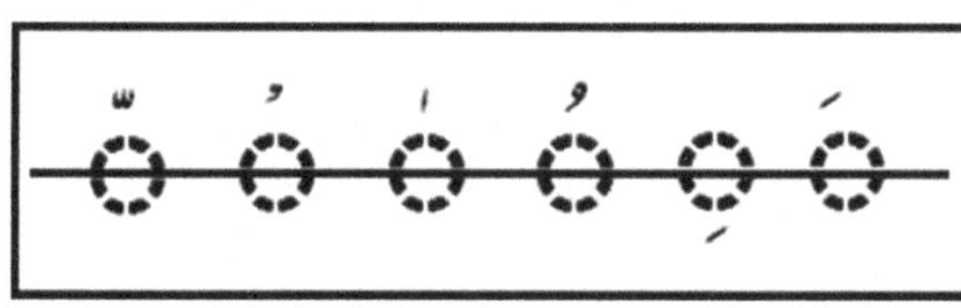

Fig. 3 Diacritics in Urdu (right to left: Zabr, Zair, Paish, Khari Zabr, Juzm, and Shud) [7]

2.3 *Cursiveness*

Characters in Nastalique style of writing are naturally cursive due to its streaming character and shapes. The character joins together cursively to form words or ligatures [2]. In Nastalique, ligature is a solitary unit of various characters bound together cursively in a liquid structure making an assortment of compound characters, and Fig. 4 demonstrates the cursiveness of character 'ye' set apart with a circle. A solitary word may in this manner be made of more than one ligature or as disconnected characters. Consequently, characteristic highlights of cursiveness make the nastalique content progressively mind boggling [8].

Fig. 4 Cursiveness in Urdu writing

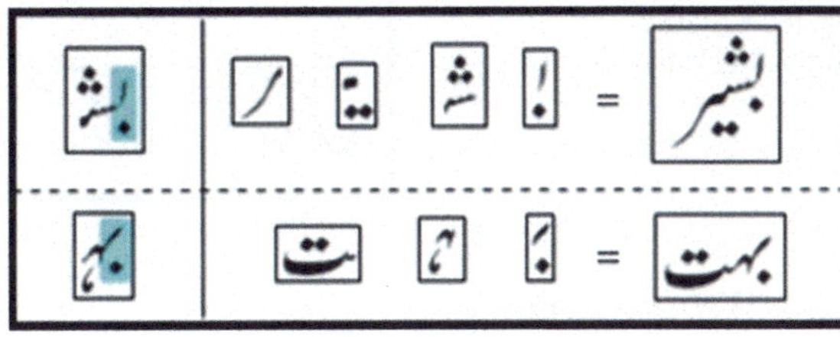

Fig. 5 Context sensitivity—two shapes of bay initial [2]

2.4 Context Sensitivity

Urdu is delicate language; during word or ligature development, state of the character changes because of the reliance of each character on the former or succeeding characters it unites with. Each character contingent on its situation in a word can frame two to four unique shapes [1]. For instance, "bay (ج)" which is the second character in the letters in order is very unique in the shape it bears as an initial, average, last, or a disengaged one. Up to 20 unique shapes, any character may frame while uniting with different characters. In some cases even the third or fourth going before or succeeding character may start an adjustment in shapes. Therefore, Bay (ج) has numerous occurrences and may have around 20 distinct shapes for its underlying structure. Figure 5 illustrates two different context sensitive shapes of 'bay (ج)' initial [2].

2.5 Dots' Position and Number

Dots are the compulsory piece of a portion of the Urdu characters; complete 17 characters out of 38 characters in Urdu letters in order to have dots, which are situated above, underneath, or inside the character. The number of dots associated with any character can be one, two, or three [7]. Diverse dots' location and counts associated with the characters discriminate the characters which are identical in shape.

2.6 Diagonality

Urdu nastalique script is composed corner to corner at various angles with the even base. This way of composing causes unpredictable tallness and width of words and ligatures and consumes less flat space [7]. Also, the number of characters joining together does not have limit.

Figure 6 shows the diagonality of characters, and the word development is appeared in five stages; in each progression, one character is included. With the expansion of a character, previous characters are heaped up slantingly and move toward the upper right corner.

Fig. 6 Nastaleeq diagonality behavior

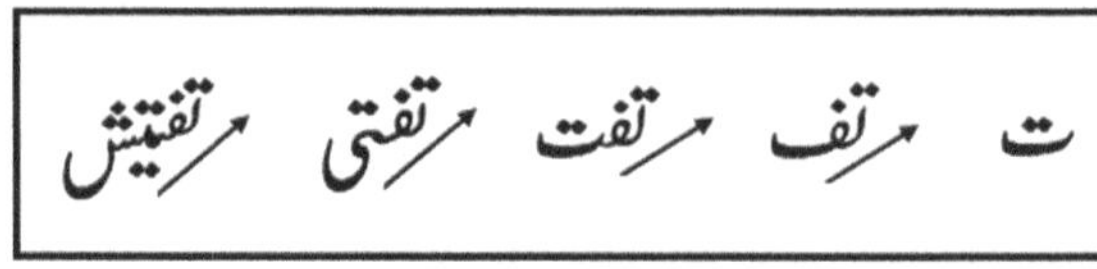

2.7 *Writing Styles*

Urdu can be inscribed in the several multiple styles such as Nashk, Nastalique, Batol, Aswad, Taleeq, Zaben, but commonly used styles are Naskh and Nastaleeq only. The Naskh style has a plane single gauge, and letter sets spread evenly along the single guage taking extensive space for composing a word or ligature [2]. The Naskh style because of its linearity recorded in writing is simpler for division contrasted with Nastalique.

Nastalique style is the conventional writing style for Urdu printings and advanced from two contents named 'Naskh' and 'Taleeq'. Nastalique is exceptionally cursive in nature, and because of various baselines and diagonality of characters, it is one of the most convoluted contents to be utilized electronically [8]. A sample of Nashk and Nastalique style is shown in Fig. 7.

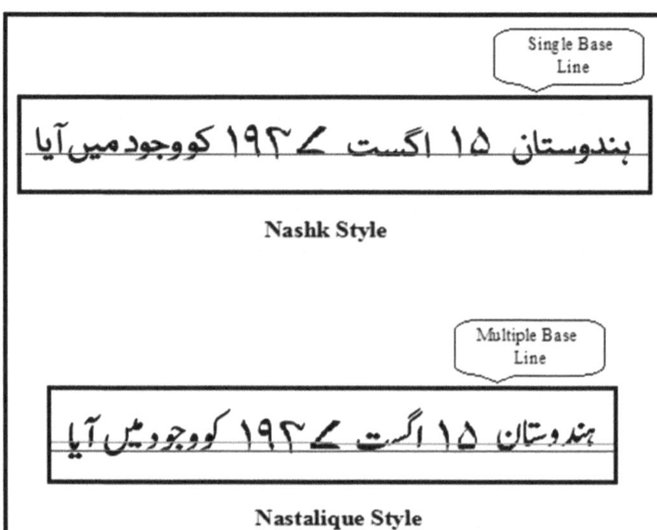

Fig. 7 Styles of Urdu writing

2.8 Overlapping

The overlapping (projection) of characters due to cursiveness and corner to corner composing conduct of nastalique contents introduces complexity in the text segmentation. Overlapping exists are of two types: One is projection between two imminent ligatures, and another is projection within a ligature.

In projection within a ligature, characters are connected with the neighboring character in a liquid structure, beginning from the correct top and moving slantingly toward the base left. As characters are heaps up vertically and push forward, it spares space for composing. Additionally, certain characters demonstrate a non-monotonic nature which is written backward heading (left to right). In some cases when characters unite with other characters, they end up on the right side of the beginning stage, rendered in Fig. 8.

Projection between two neighboring ligatures happens when a few characters having a place with various ligatures cover one another [7]. This kind of covering brings about disposing of the vertical blank area between two ligatures and making ligature detachment progressively unpredictable (Fig. 9).

3 Review and Implemented Approaches

There are large quantities of text-line segmentation (division) method accessible in the literature. Line segmentation is observed as an essential and advance step for all-normal language-handling applications. Right line division for any context archives

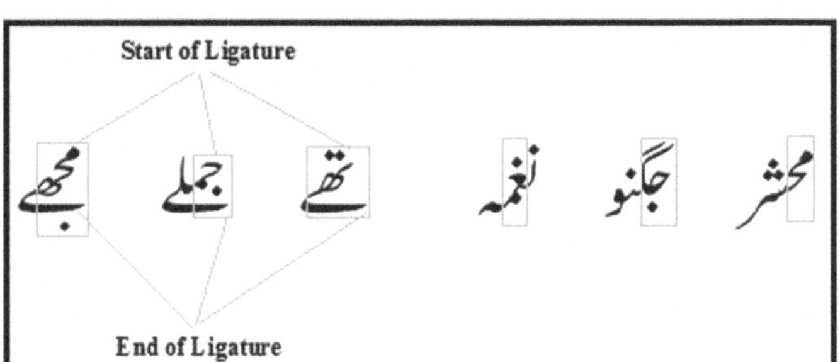

Fig. 8 Projection within a ligature

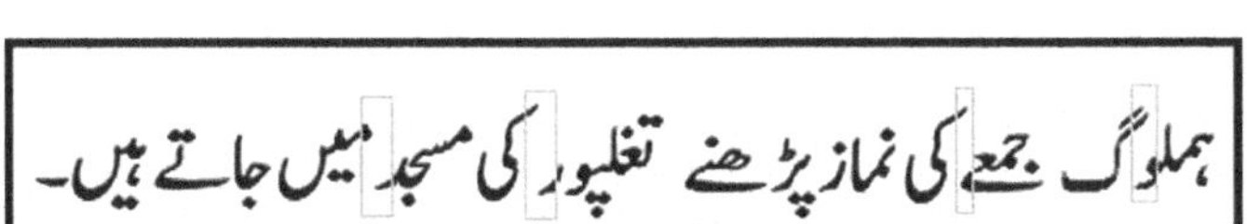

Fig. 9 Projection between neighboring ligatures

may improve the presentation of context division applications. Some of the well-known context line division strategies that have been reported in the study for various dialects are as follows: projection profile method [9–11], repulsive–attractive network method [12], smearing method [13, 14], docstrum method [15], hough transform-based method [16, 17], stochastic (probabilistic Viterbi algorithm) [18], constrained text-line detection method [19], edges' information-based segmentation method [3], etc. Table 1 summarizes the above papers related to different types of segmentation methods developed under different projects.

Table 1 Overview of text-line segmentation methods

Author	Algorithm	Functionality	Writing Type	Documents
[10]	Projection method	Line and word detection	Printed English, Hindi, and Urdu text	Newspaper, magazine, book, computer printouts, question paper, etc.
[11]	Projection method	Line detection	Latin printed	Memorial/personal records(World War II)
[12]	Projection method (Recursive X-Y cut)	Line detection	Printed English	UW English document image database
[13]	Smearing Method (Run length moothing)	Line detection	Printed English text	Mixed text image documents
[14]	Smearing method (fuzzy run length)	Line detection	Latin handwriting	Newton, Galileo manuscripts
[16]	Constrained line finding method	Column separator and line detection	US style printed text	UW3 document database
[15]	Docstrum method	Text block and line detection	Printed English text	Scanned journal pages and business cards
[16]	Hough transform, (hypothesis validation scheme)	Connected component and lines detection	Latin handwriting	Philectre/authorial manuscripts, manuscripts of the sixteenth century
[17]	Hough transform (moving window)	Clusters and lines' detection	Latin handwriting	Handwritten documents
[12]	Repulsive–attractive network	Line detection	Arabic and Latin handwriting	Ancient Ottoman documents
[18]	Stochastic(probabilistic Viterbi algorithm)	Line detection	Chinese handwriting	Handwritten documents
[3]	Edges' information-based segmentation method	Connected component and lines' detection	Printed Urdu text	Papers, books, and magazines

It was observed that projection and smearing methods are easy to implement and normally adopted for straight lines. We have selected three algorithms for the performance analysis; brief working of each method along with their potency and debility is discussed underneath.

3.1 Projection Method

Projection method is widely used to discover lines in most of the text images. A horizontal projection is applied over text image that reckon sum of all black pixels for each individual row of text image and build a histogram. The ditch between consecutive histogram peaks is identified and used to mark the boundary between the consecutive text lines [9, 20]. This method works well for printed clean text image with good interline spacing, but fails for text image which contains noise, small interline spacing, skew-angled, and multi-column formats.

3.2 Smearing Method

Smearing technique takes a shot at binary pictures where high- and low-contrast pixels of the pictures are spoken as '1' and '0,' respectively. This technique changes a twofold grouping x into y for a predefined limit esteem *C*, by following two guidelines.

a. If the number of imminent 0s is not exactly or equivalent to *C*, at that point 0s in *x* are changed to 1s in *y*.
b. 1s in *x* are unaltered in *y*.

These two stages perform connecting together neighboring dark regions that are isolated by not exactly or equivalent to *C* pixels [21].

Smearing technique works for printed clean context pictures having some interline dividing space and for twofold text format. However, this method fails in a few situations where context pictures are slant, having no interline separation and having fringe clamor.

3.3 Edge Information-Based Segmentation Method

This segmentation method comprises three modules: Very first module is "connected component detection module," which discovers the connected components (i.e., single characters, words, or diacritics) in a given input text image. It is being observed that line and section places of a segment are connected to its close-by segment or not. The second module which is "connected component edge detection module" finds out the edge's information (i.e., start, end, merge, split, and continue points) for each individual connected component. Lastly, in the segmentation module, a line is set apart between back-to-back context lines in a context picture by taking an average of begin status and end status for each associated segment in a specific content line [3].

This method works well for printed as well as handwritten text, and for text having small interline spacing, having single character or word within a text line, and having border noise, and performance is fairly low for text image of double-column format, no interline spacing, and skew-angled text images.

4 Experiments and Results

This section demonstrates the exhibition of all the three text segmentation strategies tried over the gathered informational collection, utilizing standard precision metric and recall metric. In reference to content division, the precision metric is characterized as the precise number of boundaries identified by the framework divided by the general boundaries (i.e., right or mistaken) identified by the framework. And the recall metric is characterized as the exact number of boundaries identified by framework divided by the in general real boundaries that exists in the example informational collection [21].

For the experiment purpose, we have utilized absolute 115 context examples having 1664 context lines from various papers (20 samples), books (60 samples), and magazines (35 samples); each sort of sample contains 216, 910, and 532 context lines individually. Result assessment for paper, book, and magazine samples, utilizing precision and recall measurements for three text division techniques, is exhibited in Tables 2 and 3, respectively. The overall evaluation for all 115 data set using precision and recall metrics is presented in Table 4.

Table 2 Results using precision metric

Text segmentation method	Newspaper (%)	Book (%)	Magazine (%)
Projection profile method	59.22	82.24	76.58
Smearing method	64.92	88.72	90.33
Edge information-based segmentation method	63.55	85.92	82.62

Table 3 Results using recall metric

Text segmentation method	Newspaper (%)	Book (%)	Magazine (%)
Projection profile method	59.22	82.24	76.58
Smearing method	64.92	88.72	90.33
Edge information-based segmentation method	63.55	85.92	82.62

Table 4 Results for overall data set

Text segmentation method	Precision metric (%)	Recall metric (%)
Projection profile method	72.68	70.05
Smearing method	81.32	83.61
Edge information-based segmentation method	77.36	74.75

Results show that smearing method outperforms the other two methods for precision and recall metric, for all three types of data set (i.e., newspaper, books and magazine dataset), while the performance of projection profile method lags behind other methods.

5 Conclusions

This paper represents the performance comparison of three text line segmentation methods over the gathered data set using precision and recall metrics. The evaluation results show that smearing method performs better than the edge information-based segmentation method and projection method. Performance of all the three methods is satisfactory for book and magazine samples but for newspaper samples, performance degrades due to the presence of overlapping of text between consecutive lines. It was found that certain issues related to text sample, such as "text images having less or no interline spacing, skew-angled text images, double-column text lines, and border noise text images" cannot be resolved through any of the presented methods. To achieve more accuracy, some other segmentation methods can be used, or the presented methods could be further modified to resolve them.

References

1. Mahmood A (2013) Arabic & Urdu text segmentation challenges & techniques. IJCST 4(special-1):32–34
2. Sattar SA (2009) Ph.D. thesis: A technique for the design and implementation of an OCR for printed Nastalique text. N.E.D. University of Engineering & Technology, Karachi, Pakistan
3. Mahmood A, Srivastava A (2018) A novel segmentation technique for Urdu type written text. In: IEEE international conference on recent advances in engineering, technology and computational sciences, Allahabad (U.P), India, 06–08 February, 2018
4. Rehman Z, Anwar W, Bajwa UI (2011 Nov 8) Challenges in Urdu text tokinization and sentence boundary disambiguation. IJCNLP, 40–45
5. Ahmed Z, Orakzai JK, Shamsher I, Adnan A (2007) Urdu Nastaleeq optical character recognition, 249–252
6. Nawaz T, Naqvi S, Rehman H, Faiz A (2008) Optical character recognition system for Urdu (Nashk font) using pattern matching technique IJIP 3(3):92–104
7. Muaz A (2010) MS thesis: Urdu optical character recognition system. National University of Computer & Emerging Sciences, Lahore, Pakistan

8. Iftikhar U (2011) M.Sc. thesis: Recognition of Urdu ligatures. [S.l.]: VIBOT Consortium and German Research Center for Artificial Intelligence
9. Chanda A, U Pal (2005) English, Devnagari and Urdu Text Identification. In: Proceedings of the international conference on cognition and recognition. [S.l.]: [s.n.]
10. Antonacopoulos A, Karatzas D (2004 Jan) Document image analysis for world war II personal records. In: Proceedings of the international workshop on document image analysis for libraries (DIAL2004). IEEE Computer Society Press, Palo Alto, pp 336–341
11. Ha J, Haralick RM, Phillips IT (1995) Recursive X-Y cut using bounding boxes of connected components. In: Proceeding of 3rd international conference on document analysis and recognition (ICDAR), Aug 1995. IEEE Computer Society, pp 952–955
12. Öztop E, Mülayim AY, Atalay V, Yarman-Vural F (1997) Repulsive attractive network for baseline extraction on document images. In: IEEE international conference on acoustics, speech, and signal processing. IEEE, Munich, Germany
13. Wong KY, Casey RG, Wahl FM (1982) Document analysis systems. IBM J Res Dev 26(6)
14. Shi Z, Govindaraju V (2004) Historical document image enhancement using background light intensity normalization. In: Proceedings of the 17th international conference on pattern recognition (ICPR'04). IEEE Computer Society
15. O'Gorman L (1993) The document spectrum for page layout analysis. IEEE Trans 15:1162–1173
16. Likforman-Sulem L, Hanimyan A, Faure C (1995) A hough based algorithm for extracting text lines in handwritten documents. In: Proceedings of 3rd international conference on document analysis and recognition. IEEE, pp 774–777
17. Pu Y, Shi Z (1998) A natural learning algorithm based on Hough transform for text lines extraction in handwritten documents. In: Proceedings of the 6 international workshop on frontiers in handwriting recognition, Taejon, Korea, 637–646
18. Tseng Y-H, Lee H-J (1999 Aug) Recognition-based handwritten Chinese character segmentation using a probabilistic Viterbi algorithm. Pattern Recognit Lett 20(8):791–806
19. Breuel TM (2001) Two geometric algorithms for layout analysis. In: Proceedings of the international workshop on document analysis systems, 188–199
20. Khorsheed MS (2001) Offline Arabic character recognition—a review. Pattern Anal Appl 5(1):31–45
21. Shafait F, Keysers D, Breuel TM (2006) Performance comparison of six algorithms for page segmentation. Image Understanding and Pattern Recognition (IUPR) Research Group, German Research Center for Artificial Intelligence (DFKI) and Technical University of Kaiserslautern, Kaiserslautern, Germany

Chapter 13
Improvement in Power Quality Using Ultra-Capacitor-Integrated Hybrid-Active Filter for Current Harmonic Mitigation

Soumya Ranjan Das, Prakash K. Ray, Asit Mohanty, V. P. Singh and Alok K. Mishra

1 Introduction

In recent days, there is an increasing use of power electronics device (PED) in different fields. These uses of PED in turn produce harmonics [1] and create lots of disturbances in the distribution network. Nonlinear loads pull current which is non-sinusoidal and creates drop in voltage across conductors attached to the supply system. Factors like overloading or under loading, capacitor disasters, etc., are major causes due to harmonics. Because of nonlinear features of PED, several power quality issues have been raised. Excess reactive power adversely affects the loss in the transmission network and in addition to that improves the generating capacity of generating stations. Generally, the load side requires large amount of reactive power because losses in transmission network completely depend on reactive power. To overcome from this PQ issues conventionally, passive filters were used for reducing the harmonics, but due to its inherent characteristics of bulky size, resonance problems, etc., they are used very limited. Subsequently, hybrid filters [2, 3] are utilized for management of reactive power and reduction in distortion of harmonics at load end. In the power system network, harmonics play a major role, create unbalance at load end and produce high neutral currents. Due to the presence of harmonics,

S. R. Das
IIIT, Bhubaneswar, India

P. K. Ray (✉) · A. Mohanty
CET, Bhubaneswar, India

V. P. Singh
Rajkiya Engineering College, Sonbhadra, India

A. K. Mishra
SOA University, Bhubaneswar, India

V. K. Giri et al. (eds.), *Computing Algorithms with Applications in Engineering*, Algorithms for Intelligent Systems, https://doi.org/10.1007/978-981-15-2369-4_13

the system gets low power factor with less efficiency. In the same time, voltages at different level get highly affected due to existence of harmonics.

The basic design of hybrid filter reduces the restrictions of compensating devices. Production of renewable source is increasing quickly and is suitably active in energy storage resources like ultra-capacitor (UCAP) [4–6] for providing real and reactive power. Compared to conventional storage battery, UCAP has higher number of charge/discharge cycles and is cost-effective. The prime motto is to develop incorporation of SHAF with UCAP which provides the dynamic power ability to the utility network system. In this paper, incorporation of SHAF with UCAP via a bidirectional DC-DC boost converter [7] is suggested. In this paper, simulation model of integrated UCAP-based SHAF is performed with and without DC-DC converter. The results are compared and analysed using MATLAB/ Simulink tool.

2 Model Configurations

Figure 1 describes the circuit composition of SHAF with UCAP feed through DC-DC boost converter. The UCAP supported with three-phase, three-wire SHAF is used as prototype. Generally, the function of HAPF is to improve significantly the filtering performance of harmonics of higher order. The DC link of the HAPF is maintained by the UCAP power-generating system to provide a long-lasting compensation against current-based disturbances in the power distribution system. A DC-DC converter is employed to act as an interface between UCAP and loads. The DC-DC boost converter maintains UCAP output voltage. Such an arrangement of SHAF with the

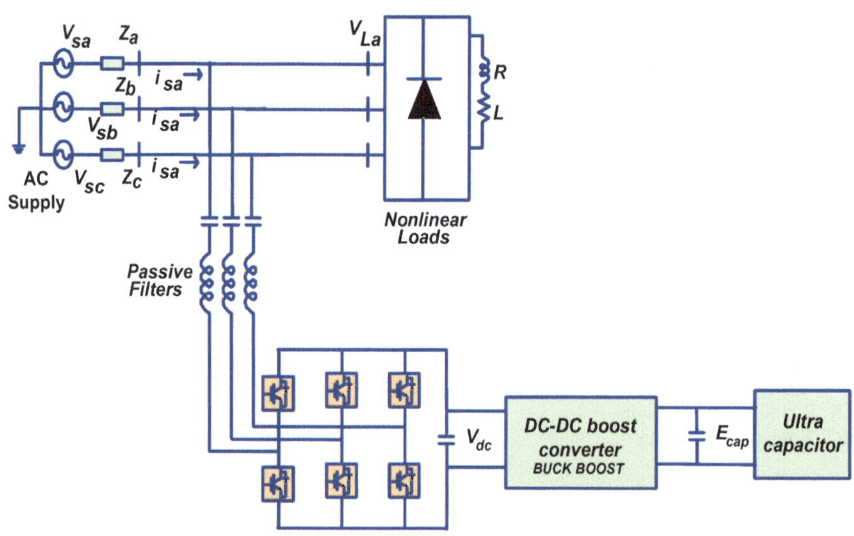

Fig. 1 Configuration of SHAF with UCAP-tied DC-DC boost converter

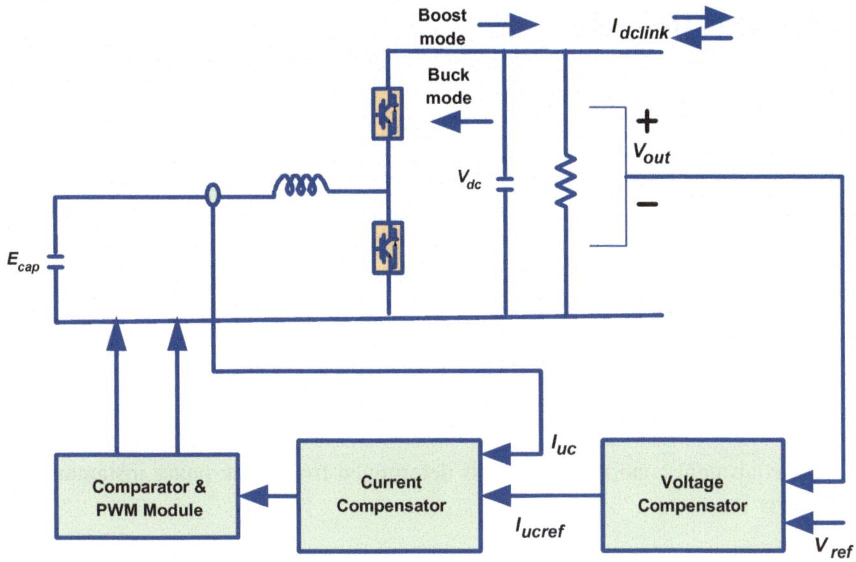

Fig. 2 Block diagram of bidirectional DC-DC converter

passive filter decreases the active filter ratings. The main objective of an SHAF is relatively to achieve isolation of harmonics between source and load. Two different control schemes are involved in SHAPF: one is controlling with boost converter, and other is running the SHAF with UCAP without using boost converter. During the energy discharge, the value of voltage of UCAP changes simultaneously. Hence, feeding UCAP with VSI of SHAP is not convenient. In order to overcome from these issues, a bidirectional DC-DC converter is connected in between VSI and UCAP. Bidirectional DC-DC converter controls the DC-link voltage during decrease and increase of UCAP voltage with corresponding discharging and charging.

Figure 2 illustrates the configuration model of bidirectional converter. The bidirectional converter is capable to holdout power on discharge mode during the occurrence of voltage sag. Simultaneously, bidirectional converter is capable of charging or absorbs extra power from the supply system during the occurrence of voltage swell. This bidirectional converter operates as boost and buck converter during discharge and charge of power from UCAP. UCAP provides very high power within short period, higher power density and lower energy density. UCAP has less loss while charging and discharging.

3 Control Strategies

3.1 Fryze Power Theory

This technique describes a minimum rms value as such three-phase average real power is received from the supply as the real load current as depicted in Fig. 3. Implementing the technique, the resistance loss gets reduced in the grid network which maintains a linear link among the source voltage and compensation current. The three-phase instantaneous real power is given as

$$\begin{aligned} P(t) &= v_{sa}(t)i_{sa}(t) + v_{sb}(t)i_{sb}(t) + v_{sc}(t)i_{sc}(t) \\ &= P_{sa}(t) + P_{sb}(t) + P_{sc}(t) \end{aligned} \tag{1}$$

And equivalent conductance (G_e) is determined from three-phase instantaneous real power [8]

$$G_e = \frac{v_{a1}i_{La} + v_{b1}i_{Lb} + v_{c1}i_{Lc}}{v_{a1}^2 + v_{b1}^2 + v_{c1}^2} \tag{2}$$

where v_{a1}, v_{b1} and v_{c1} are the three-phase instantaneous phase voltages and i_{La}, i_{Lb} and i_{Lc} are the three-phase load currents.

The corresponding reference voltage is given by load currents.

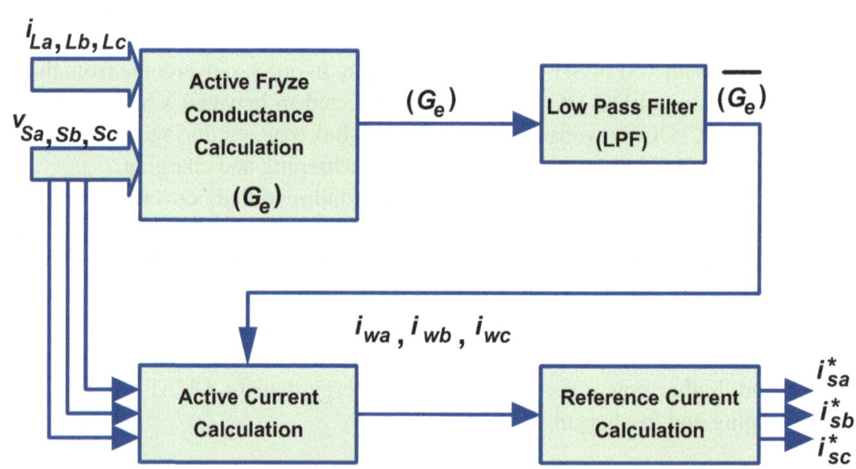

Fig. 3 Configuration of calculation of Fryze power theory

$$
\begin{aligned}
i_{\mathrm{sa}}^{*} &= i_{\mathrm{pa}} = I_{\mathrm{rms}} \sin \omega t \\
i_{\mathrm{sb}}^{*} &= i_{\mathrm{pb}} = I_{\mathrm{rms}} \sin(\omega t - 120^{\circ}) \\
i_{\mathrm{sc}}^{*} &= i_{\mathrm{pc}} = I_{\mathrm{rms}} \sin(\omega t + 120^{\circ})
\end{aligned} \tag{3}
$$

where i_{pa}, i_{pb} and i_{pc} are the instantaneous active currents and expressed as

$$
\begin{aligned}
i_{\mathrm{pa}} &= \overline{G_{\mathrm{e}}} v_{\mathrm{a}} \\
i_{\mathrm{pb}} &= \overline{G_{\mathrm{e}}} v_{\mathrm{b}} \\
i_{\mathrm{pc}} &= \overline{G_{\mathrm{e}}} v_{\mathrm{c}}
\end{aligned} \tag{4}
$$

3.2 Hysteresis Current Controller

For generating the gating signal to the VSI, the obtained reference current signal ($i_{\mathrm{ref}}(t)$) is related to actual current ($i_{\mathrm{actual}}(t)$) and the error between them is determined to HCC. This controller is employed individually for individual phase and straight produces the gating signals for the PWM-VSI.

Compared to other controlling techniques, hysteresis current controller is one of the simple and uncomplicated technologies. Performance of this theory is better with good dynamic response. The switching performance is shown in Fig. 4 and defined as

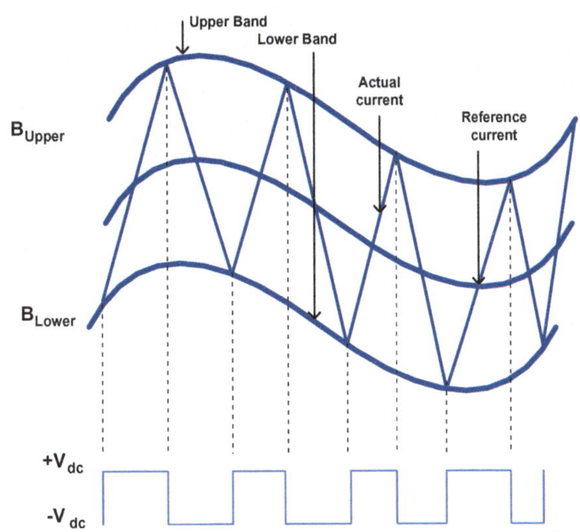

Fig. 4 Diagram of hysteresis current control

$$S = \begin{Bmatrix} \text{OFF} & \text{if} & i_{\text{actual}}(t) > i_{\text{ref}}(t) + H \\ \text{ON} & \text{if} & i_{\text{actual}}(t) < i_{\text{ref}}(t) - H \end{Bmatrix} \quad (5)$$

4 Results and Analysis

The overall model of the proposed system is performed using MATLAB/Simulink tool. Initially, the proposed system is performed without using SHAF, and the output results are illustrated in Fig. 5a. It has been found that due to the presence of harmonics the signals of load current get distorted, and the corresponding THD values are shown in Fig. 5b. THD value is found to be 47.94%. Subsequently, the system is operated with using UCAP and with absence of DC-DC boost converter as depicted in Fig. 6a, and the THD results are analysed from Fig. 6b with 3.78%. The simulated results are compared with performing the filter using UCAP tied with DC-DC boost converter and found to get improved results with THD value of 2.24% (Table 1). The corresponding output and THD value are depicted in Fig. 7a and b, respectively. The parameters of the system are shown in Table 2.

5 Conclusions

In this paper, proposed model is operated using MATLAB/Simulink software environment. Proposed model is analysed under two different schemes. In one scheme, the complete model including SHAF and UCAP is performed without utilizing DC-DC boost converter. And in second case, the proposed model including SHAF and UCAP is operated with DC-DC boost converter. Simulation outcome indicates that by connecting SHAF connected with UCAP and DC-DC boost converter, the harmonic current content and reactive power management get reduced, and with implementing Fryze theory, the advancement in the current signal is detected with substantial progress in the THD values.

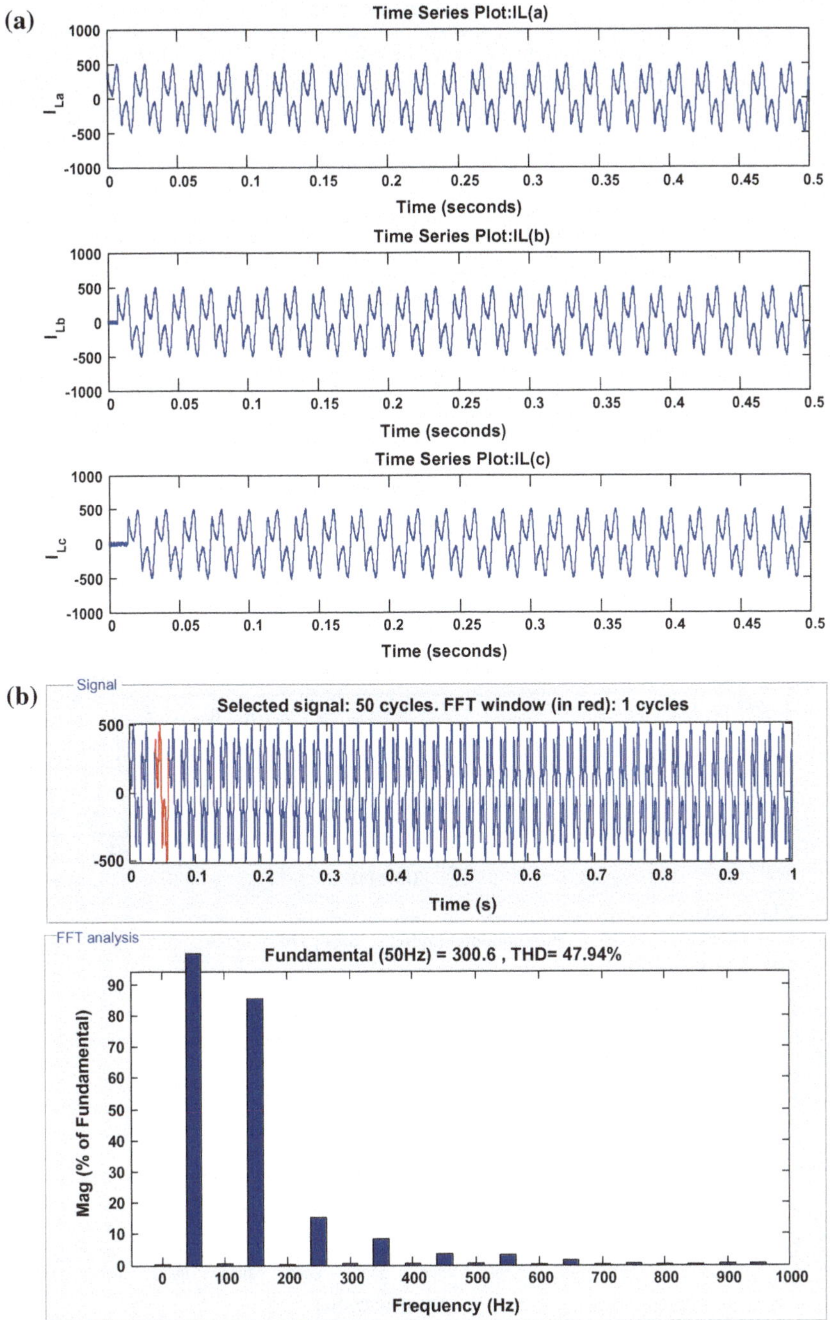

Fig. 5 **a** Load current before using SHAF with UCAP. **b** Total harmonic distortion value without UCAP

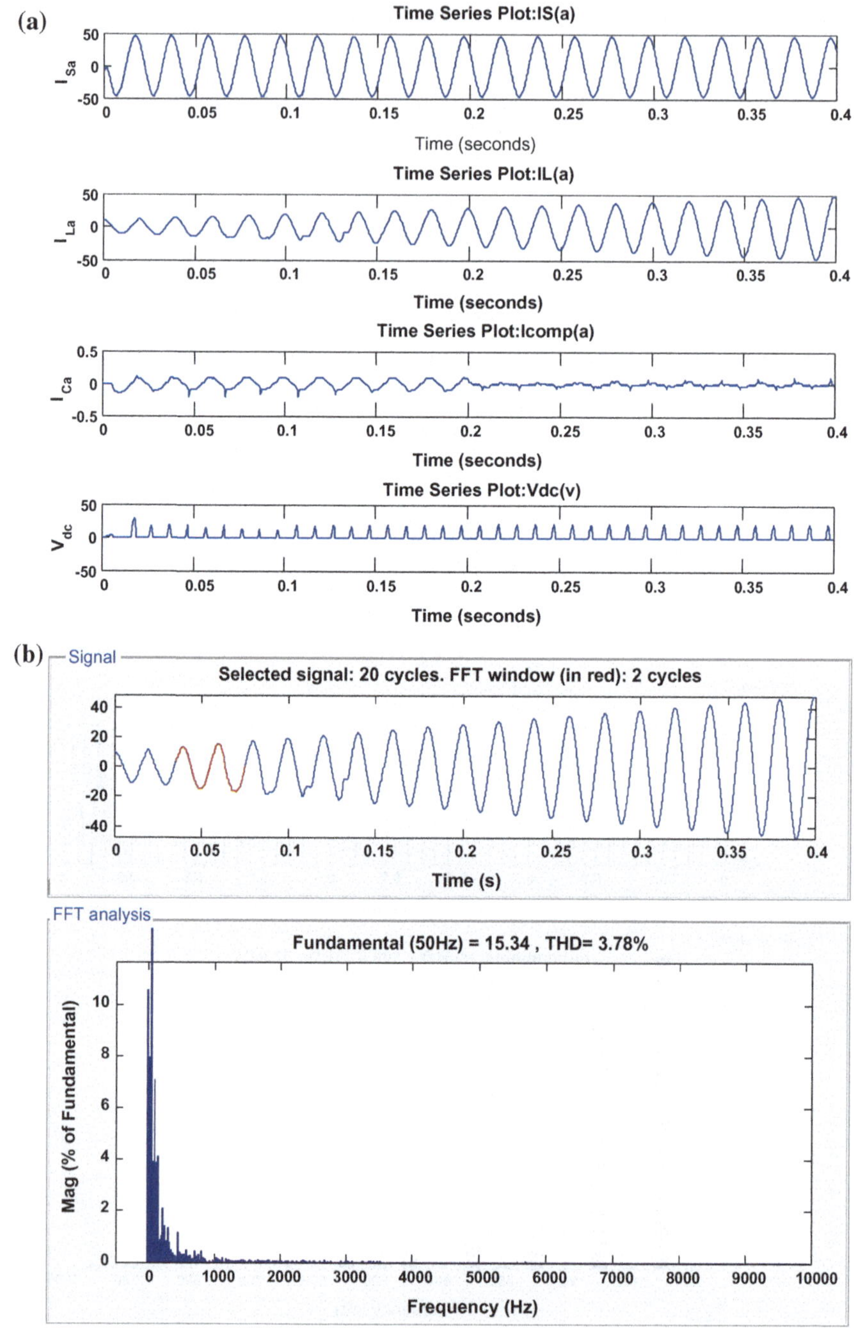

Fig. 6 **a** Load current after compensation without DC-DC converter. **b** Corresponding total harmonic distortion value

Table 1 THD analysis of load current

THD analysis			
	Without filter	Without DC-DC boost converter	With DC-DC boost converter
THD	47.94%	3.78%	2.24%

(a)

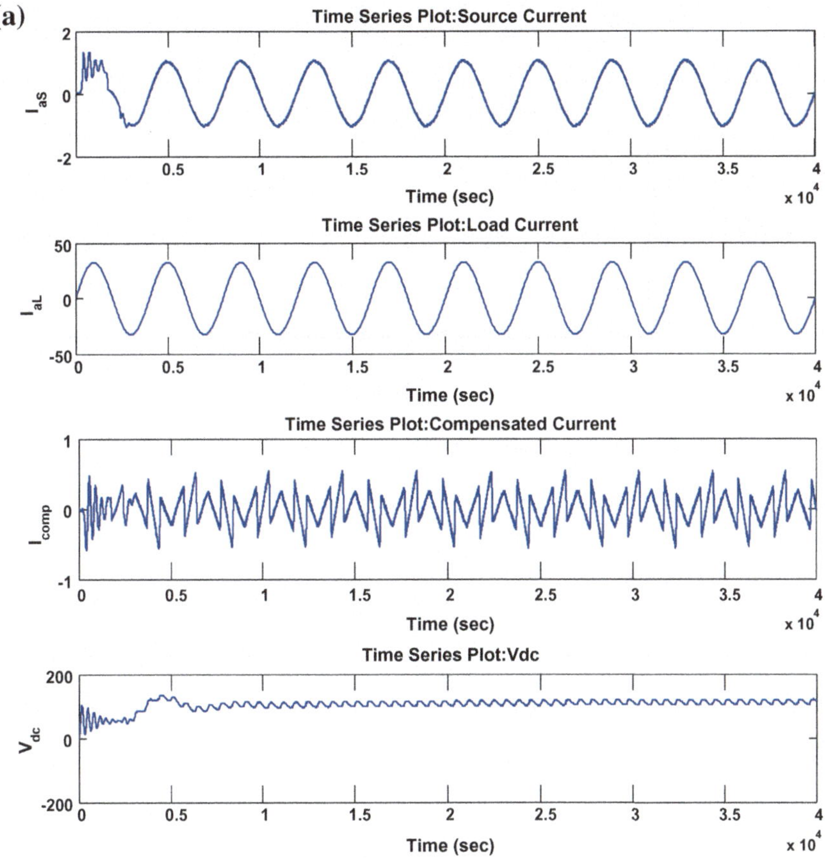

Fig. 7 **a** Load current after compensation with DC-DC converter. **b** Total harmonic distortion

(b)

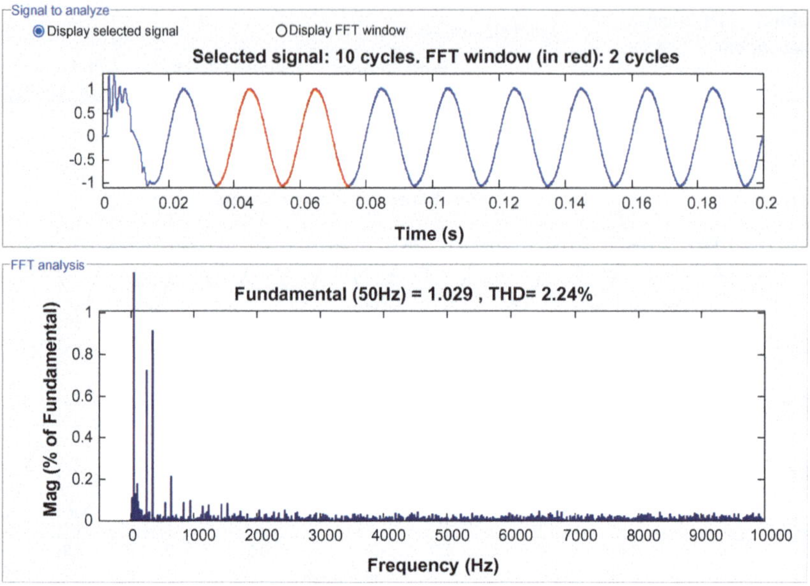

Fig. 7 (continued)

Table 2 Parameters of the systems

Parameter	Values
Line to line voltage, frequency	4405 V, 50 Hz
Line, load impedance	$L_s = 0.35$ mH, $L_{ac} = 1.25$ mH, $C_{DC} = 1500$ μF, $R_L = 60$ Ω, $L = 20$ μF; ($R_1 = 6$ Ω, $R_2 = 8$ Ω, $R_3 = 10$ Ω)
Active filter parameter	$C_D = 1200$ μF $V_D = 800$ V
Coupling inductance	4.5 mH
PI controller	$K_P = 24$, $K_I = 1.2$

References

1. Singh B, Bhuvaneswari G, Arya SR (2012) Review on power quality solution technology. Asian Power Electron 6(2)
2. Rahmania S, Haddad KA, Kanaan HY (2006) A comparative study of shunt hybrid and shunt active power filters for single-phase applications—simulation and experimental validation. Math Comput Simul 71:345–359
3. Kumar C, Mishra MK (2014) An improved hybrid DSTATCOM topology to compensate reactive and nonlinear loads. IEEE Trans Industr Electron 61:6517–6527
4. Ortuzar ME, Carmi RE, Dixon JW, Moran L (2006) Voltage-source active power filter based on multilevel converter and ultra-capacitor DC link". IEEE Trans Ind Electron 53(2):477–485
5. Casadei D, Grandi G, Rossi C (2002) A super capacitor-based power conditioning system for power quality improvement and uninterruptible power supply. Proc IEEE Conf Ind Electron 4:1247–1252
6. Somayajula D, Crow ML (2014) An integrated active power filter–ultra capacitor design to provide intermittency smoothing and reactive power support to the distribution grid. IEEE Trans Sustain Energy 5 (2014)
7. Bottu MB, Crow ML, Elmore AC, Atcitty S (2013) A power electronic conditioner using ultracapacitors to improve wind turbine power quality. Smart Grid Renew Energy 4:69–75
8. Karupannan P, Mahapatra KK, Jeyaraman K, Viji J (2011) Fryze power theory with adaptive HCC based active power line conditioner. In: International conference on power and energy system, 1–6

Chapter 14
Ambiguity Function Analysis of Polyphase Codes in Pulse Compression Radars

Ankur Thakur and Davinder Singh Saini

1 Introduction

Radar system is used to determine the characteristics of the detected target like range, location and velocity, etc. Long duration pulses are transmitted for better and long-range detection because these pulses have more energy which cannot be easily attenuated in the environment [1]. Short duration pulses are used for better range resolution. Resolution is the ability of the system to distinguish two or more objects in terms of their range and velocity. Range resolution is calculated as $R = \frac{c}{2\beta}$, where R is the range resolution, β is the pulse bandwidth and c is the light velocity. In practical applications, these resolutions must be small as much as possible [2, 3]. In the radar signal processing, transmitting signal is communicated toward target and then echo signal is captured at radar receiver. Using signal processing techniques, this echo signal is processed to get the characteristics of the target.

Transmitting pulses duration cannot be reduced in a greater extend because these pulses leads to large transmission power which increase the overall system cost. To overcome this issue, long duration pulses are preferred for transmission purposes [1, 4]. In pulse compression, an effective short duration pulse is attained by communicating long duration pulse with an angle modulation. The matched filter is used in the radar receiver to observe the features of the target filter. Echo signal is passed through the matched filter which maximizes the detect-ability of an object. Usually, matched filter output has main response as well as some side-lobes. These side-lobes are unwanted because these might have information related to the nearby target. This will create the problem of ambiguity during detection process [5, 6]. So, for accurate detection side-lobes must be less as much as possible.

Using ambiguity function, the range and velocity resolution through delay-Doppler plane can be determined. Ambiguity function consists time-delay and

A. Thakur (✉) · D. S. Saini
Department of Electronics and Communication Engineering, Chandigarh College of Engineering and Technology, Chandigarh 160019, India

V. K. Giri et al. (eds.), *Computing Algorithms with Applications in Engineering*, Algorithms for Intelligent Systems, https://doi.org/10.1007/978-981-15-2369-4_14

Doppler shift as two measuring parameters through which range and velocity of the target can be determined, respectively. Doppler frequency shift is calculated as $f_d = \frac{2v_r}{\lambda}$, where v_r is the radial velocity of the target and λ is the transmitting signal wavelength.

1.1 Related Work

There are many pulse compression techniques in the literature for the reduction of side-lobes. In [4], non-convex problems are handled to minimize side-lobes but main-lobe width is disturbed in this method. In [5], time-bandwidth is observed by different classical orthogonal polynomials. In [6], a correlation vector matrix is used for an angle estimation of nearby targets in non-stationary noise. In [7–9], linear frequency modulated waveforms are presented for the side-lobe reduction. For practical requirement, this reduction in side-lobe is not more to avoid target masking. In [10–13], nonlinear frequency modulated waveforms are designed for more reduction in side-lobes. These waveforms give better reduction in side-lobes but increase the complexity for designing of the waveforms. Different polyphase codes are analyzed in [14–16]. Polyphase code-based Woo filter techniques are much good choices for side-lobes reduction. Different windowing techniques can be used to reduce the side-lobes further but this approach increases the relative main-lobe width which disturbs the range resolution [17–19].

In this paper, P3 and P4 codes are observed for side-lobes reduction in pulse compression for radar system. Using ambiguity function, Doppler tolerant properties are also observed. This paper has been organized as: Sect. 2 describes the polyphase codes; Sect. 3 gives the simulation and results. Lastly, Sect. 4 concludes the research work.

2 Polyphase Codes

Many polyphase codes are existing in literature like Barker code, Frank code, P1, P2, P3 and P4 codes. These polyphase codes have harmonically correlated phases based on an assured phase rise. These codes are consequent logically distinguishing a frequency modulation pulse compression waveform. Frank projected a polyphase code having non-periodic connection properties. Frank, P1, P2, P3 and P4 codes have been resulting from step estimate to linear frequency modulation waveforms [6, 14]. These codes are achieved by separating the given polyphase codes into sub codes having equal length. With the help of each sub codes phase values finest matches the complete phase trajectory of the underlying waveform. Phases of the P3 and P4 codes are calculated using Eqs. (1) and (2) [18], where N represents the code length and $i = 1, 2, 3 \ldots N$.

$$P3: \phi(i) = \pi(i-1)^2/N \tag{1}$$

$$P4: \phi(i) = \pi(i-1)(i-1-N)/N \tag{2}$$

Figure 1 represents the phase variation of first 100 elements of P3 and P4 codes. It is clear that P4 codes are more symmetrical around the center which gives good Doppler tolerant properties than P3 codes. Elements are calculated using exponential of the phases as $x(i) = \exp(j\phi(i))$ and are shown in Figs. 2 and 3.

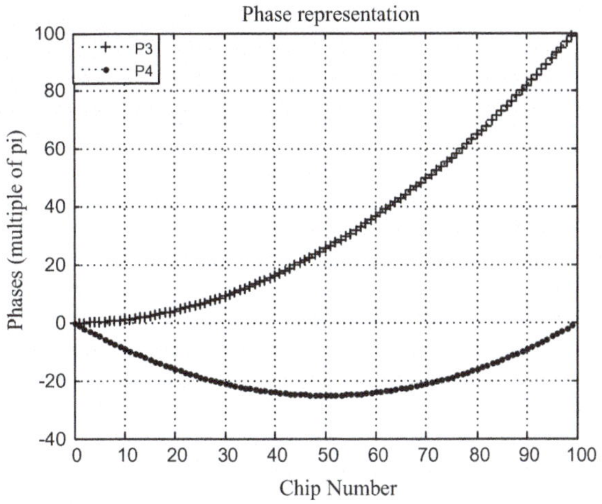

Fig. 1 Phase variations of P3 and P4 codes for first 100 elements

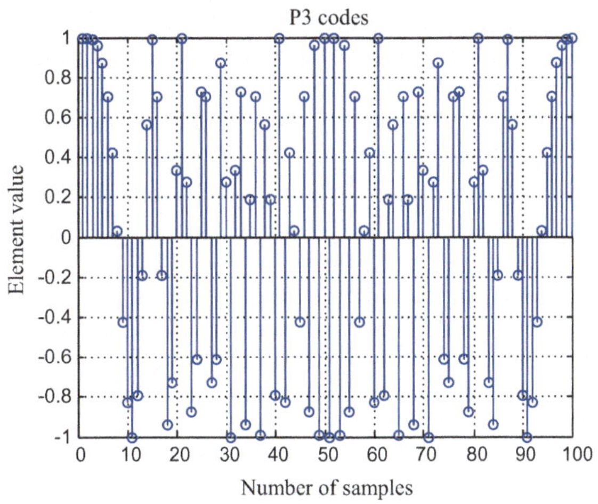

Fig. 2 P3 elements for $N = 100$

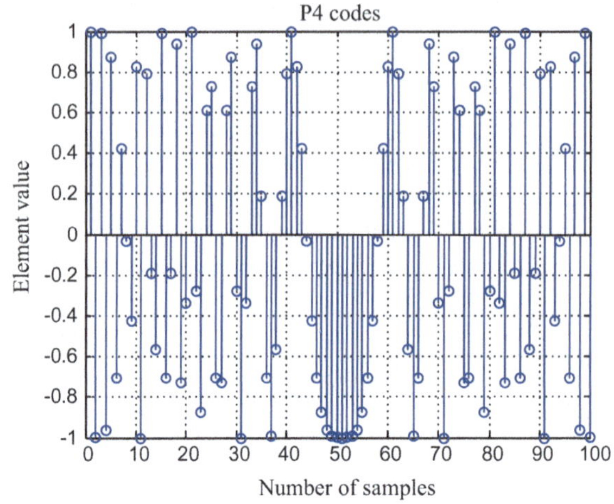

Fig. 3 P4 elements for $N = 100$

3 Simulations and Results

In this paper, side-lobes reduction is observed for the first 100 elements of P3 and P4 codes. In order to examine designed codes suitability for range and velocity resolution, the ambiguity function has been observed and this is represented as

$$X(t_d, f_d) = \int_{t=-\infty}^{\infty} x(t)x^*(t - t_d)e^{j2\pi f_d t}dt \tag{3}$$

$X(t_d, f_d)$ is an ambiguity function, t_d and f_d are the time-delay and Doppler shift, respectively. It behaves like a matched filter if no Doppler shifts exist and its response is at all times symmetrical around the origin [17–19]. In most cases, ambiguity function offers single central peak at the origin while the rest energy can be spread uniformly in the delay-Doppler plane. The narrow peak which lies at the origin implies good resolution in each range and velocity. At the receiving end, when filter is matched in Doppler to the reflected signal and is sampled at a delay similar to the precise range of the goal, then the matched filter response will likely be maximum on the origin which gives good decision property.

The skew in the ambiguity function plots gives information about the Doppler mismatch. Usually, for practical application matched filter response main peak must lie on the origin of the delay-Doppler plane. For a pulse having τ duration gives $1/\tau$ Hz Doppler resolution. If Doppler mismatch is more, then output of the matched filter will be highly attenuated which leads to miss the target detection.

On comparing Figs. 4 and 5, it is clear that P4 codes are more Doppler tolerant than P3. Contour plot of P4 code is narrower than P3 code for code of length 100 which is responsible for less ambiguity during target detection. As x-axis are responsible

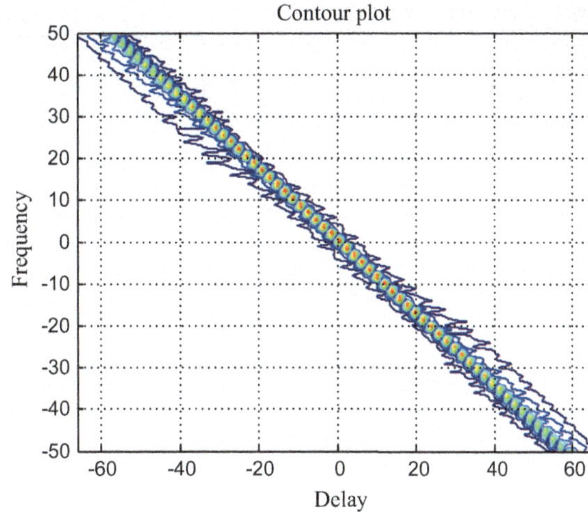

Fig. 4 P3 codes contour plot

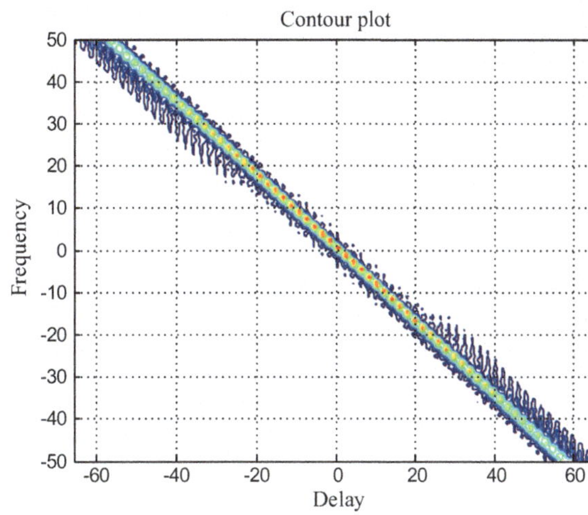

Fig. 5 P4 codes contour plot

for range resolution and y-axis responsible for frequency resolution. It is observed that P4 code has more range resolution.

Peak side-lobe level measures the ratio of maximum side-lobe magnitude to the in phase value of the matched filter response and it is calculated as

$$\text{Peak Sidelobe Level} = 20\log_{10}\frac{\max_{1\leq l\leq N}|C(l)|}{|C(0)|} \tag{4}$$

where N is number of side-lobes, $C(l)$ is output of matched filter. As zero Doppler cut of ambiguity function is similar to the matched filter response.

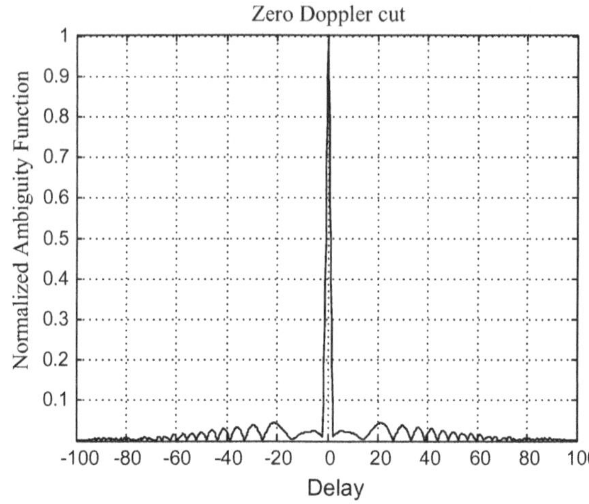

Fig. 6 Zero Doppler cut for P3 codes

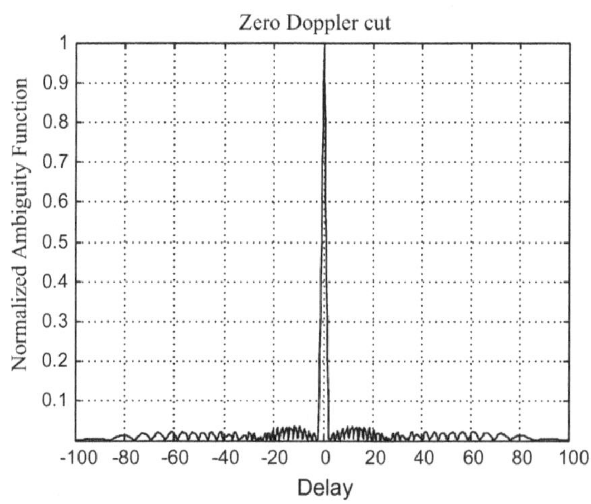

Fig. 7 Zero Doppler cut for P4 codes

Here, in Figs. 6 and 7, it is observed that less side-lobe exist for P4 codes than P3 codes. −29.11 and −26.37 dB reduction in side-lobe exists for P4 and P3 codes, respectively. So, P4 codes are better for side-lobe reduction as well as for better Doppler tolerant properties.

4 Conclusions

This paper has demonstrated polyphase codes-based study for side-lobe reduction having applicability inside the pulse compression radar. It is observed that P4 codes are better for side-lobe reduction than other polyphase codes. The ambiguity function result shows that P4 codes have better side-lobe reduction and range resolution performance with respect to the pulse compression. −29.11 and −26.37 dB reductions in side-lobe exist for P4 and P3 codes, respectively. More analysis can be done for the design of new codes which gives superior resolutions in pulse compression radars.

Acknowledgements Authors are thankful to the Department of Science and Technology (DST) for sanctioning a INSPIRE Fellowship with Registration No. IF-180847 for Ph.D. Programme under which this paper has been accomplished.

References

1. Skolnik MI (2001) Introduction to radar system, 3rd edn. McGraw-Hill, New York
2. Richards MA (2014) Fundamentals of radar signal processing. McGraw-Hill, New York
3. Saini DS, Upadhyay M (2009) Multiple rake combiners and performance improvement in 3G and beyond WCDMA systems. IEEE Trans Veh Technol 58(7):3361–3370
4. Leilei X, Shenghua Z, Hongwei L (2018) Simultaneous optimization of radar waveform and mismatched filter with range and delay-Doppler sidelobes suppression. Digit Signal Proc 83:346–358
5. Thakur A, Talluri SR (2018) Comparative analysis on pulse compression with classical orthogonal polynomials for optimized time-bandwidth product. Ain Shams Eng J 9(4):1791–1797
6. Mahafza BR (2013) Radar systems analysis and design using MATLAB. Chapman and Hall/CRC CRC Press LLC
7. Atilio G, Daniel C, Jessica S, Monteiro PP (2018) Research challenges, trends and applications for future joint radar communications systems. Wirel Pers Commun 100(1):81–96
8. Chengjie L, Lidong Z, Anhong X, Zhongqiang L (2017) Blind separation of weak object signals against the unknown strong jamming in communication systems. Wirel Pers Commun 97(3):4265–4283
9. Guodong J, Yunkai D, Robert W, Wei W, Yongwei Z, Yajun L, Liang D (2019) Mitigating range ambiguities with advanced nonlinear frequency modulation waveform. IEEE Geosci Remote Sens Lett, 1–5
10. Guodong J, Yunkai D, Robert W, Wei W, Pei W, Yajun L, Zhimin Z, Zhang Y (2019) An advanced nonlinear frequency modulation waveform for radar imaging with low sidelobe. IEEE Trans Geosci Remote Sens, 1–14
11. Thakur A, Talluri SR, Saini DS (2017) Signal generation employing Chebyshev polynomial for pulse compression with small relative side-lobe level. In: IEEE international conference communication on signal processing, 114–118. https://doi.org/10.1109/iccsp.2017.8286543
12. Joshi A, Saini DS (2017) GA-PTS using novel mapping scheme for PAPR reduction of OFDM signals without SI. Wirel Pers Commun 92(2):639–651
13. Saini DS, Balyan V (2016) An efficient multicode design for real time QoS support in OVSF based CDMA networks. Wirel Pers Commun 90(4):1799–1810
14. Thakur A, Talluri SR (2017) A novel pulse compression technique for side-lobe reduction using woo filter concepts. In: IEEE international conference communication on signal processing, 1086–1090. https://doi.org/10.1109/iccsp.2017.8286543
15. Zakeri B, Zahabi M, Alighale S (2012) Sidelobes level improvement by using a new scheme used in microwave pulse compression radars. Prog Electromagn Res Lett 30:81–90
16. Alighale S, Zakeri B (2014) An excellent reduction in sidelobe level for P4 code by using of a new pulse compression scheme. Int J Electron 101(10):1458–1466

17. Tao Y, Bo Q, Ying W, Guoyong W, Wenying L, Lang B, Yansong M (2018) Weighted discriminator function based unambiguous tracking method for dual-frequency constant envelope modulated signals. Wirel Pers Commun 103(3):1895–1925
18. Thakur A, Talluri SR, Panigrahi RK (2019) Sidelobe reduction in pulse compression having better range resolution. Comput Electr Eng 74:520–532
19. Thakur A, Saini DS (2019) Signal compression method based on discrete linear chirp transform. In: IEEE International conference communication on signal processing, 17–21. https://doi.org/10.1109/iccsp.2019.8697962

Chapter 15
Early, Demagnetization Diagnosis in Multiphase PMSM Machine by Advanced MCSA Technique

Khadim M. Siddiqui, Rafik Ahmad, V. K. Giri and Kuldeep Sahay

1 Introduction

The rotor is made by permanent magnet (PM) material in permanent magnet synchronous motor. Therefore, the power factor of this motor is improved, and also, it becomes easy and supple. But, the magnetic demagnetization in rotor PM material is the largest threat in the PMSM machine beneath real operating conditions [1–6].

In the PMSM, exciting winding is absent; therefore, characteristics are always chosen by the PM material. Thus, the stability and efficiency improve due to the absence of electromagnet and slip rings. In recent times, the popularity of PMSMs is being increased because of lightweight, small volume with better control performance and economic profit. Some main applications of this motor are electric vehicle, wind power system and air systems, etc. [7–11].

In the PMSM operation, partial and irreversible magnetic demagnetization problems occur due to the pressure of high temperature and vibration. Hence, the operating reliability, output efficiency and life of the motor are severely affected. So, the permanent magnet's flux monitoring is highly essential [12–16].

Nowadays, the resolution normally adopted is to optimize the magnetic circuit structure when the motor is not dependable so as to reduce the risk of demagnetization, but this solution increases manufacturing costs with limited effects. The demagnetization of motor during operation can only be detected after observable failures which cause the motor to stop whereas the demagnetization has been very severe at this time [13, 14].

K. M. Siddiqui (✉) · R. Ahmad
Babu Banarsi Das National Institute of Technology and Management, Lucknow, India

V. K. Giri
Rajkiya Engineering College, Sonbhadra, India

K. Sahay
Institute of Engineering & Technology, Lucknow, India

V. K. Giri et al. (eds.), *Computing Algorithms with Applications in Engineering*,
Algorithms for Intelligent Systems, https://doi.org/10.1007/978-981-15-2369-4_15

The magnetization is reduced of the PM material if the temperature exceeds the Curie temperature. Clearly, for temperatures under this limit, the B-H curve is affected in a way that residual flux density, coercive force and energy product decrease (except for ceramic ferrite) leading to potential demagnetization even for nominal operating conditions and not only in case of large currents due to overloading or to faults in inverter or stator windings [6, 12].

If the temperature rises, then the flux linkage decreases during the motor operation for permanent magnet rotor. Though, the flux linkage is usually considered constant for permanent magnet rotor in a PMSM controller design. If the flux linkage fluctuation occurs, then it leads to unstable output torque. This unstable output torque affects the electric vehicle (EV) performance. Moreover, if the motor internal temperature exceeds the material critical point, then the permanent magnet flux density cannot be restored to the actual remainder flux density and results in permanent demagnetization. This permanent demagnetization leads to motor failure and inferior performance [15, 17, 18]. Therefore, online monitoring of PM flux linkage and demagnetization diagnosis is critical to make sure better performance from a PMSM drive.

For demagnetization diagnosis in the machine, an accurate analysis of demagnetization characteristics is needed. Many researchers already studied demagnetization characteristics in diverse perspectives. The speed harmonics analysis is reported in [19] for the PM flux linkage estimation. The position offset-based parameter estimation is reported in [20] to attain the accurate PM flux linkage. The Kalman filter analysis is reported in [21] for observation of PM flux linkage/temperature using *dq*-axis equations. Torque with inductance is reported in [22] and studied in detail to understand the flux linkage reduction in PMSM. However, most of the demagnetization detection techniques are based on costly sensor-based diagnosis such as acoustic noise [23], output torque and temperature [16].

Presently, five-phase variable speed drives are being popular in many industrial applications. The five-phase PMSM machine is having advantages over three-phase PMSM machines. In the five-phase machines, the per phase stator current reduces, and the reliability is also improved. The pulsating torque is reduced in five-phase machines, and these machines have more fault-tolerant capability. The mathematical modeling of PMSM machine is given in [4] in which the performance and simulation analysis of PWM inverter-fed five-phase PMSM drive have been discussed.

In this research paper, the demagnetization diagnosis has been done for multiphase PMSM machine by wavelet transform-based advanced MCSA technique. From this proposed approach, one can do permanent magnetic demagnetization diagnosis in the transient condition precisely with the help of time–frequency information. The encouraging results obtained by this proposed approach have been given the effective solution for diagnosing of this kind of problem.

2 Proposed Five-Phase Machine Setup

The block diagram of the proposed setup is shown in Fig. 1. The simulation analysis of this PMSM drive has been given in [6]. The five-phase 6 HP motor has been used for analysis purpose. The back emf and rotor are chosen sinusoidal and round, respectively.

The block diagram of the experimental setup is shown in Fig. 1. From this experimental setup, permanent magnet synchronous motor demagnetization status has been examined by advanced MCSA technique. The processes of demagnetization diagnosis through motor current signal may be understood from Fig. 1.

Firstly, the pulse width modulated inverter gives the supply to PMSM for operation. The current transducer senses the motor current, and this current will be used for differentiating healthy and magnetic demagnetization conditions of the motor. For fast and high accuracy response, the closed-loop sensor has been used. The anti-aliasing filter is used before a signal sampler to limit the bandwidth of a signal to approximately or completely satisfy the sampling theorem over the band of interest. Now, the analog stator current signal is processed through the analog-to-digital

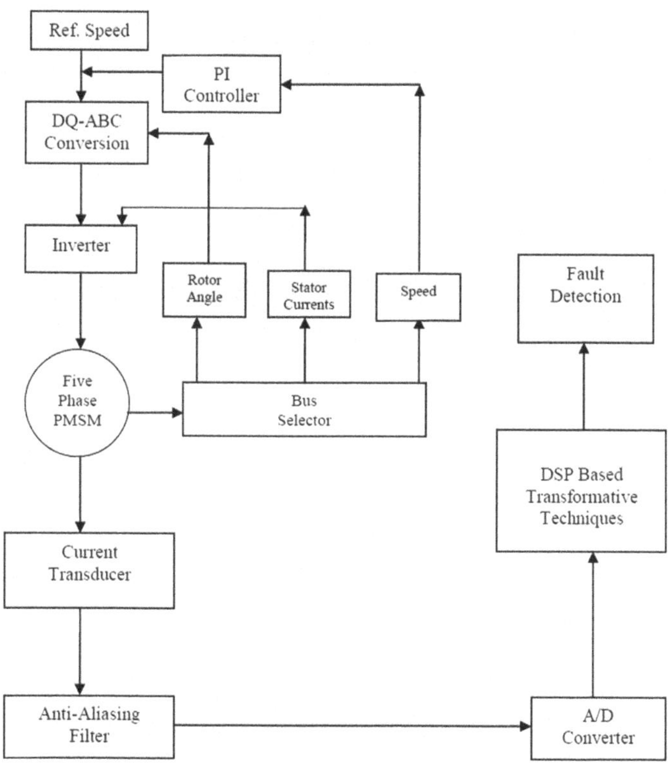

Fig. 1 Proposed setup of five-phase permanent magnet synchronous machine

converter, and it gives the required digital signal. This stator current digital signal is processed through the wavelet transform-based advanced MCSA technique. This approach decomposed the stator current signal for removing the strong fundamental component. The final level high-frequency band detailed signal has been used for demagnetization diagnosis purpose in the transient condition.

3 Time-Domain Analysis

3.1 Healthy Condition

In the proposed setup (shown in Fig. 1), effective speed regulation of the motor has been achieved by using two control loops. The inner loop is used to regulate the stator currents whereas the outer loop is used to regulate the speed of the motor in PMSM machine. In the healthy condition of the motor, rated values have been applied. The rated torque of the motor is 7 N m., and the motor starts under all healthy phases, viz. symmetrical or normal operation.

The obtained results of healthy condition of the motor are shown in Fig. 2a–d. Four signals have been used for healthy and demagnetization diagnosis purpose in time domain. These signals are stator current in phase *a*, stator currents in phase *abcde*,

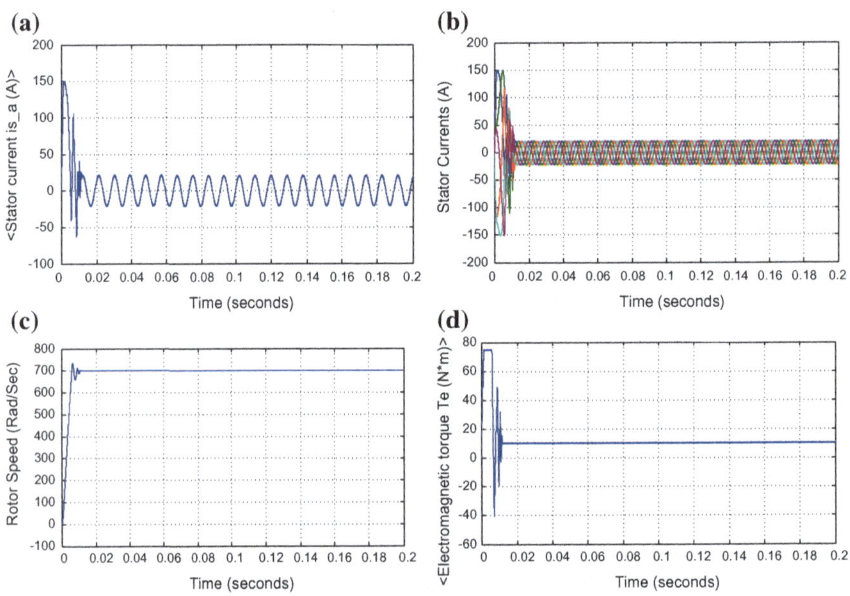

Fig. 2 Results in healthy condition, **a** stator current in phase *a*, **b** stator current in phase *abcde*, **c** rotor speed, **d** electromagnetic torque

rotor speed and developed electromagnetic torque. The rated speed of the motor is 700 rad/s, and it is achieved efficiently in the healthy condition. The developed electromagnetic torque is shown in Fig. 2d. All signals are reached in steady state before 0.02 s.

3.2 Demagnetization Diagnosis of Permanent Magnet Synchronous Motor

Diagnosis of demagnetization in the motor has been implemented by varying load torque largely. In this case, ten times load torque has been applied ($T_L = 10 \times T_L$) for effective diagnosis of demagnetization in five-phase PMSM machine. This applied load torque disturbs the magnetization characteristic of the motor in PMSM machine. The main scope of this paper is the early diagnosis of demagnetization of the motor.

The fundamental component of the stator voltage and stator current of the motor with radial demagnetization will take place when load torque exceeded largely. For large load torque, the fundamental component of the stator current becomes large, and if the applied load torque is small, then the stator voltage of the motor will be demagnetized and will achieve the lowest value. But, in this work, our concentration is toward the analysis for increased load torque not the decreased load torque. In the future, investigations with low load torque could be done by the same technique for verification purpose.

The stator voltage of the motor demagnetizes largely for largely increased load torque compared to its rated value. In this analysis, space harmonics has been ignored. However, the actual values of inductances were changed according to the stator current and were assumed to be constant.

When the load torque is applied ten times as compared to its rated load torque means $T_L = 10 \times T_L$, consequently all the parameters of the transient state will be disturbed. If one observes the rotor speed and electromagnetic torque waveforms, he/she may understand clearly how parameters are being changed. The results in demagnetization condition for ten-time load torque are shown in Fig. 3a–d.

For clear verification purpose by time-domain approach, the load torque was further increased. But, for remaining load torque conditions, only stator current signal has been used for comparison purpose. Since, phase '*a*' stator current parameter has been used for comparison purpose because of the advantages of Motor Current Signature Analysis (MCSA) technique over other techniques. Recently, the MCSA technique has emerged very popular for diagnosing electrical machine faults with competitive cost. Therefore, only stator current parameter has been used for demagnetization assessment purpose.

In the simulation model, for effective demagnetization diagnosis purpose, the following load torque values are used.

$$\text{Load Torque } (T_L = 10 \times T_L)$$

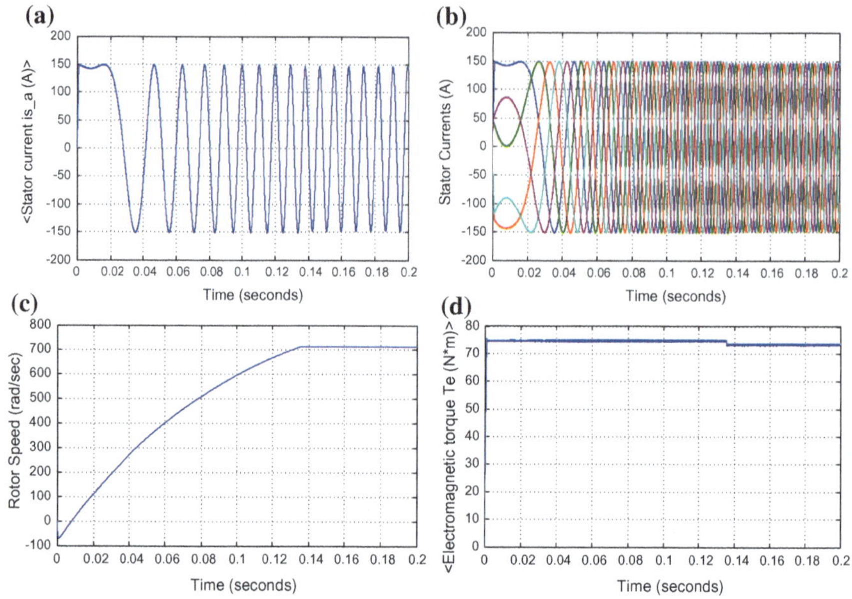

Fig. 3 Results in demagnetization condition for load torque ($T_{\text{L}} = 10 \times T_{\text{L}}$), **a** stator current in phase *a*, **b** three-phase current in phase *abcde*, **c** rotor speed, **d** electromagnetic torque

$$\text{Load Torque } (T_{\text{L}} = 10 \times T_{\text{L}} + 10)$$
$$\text{Load Torque } (T_{\text{L}} = 10 \times T_{\text{L}} + 15)$$

From Fig. 4a–d, one may observe changes in the stator current signal as per changes of the load torque. If obtained results by increased load torque conditions from healthy condition are compared, then one can observe that all signals are different. So, demagnetization diagnosis has been possible of PM synchronous motor by the time-domain analysis approach.

However, demagnetization diagnosis of the motor by time-domain approach is a tedious process, and major information of the problem is not visualized in the effective way. Therefore, in the next section, discrete wavelet transform (time–frequency domain) theory has been applied for healthy condition as well as various demagnetization diagnosis purposes at the early stages.

4 Demagnetization Diagnosis by Advanced MCSA Technique

In this section, the magnetic demagnetization diagnosis has been done for five-phase motor in PMSM drive by advanced MCSA technique. The wavelet transform has

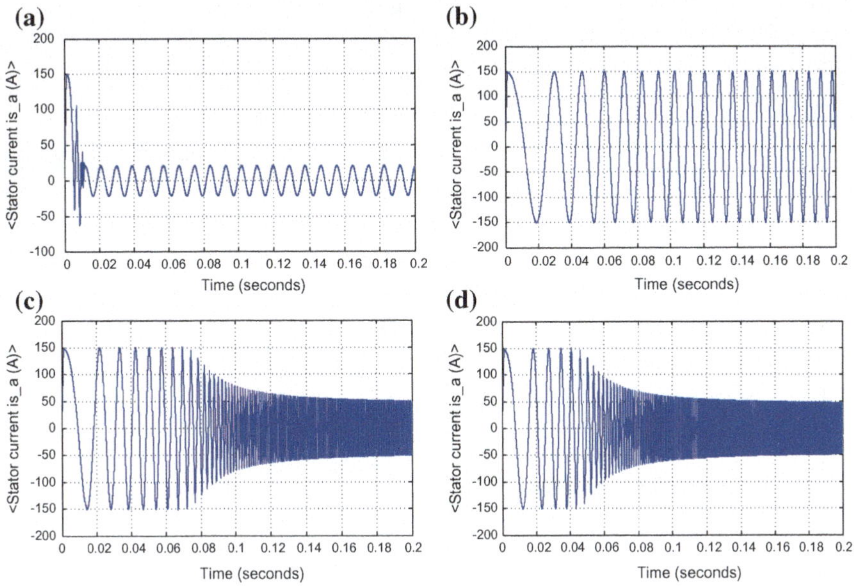

Fig. 4 Stator current signal, **a** healthy condition, **b** load torque $T_{\mathrm{L}} = 10 \times T_{\mathrm{L}}$, **c** load torque ($T_{\mathrm{L}} = 10 \times T_{\mathrm{L}} + 10$), **d** load torque ($T_{\mathrm{L}} = 10 \times T_{\mathrm{L}} + 15$)

been used as advanced MCSA technique, and by using this approach, one may extract time–frequency information by time-domain signal. The applied theory in this application could be understood in [6].

The wavelet algorithm has been applied for demagnetization diagnosis purpose in the early stage. In this analysis, the sensed stator current signal (time domain) is decomposed by using empirical decomposition method. In the empirical decomposition analysis, the low- and high-pass filters are used. The low-frequency band approximation signal and high-frequency band detailed signals are generated by the empirical decomposition.

In this time–frequency domain analysis, original stator current signal has been decomposed up to nine levels for extraction of relevant information concerning magnetic demagnetization. The Daubechies mother wavelet has been used in this application. Since the mother wavelet is located more in time and oscillates faster in a short period of time, high decomposition level is preferred for accurate fault demagnetization diagnosis purpose [7]. In this paper, the Debauchees (db-10) mother wavelet has been used for demagnetization diagnosis purpose because of its advantages.

Figure 5a–d demonstrates results of healthy condition as well as magnetic demagnetization conditions of the motor in PMSM drive. The noninvasive technique has been used in this application and that means the demagnetized results have been compared with the healthy condition results of the motor. The ninth level high-frequency band detailed signal has been used for comparison purpose. For better understanding

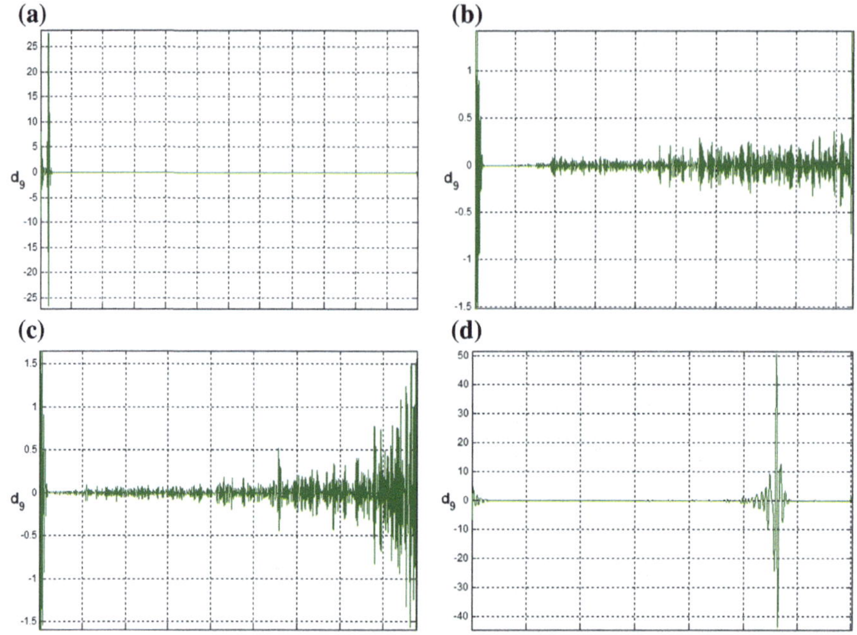

Fig. 5 Stator current high-frequency band detailed signal, **a** healthy condition, **b** load torque $T_L = 10 \times T_L$, **c** load torque ($T_L = 10 \times T_L + 10$), **d** load torque ($T_L = 10 \times T_L + 15$)

of the results, the wavelet coefficients have been given in different conditions and are shown in Table 1.

From Fig. 5a–d, one may observe the changes in the signal in healthy as well as magnetic demagnetization conditions. If one increases the load torque largely and that exceeded the limit, then harmonics will be produced in the stator current signal. Due to this variation, harmonics are not easily detectable in the time-domain analysis, but in the time–frequency domain analysis, one can observe these harmonics efficiently as per the nature of the results. Therefore, it is concluded that an efficient technique has been attempted to solve the demagnetization problem.

Table 1 Wavelet coefficients in different conditions

Healthy		$T_L = 10 \times T_L$		$T_L = 10 \times T_L + 10$		$T_L = 10 \times T_L + 15$	
A_9	D_9	A_9	D_9	A_9	D_9	A_9	D_9
485	485	387	387	362	362	156	156

A_9: Low-frequency band approximation signal coefficients, D_9: high-frequency band detailed signal coefficients

5 Conclusions

In the present article, the demagnetization of five-phase PM magnet synchronous motor has been diagnosed precisely in multiphase PMSM drive. The nonintrusive technique has been used to differentiate healthy and magnetic demagnetized signals. The motor current signal has been measured and preprocessed. Then, advanced motor current signature analysis is employed to diagnose PM demagnetization. From obtained results, we can say that the proposed approach is proficient to recognize and detect the demagnetization stages. The proposed method achieved better results as compared to conventional techniques for different operating conditions. By this method, the system reliability and accuracy increased whereas insecurity decreased.

References

1. Wang J, Wang W, Atallah K, Howe D (2008 Sept) Demagnetization assessment for three-phase tubular brushless permanent-magnet machines. IEEE Trans Magn 44(9)
2. Fu WN, Ho SL (2010 June) Dynamic demagnetization computation of permanent magnet motors using finite element method with normal magnetization curves. IEEE Trans Appl Supercond 20(3)
3. Sarikhani A, Mohammed O (2012 Sept) Real-time demagnetization assessment of PM synchronous machine. In: XXth IEEE international conference on electrical machines
4. Siddiqui KM, Upadhyay SK, Singh S, Srivastava RK, Babu R (2017) Performance and analysis of PWM inverter fed 3-Phase PMSM drive. i-manager's J Electron Eng 8(1):9–18
5. Ruoho S, Kolehmainen J, Ikäheimo J, Arkkio A (2010 Mar) Interdependence of demagnetization, loading, and temperature rise in a permanent-magnet synchronous motor. IEEE Trans Magn 46(3)
6. Siddiqui KM, Chatterji S (2018) Early, demagnetization assessment of PMSM machine by discrete wavelet transform. i-manager's J Circuits Syst 6(4):1–10
7. Ding X, Bhattacharya M, Mi C (2010 June) Simplified thermal model of pm motors in hybrid vehicle applications taking into account eddy current loss in magnets. J Asian Electr Veh 8(1)
8. Refaie AME, Harris NC, Jahns TM, Rahman KM (2004 June) Thermal analysis of multibarrier interior PM synchronous machine using lumped parameter model. IEEE Trans Energy Conv 19(2)
9. Demetriades GD, De La Parra HZ, Andersson E, Olsson H (2010 Feb) A real-time thermal model of a permanent magnet synchronous motor. IEEE Trans Power Electron 25(2)
10. Liu S, Mohammed OA, Liu Z (2007 Apr) An improved FE-based phase variable model of PM synchronous machines including dynamic core losses. IEEE Trans Magn 43(4)
11. Morimoto S, Sanada M, Takeda K (1994) Wide-speed operation of interior permanent magnet synchronous motors with high performance current regulator. IEEE Trans Ind Appl 30:920–926
12. Siddiqui KM, Sahay K, Giri VK (2017 Aug–Oct) Modelling and open phase fault analysis in three and five phase permanent magnet synchronous motor machine. i-manager's J Instrum Control Eng 5(4)
13. Sudhoff SD, Corzine KA, Hegner HJ (1995) A flux-weakening strategy for current-regulated surface-mounted permanent-magnet machine drives. IEEE Trans Energy Conv 10:431–437
14. Zhu M, Hu W, Mukundan S, Kar NC (Oct 2018) Multi-sensor fusion based permanent magnet demagnetization detection in permanent magnet synchronous machines. IEEE Int Magn Conf
15. Siddiqui KM, Sahay K, Giri VK (2014) Health monitoring and fault diagnosis in induction motor—a review. Int J Adv Res Electric Electron Instrum Eng 3(1):6549–6565

16. Zhu M, Hu W, Kar NC (2017 May–June) Torque ripple based interior permanent magnet synchronous machines rotor demagnetization fault detection and current regulation. IEEE Trans Ind Appl 53(3)
17. Qiao W, Lu D (2015) A survey on wind turbine condition monitoring and fault diagnosis—Part I: Components and subsystems. IEEE Trans Ind Electron 62(10):6536–6545
18. Simani S, Farsoni S, Castaldi P (2015) Fault diagnosis of a wind turbine benchmark via identified fuzzy models. IEEE Trans Ind Electron 62(6):3775–3782
19. Feng G, Lai C, Kar NC (2017) Particle-filter-based magnet flux linkage estimation for PMSM magnet condition monitoring using harmonics in machine speed. IEEE Trans Ind Inf 13(13):1280–1290
20. Liu K, Zhu ZQ (2015) Mechanical parameter estimation of permanent magnet synchronous machines with aiding from estimation of rotor PM flux linkage. IEEE Trans Ind Appl 51(4):3115–3125
21. Feng G, Lai C, Kar NC, Noninvasive Kalman filter based permanent magnet temperature estimation for permanent magnet synchronous machines. IEEE Trans Power Electron, to be published
22. Lai C, Iyer KLV, Mukherjee K, Kar NC (2015) Analysis of electromagnetic torque and effective winding inductance in a surface mounted PMSM during integrated battery charging operation. IEEE Trans Magn 51(11):1–4
23. Zhu M, Hu W, Kar NC (2018) Acoustic noise-based uniform permanent-magnet demagnetization detection in SPMSM for high performance PMSM drive. IEEE Trans Transp Electron 4(1):303–313

Chapter 16
Color Sensor-Based Object Sorting Robotic Arm

Shantani Sinha, Santosh Kumar Suman and Awadhesh Kumar

1 Introduction

The twenty-first century is considered to be the Industrial Revolution in the field of robotics. This era deals with the robotics technologies which interact with the real world and thus convert data in them to useful information. The application ranges from industrial, agricultural, medical, military equipments [1]. Robotic arm also employs its usage in artificial limbs to assist the handicapped person [2–7]. By introducing the color capture functionality in the robotic arm, it can intelligently perform tasks such as differentiate, sort and organize, thus reducing the human labor. It also enhances the precision and the accuracy of the task. The robots can also be programmed to perform recurrent tasks.

A robotic arm is an electromechanical device which consists of motors, sensors, actuators and controllers. It is guided by motors and sensors and further programmed through software hosted by a microcontroller. It is used to perform many important tasks in a real-world environment. Due to mechanical structure and automatic control, it is used to perform numerous tasks with the utmost reliability and flexibility. The use of a conventional type of wheels on the platform of the robotic arm enables it to have a free movement. It possesses a single rotational axis in which the movement is restricted to an axis shaft which is attached to it. The base of the robotic arm constitutes DC gear motor and castor wheels to ensure the free movement of the wheels. The motors attached to the wheels are thereby controlled by the Android application. Android app is installed on the phone which has keys to move the motors in all the four directions.

S. Sinha
Solinas Integrity Pvt. Ltd., IIT Madras Research Park, Kanagam Road Taramani, Chennai, India

S. K. Suman (✉) · A. Kumar
Department of Electrical Engineering, Madan Mohan Malaviya University of Technology, Gorakhpur, India

V. K. Giri et al. (eds.), *Computing Algorithms with Applications in Engineering*, Algorithms for Intelligent Systems, https://doi.org/10.1007/978-981-15-2369-4_16

The robotic arm is equipped with the motor at each joint. Furthermore, the calculation of the DOF is done by identifying the number of joints. The number of degrees of the robotic arm is directly proportional to the number of joints. This work comprises of color sensor TCS3200, PIR sensor, servo motors (MG995R, SG90) and Arduino Uno microcontroller [4–12]. Servo motor is used to provide feedback. PIR sensor is attached to the arm of the robot. Motors are programmed to move in the given direction to detect the presence of the object. PIR sensor detects the object within approximately 10 m from the sensor. As soon as the PIR sensor detects the object, interrupt is generated to perform the further actions. After the recognition of the object, the command is sent to the color sensor TCS3200 to distinguish the RGB color of the object [5]. The program is written to recognize the RGB color of the nearby object. After the color is detected, it further sends the signal to the input ports of microcontroller. The microcontroller conveys its decision to the station that has been programmed. If the color of the object lies within RGB (255, 99, 71) [6], the Arduino Uno microcontroller further adjusts the servo motor according to the programming code and then moves an end effector arm to pick up the object. Subsequently, after the color is detected, the motors attached to the joints are triggered to rotate, thus grab the object and place it at a specific location [8]. It is further programmed to return to its default position once the task has been performed. It is programmed to perform the current task again and again. The remote control is used to move the robotic platform which has two wheels along with castor wheel attached to it. The remote control is used through the application installed on our Android phones, whose implementation provided multi-directional and sideways movement to the wheel. The wheel driving concept of the robot plays a vital role to navigate along many directions depending on its wheel architecture. This research work plays an indomitable role in understanding the implementation of the color sensor in the robotic arm, dynamic motion control and the role played by the PWM in the rotation of servo motors in a different direction to perform the required tasks.

The objectives of this work are as follows:

To design and develop a 3DOF robotic arm by using Arduino microcontroller.
To detect the presence of the object using PIR sensor.
To write a program of PWM to provide control to servo motors.
To write a color recognizing and color sorting codes.
To interface servo motor, color sensor, PIR sensor with Arduino microcontroller.
To test the codes and troubleshoot with the efficiency of the system.
To sort the objects according to their respective colors using RGB color sensor.

2 Methodology

Color sensing technique is used to recognize the specified color of the object and place the object at the desired location. The aim of PIR sensor is to detect the presence of the object. They are electronic sensors used to detect the infrared waves emitting

Table 1 Different types of the photodiode

S2	S3	Photodiode type
L	L	Red
L	H	Blue
H	L	Clear (no filter)
H	H	Green

from a particular object. It is interfaced with Arduino Uno microcontroller. The three different colors, red, green and blue, are recognized by the color sensor, and thus comparing the value of the color reflected in the RGB filter of the sensor, the arm gradually rotates to grab that object. TCS3200 color sensor is the complete color detector. It also includes the TAOS TCS3200 RGB sensor chip and 4 LEDs. The TCS3200 RGB color sensor has an array of photodetectors, each with a red, green or blue filter or no filter. The filter of each color is distributed throughout the array to eliminate location bias among colors. The oscillator is also present which produces the square wave output whose frequency is proportional to the color of the object. It contains four pins S0, S1, S2 and S3 which directly communicate with the microcontroller. TCS3200 has four photodiodes: red, green, blue and clear.

2.1 Color Detection

TCS3200 can output the frequency of different square waves, based on its color. There is a direct correlation between the output and the light intensity. The range of the typical output frequency is 2 Hz–500 kHz. We can get different scaling factors by a different combination of S0 and S1 (Table 1).

2.2 Frequency Selection

To detect various colors, filters of different colors are used such as red, blue and green and compare the value of the light reflected on it. Furthermore, it analyzes the color of the object. The color sorting robot is programmed using Arduino software (Table 2).

Table 2 Different scaling factors

S0	S1	Output frequency scaling
L	L	Power down
L	H	2%
H	L	10%
H	H	100%

Programming code is written to determine the color to perform sorting mechanism. The color sensor is interfaced with Arduino board. The Arduino then receives signals and further establishes communication with the servo motors through the input ports of Arduino. The servo motor then performs its task of picking and placing the object according to the programming of the microcontroller. Four servo motors are also interfaced on the input pins of the Arduino Uno board. The servo motor attached to the base, shoulder, elbow and wrist of the arm rotates along an angle of 180°. After performing the stated action, it then returns back to its default value. The hardware consists of color sensor TCS3200, PIR sensor, robot body and servo motor. The software is programmed through Arduino and simulated on Proteus software.

2.3 Flowchart Description

The flowchart depicts that initially the robot will remain in its initial position. Flowchart description algorithm as shown in Fig. 1 describes all the steps for color sensor-based sorting process. When the object is detected using a PIR sensor, the color sensor will initiate its property of color recognition and detects the color of the object. If the detected color does not fall in the range of RGB (255, 99, 71), then the robotic arm will remain in its initial position. If the detected color is true, the robots start its functioning of picking and placing the object and drop it to its specified place. After performing the task, it then returns back to its initial position [9].

3 Design of Robotic Arm

3.1 Hardware Design

See Fig. 2.

3.2 Simulation

The simulation is performed on Proteus software. The interfacing of a servo motor with Arduino microcontroller is shown in Fig. 3.

The Arduino Uno is a microcontroller board based on the ATmega328 connected with I/O pins shown in Fig. 4. It contains 14 digital input/output pins (of which 6 can be used as PWM outputs), 6 analog inputs, a 16 MHz crystal oscillator, a USB connection, a power jack, an ICSP header and a reset button. The operating voltage is 5 V, and the clock speed is 16 MHz.

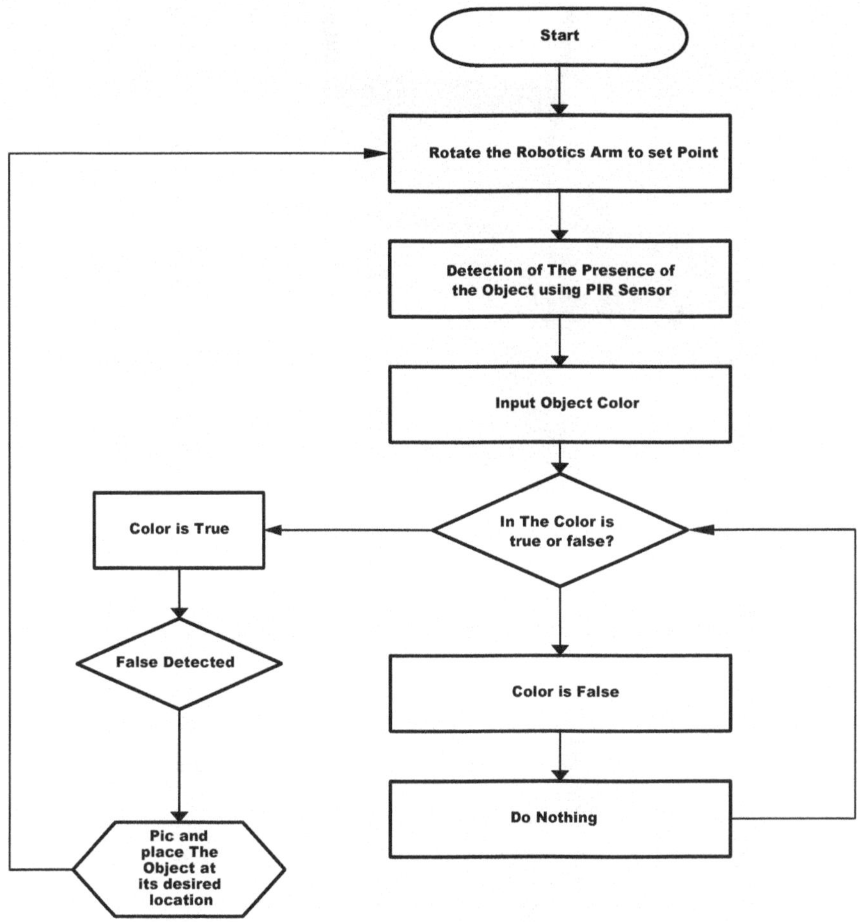

Fig. 1 Flowchart for the color sensor-based sorting process

4 Components of the System

4.1 Main Control Unit

The main control unit consists of Arduino microcontroller, 4 servo motors and motor driver IC (L239D). The MG996R servo motor is attached to the base of the arm. MG996R is a high torque featuring metal gear with a stall torque of 10 KG cm. It has a current of 2.5A and an operating voltage of 4.8–5.0 V. It can rotate along 180°. The SG90 metal gear servo motor is equipped at the elbow, shoulder and wrist. It provides the torque of 2.5 kg cm. Crystal oscillator of 16 MHz is connected to the Arduino Uno. It produces a clock pulse of 16 MHz. Four stepper motors are used for the controlling of robotic arm operations. Each servo motor is controlled by the

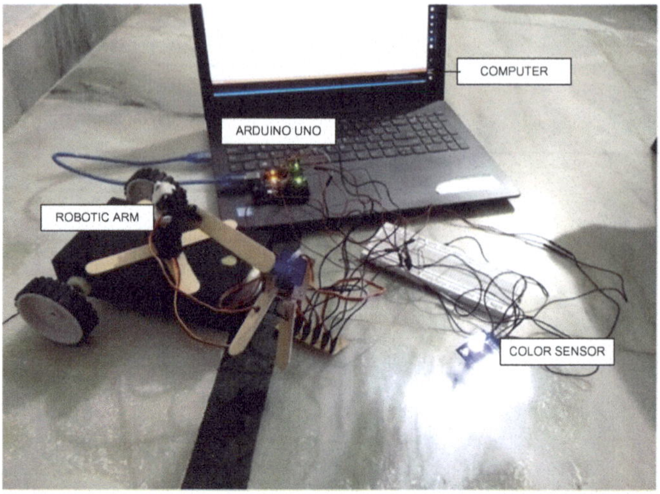

Fig. 2 Interfacing of components with Arduino

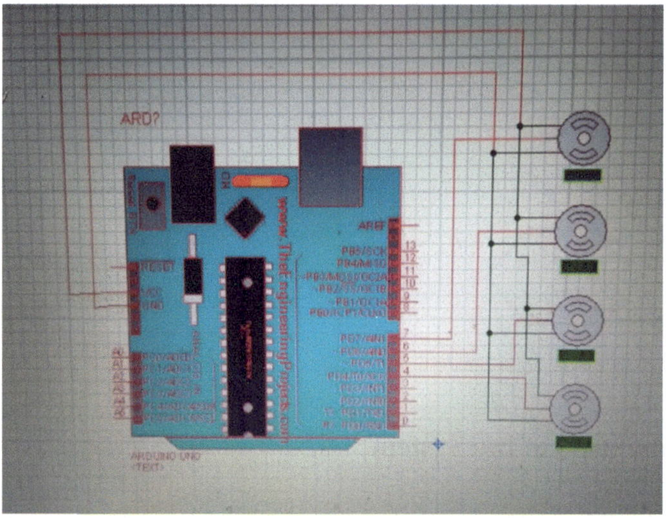

Fig. 3 Software simulation layout of interfacing of servo motor using Arduino microcontroller system architecture

motor driver IC. Every servo motor has three input pins to control the motor. The input pins are Vcc, GND and PWM. PWM ports of the Arduino microcontroller are connected to the input pins of motor driver IC. PWM pins of the stepper motor are connected to the output pins of the motor driver IC. In other words, it connotes that the input pins of the Arduino Uno microcontroller pins are connected to the output

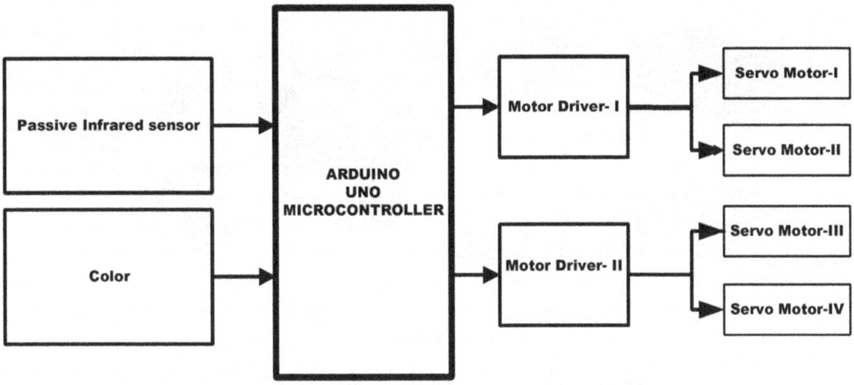

Fig. 4 Block diagram of Arduino Uno connected I/O pins

pin of the servo motor. The remaining two pins Vcc and the ground pin are directly connected to the Vcc (5 V) and the ground pins of the Arduino Uno.

Servo motors can rotate up to 180°. In this work, each servo motor is programmed to rotate in the anticlockwise and the clockwise direction. Pulses are given to rotating the motor in the particular direction. Low to high pulses are given to rotating motors [10]. The motor attached to the base will move left and right, whereas the motor attached to the shoulder will move upward and downward. The motor attached on elbow will move in the left and right direction. Furthermore, the motor implemented at the end effector will close and open to pick and place the object.

The pulse of a specific width is applied to the servo motor. The duty cycle of the pulse varies from 1 to 2 ms. The pulse should occur within each 20 ms. Pulse with 0.6 ms duration corresponds to 0° position, 1.45 ms to the 90° and 2.3 ms to the 180° (Fig. 5).

4.2 *Motor Driver*

Motor driver is used for positioning the arm in robot work, PWM programming of each servo motor is a vital task. Implementation of the L239D motor driver is done to enhance the overall functioning of the motor. It is a dual H-bridge motor driver. It is basically current amplifiers which accept low current signals from the controller and thus convert it to the high-level signal in order to drive the motor. L239D receives its signal from the Arduino Uno microcontroller and then directly implements it on the PWM pins of the servo motor. Each PWM pin of the servo motor is connected to the output pins of the motor driver. Each L239D motor driver IC can drive two motors simultaneously.

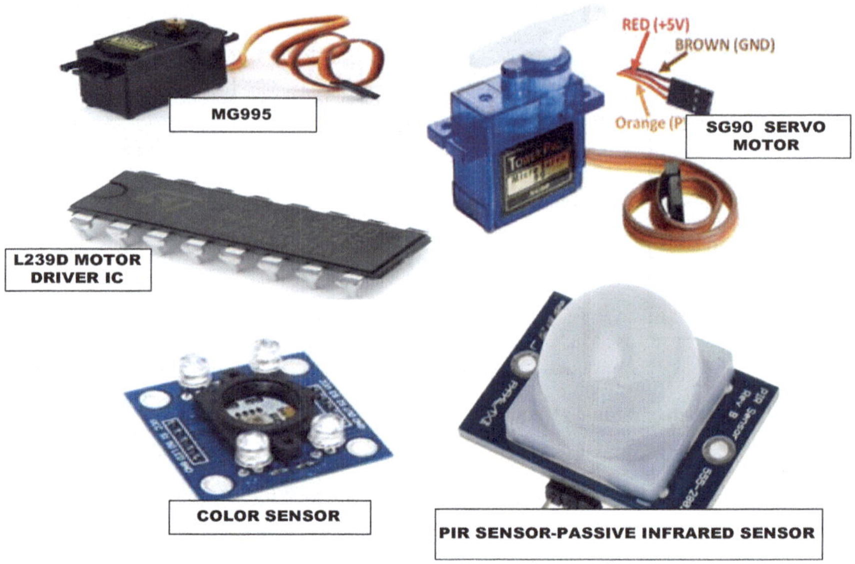

Fig. 5 Main component of system

4.3 Color Sensing Operation

This unit includes the use of TCS3200 color sensor. The function this unit performs is the sorting mechanism using color detection approach. The color sensor will detect the color of the object and then subsequently send a signal to the microcontroller to perform the further action as defined by the programming. It is also integrated with four LEDs that light up the object in front of it and detects the color within an array of 64 photodiodes, each with a red, green and blue, or no filter. The microcontroller will detect the color of the object based on the intensity of the color reflected. If the intensity of the red color is greater than that of green and blue, then the object's color is red. Moreover, if the intensity of the green color is greater than the respective red and blue color, then the object's color is green [11]. In the same way, if the intensity of the blue color is greater than the red and blue color, then the object's color is blue. The microcontroller will then receive the signal from the color sensor, and that corresponding pin switches from low to high with a delay of 100 ms. After the appropriate delay, the pin of the corresponding color goes low. In addition to it, it can also detect purple, orange and pink colors through appropriate programming.

4.4 Hardware Description

Different components are used in this work. List of components used in this work is described below.

4.5 Servo Motor

Servo motor is an error sensing feedback control device which is used for precise control of performance of the system. It is a DC motor which is used to control the angular position in steps. The servo motors usually have a revolution from 90 to 180°. Servo motor is controlled by the pulse width modulation technique. The width of the pulse determines the angular position of the motor [12]. PWM signal for servo control generally has a period of 20 ms. The width of the pulse varies from 1 to 2 ms. The length of the pulse determines how far the motor moves. For example, a pulse of duty cycle of 1.5 ms will turn the motor to 90° position (neutral position). When the pulse that is sent is less than 1.5 ms, then the motor will rotate its output shaft in an anticlockwise direction from its initial neutral position. When the pulse is greater than 1.5 ms, then the motor will rotate in clockwise direction from its position. The pulse time for various angular positions of servo motor is shown in Table 3. The minimum pulse width is about 1 ms wide (for some servo it is 0.5 ms), and the maximum pulse width is about 2 ms (for some servo it is 2.5 ms). PWM pins of all the four servo motors are connected to the respective pins of Arduino Uno microcontroller. The input is given to the microcontroller which further generates a control signal to control the rotation pins of servo motors, and the corresponding pulse is generated (Fig. 6).

4.5.1 The Relationship Between Pulse and Rotation of Servo Motor

See Fig. 7.

4.5.2 High Pulse Time for Various Angular Positions of Servo Motor

See Table 3.

5 Results and Discussion

This work has been effectively designed to handle the required task. The work is working automatically on the commands that are received from the control unit. The

Table 3 High pulse time for various angular positions of servo motor

Angular position (in degrees)	High period (in millisecond)
0	0.560
10	0.64
20	0.72
30	0.88
40	1.04
50	1.12
60	1.20
70	1.28
80	1.36
90	1.52
100	1.60
110	1.68
120	1.76
130	1.92
140	2.00
150	2.16
160	2.24
170	2.32
180	2.40

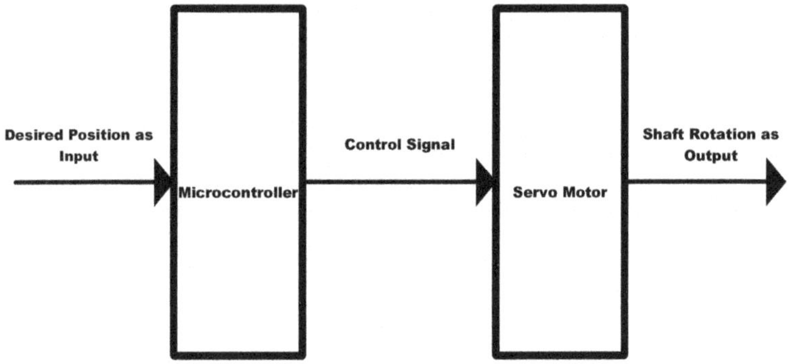

Fig. 6 Arduino Uno microcontroller with servo motor

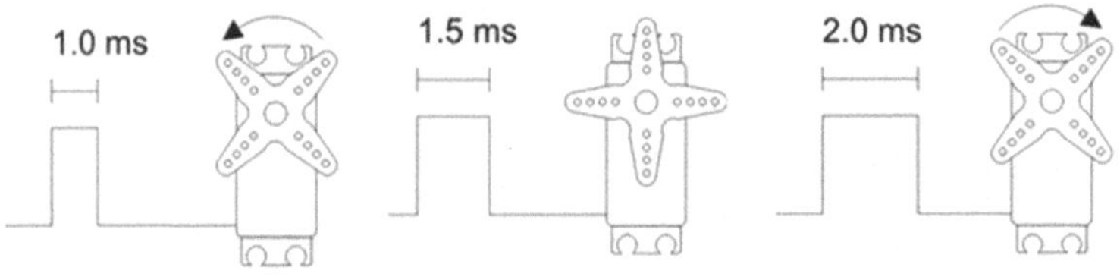

Fig. 7 Pulse width with varying duty cycle

Table 4 Robotic arms operated by servo motors

Maximum angle (°)	Axis capabilities: mechanical assembly
90	Clockwise direction
90	Anticlockwise direction
180	Anticlockwise direction

color sensor is programmed properly to perform the action of sorting of objects based on its color. The color sensor recognizes the RGB color. The value of the RGB (255, 99, 71) color sensor ranges from 0 to 255. Commands are given to rotate the arm in a particular direction to grab the object by recognizing its color and further place it at its desired location. Table 4 depicts the direction of rotation of robotic arms operated by servo motors. By considering the initial position of the robotic arm, the following conclusions are derived.

6 Future Scope and Conclusion

This system has been successfully designed and has a capability to rotate in order to handle the desired task. It can identify the presence of the object using PIR sensor and thereby generates an interrupt to perform the further operations.

Color sensor recognizes the color of the object and thereby rotates the arm to grab the object and place it at its specific location.

The main objective of this work is:

1. Detection of object.
2. Recognition of color.

The system consists of Arduino powered control unit which sorts the objects based on its color and thereby picks and places it at its desired location. The real contribution of this system is that it saves time while sorting, thus performing a large number of tasks in minimum time. It also reduces human labor to perform the recurrent task again and again. The advantage of this system is to enhance the reliability and flexibility of the system. It is a cost-effective system and performs the operations without any human errors.

The robotic arm with color recognition and sorting process has been achieved successfully. This system is useful in industrial, agricultural, military, medical, household operations and many more.

References

1. Beyer C, Pretz T (2004) Special requirements for material characterization regarding the simulation of solid waste material processing. Resour Conserv Recycl 42:121–131
2. Pillajo C, Sierra JE (2013) Human Machine Interface HMI using Kinect sensor to control a SCARA Robot. In: 2013 IEEE Colombian conference on communications and computing (COLCOM), pp 1–5
3. Singh A, Balam N (2016) An intelligent color sensing system for building a wall. In: International conference on emerging trends in communication technologies (ETCT), pp 1–4
4. Hu C-Y et al (2016) Visual servoing spanner picking and placement with a SCARA manipulator. In: 2016 IEEE international conference on industrial technology (ICIT), pp 1632–1637
5. Jawahar M et al (2017) Sensor-based color sorting system for leather shoe components. In: 2017 Third international conference on sensing signal processing and security (ICSSS), pp 296–300
6. Kwon Y, Kim W (2014) Development of a new high-resolution angle-sensing mechanism using an RGB sensor. IEEE/ASME Trans Mechatron 19(5):1707–1715
7. Su S-F, Wang K-J, Chen M-C, Rudas IJ, Tsai C-C (2015) Adaptive PD fuzzy control with dynamic learning rate for two-wheeled balancing six degrees of freedom Robotic Arm. In: 2015 IEEE international conference on automation science and engineering (CASE), pp 1258–1261
8. Feng K, Zhang X, Li H, Huang Y (2017) In: Intelligent robotics and applications, vol 10463, p 576
9. Munich ME et al (2006) SIFT-ing through features with ViPR. IEEE Robot Autom Mag 13(3):72–77
10. Angeles J, Morozov A, Navarro O (2000) A novel manipulator architecture for the production of SCARA motions. In: Proceedings of the 2000 IEEE international conference on robotics and automation, San Francisco, CA, Apr 2000
11. Getman A, Kim J, Kim T-C (2010) Imaging system having White-RGB color filter array. In: Proceedings of 2010 IEEE 17th international conference on image processing, Hong Kong, 26–29 Sept 2010
12. Farman M et al (2018) Design of three degrees of freedom robotic arm. Int J Comput Appl 179:12–17

Chapter 17
A State-of-the-Art Literature Review on Microelectromechanical Systems

Shivam Hemant Dandgavhal, Ashish Ravindra Lande and Akbar Ahmad

1 Introduction

Great potential and performance displayed by microelectromechanical systems over the conventional devices have motivated researchers in recent years to work on research for mass production and improvement of these products. Significant advantages of MEMS over the conventional devices are seen which include small form factor, small aspect ratio, downscaling IC architecture, admirable single crystal silicon mechanical property and good on-chip electronic control integration potential.

1.1 Micro-machining

Miniaturization is vital in most of the contemporary technologies. Production of the state-of-the-art micro-products for microelectronics, biotechnology, telecommunication, medical applications and MEMS demands micro-machining. Areas like electronics, optics, medicine, avionics, biotechnology, communication sector are heavily demanding manufacturing of mechanical components with the aspects of micro- to nanoscale. Deep X-ray lithography masks, micro-fluidic systems, micro molds, microscale fuel cells, valves and mixing devices, micro-holes for fiber optics, micro-nozzles for high-temperature jets, fluidic micro-chemical reactors requiring microscale pumps are some of the featured applications.

S. H. Dandgavhal (✉)
Department of Mechanical Engineering, KKWIEER, Nashik, Maharashtra, India

A. R. Lande
Department of Mechanical Engineering, KKWP, Nashik, Maharashtra, India

A. Ahmad
Department of Electrical Engineering, KKWIEER, Nashik, Maharashtra, India

V. K. Giri et al. (eds.), *Computing Algorithms with Applications in Engineering*,
Algorithms for Intelligent Systems, https://doi.org/10.1007/978-981-15-2369-4_17

Laser machining and computer numerical controlled (CNC) micro-machining, deep UV lithography, deep reactive ion etching, electrical discharge machining, X-ray lithography electro-deposition molding (LIGA) are the state-of-the-art production techniques to meet the requirement. Mechanical micro-machining is a feasible micro-machining technique for the fabrication of 3D features on composites, polymers, metals and ceramics. Micro-machining facilitates the micro-scaled features with aid of miniature drilling, milling and turning tools of nearly 10 μm in diameter. Scarcity of experience and information on the micro-machinability of materials defines numerous problems for research in micro-machining.

1.2 Morphological Analysis

State-of-the-art morphological analysis techniques like transmission electron microscopy analysis, white light interferometry, scanning electron microscopy, porosity testing, thermal gravimetric analysis, optical light microscopy, differential scanning calorimeter analysis, particle surface area analysis, X-ray diffraction, dynamic light scattering, porosity meter and energy-dispersive X-ray analysis have been of great use to the researchers in analyzing the dissertation of synthesis, manufacturing, assembly, active monitoring and testing of MEMS devices.

1.3 Smart Materials

Precise and accurate actuation of MEMS devices can be facilitated by use of smart materials. Shape memory alloy material exhibits excellent superelasticity and shape memory effect properties. If SMA is deformed under limited external force and then subjected to direct/indirect thermal actuation, then SMAs regain their original shape. Different types of shape memory alloy, their behavior, cost analysis, optimization and applications have been studied in this paper, and further areas of research work have been depicted.

1.4 Micro-laser Beam Machining

Material removal by the laser ablation depends upon powerful absorption of photons; hence, appropriate laser wavelength is profitable. Ability to machine the materials transparent to the laser wavelength is gained by the use of ultrafast lasers which trigger ablation because of the multi-photon absorption at huge peak intensities.

In this paper, detailed literature in the areas of micro-machining, morphological analysis, smart material behavior and applications has been reviewed. The gray areas in these fields have been identified, and future research scope has been concluded.

2 Literature Review

This section provides an overview of the current state-of-the-art, wide and complex field of micro-machining, morphological analysis, smart materials, MEMS and their wide range of applications. Detailed state-of-the-art literature has been reviewed and presented in organized tabular format in Table 1. Graphical representation of yearwise percentage of papers referred in this review paper has been shown in Fig. 1.

2.1 Inference from the Literature Review

From the literature review as elaborated in Table 1, we have realized that micro-machining and synthesis techniques are need of hour to satisfy the production requirements of MEMS, and various researchers have focused on various areas to research on optimized MEMS production process, implementation of smart materials and suitable morphological analysis techniques. Research gap and gray areas have been identified from the literature review as discussed in Table 1, and the following research topics will be of great use to researchers in their research work.

1. Process parameter optimization of micro-fabrication process like FIB, micro-milling, micro-drilling, ion beam machining, micro-EDM, chemical vapor deposition, sputtering, micro-laser beam machining, micro-electrochemical machining, lithography, etc.
2. Design and development of test rig for micro-actuators, micro-robotics and mechatronics application.
3. Development of micro-UTM for smart materials like SMA wires.
4. Development of nanostructure-based electronic device like metal–oxide–semiconductor field-effect transistors.
5. Heat transfer study of nanostructure-based electronic devices (phase-change memory CPU).
6. Comparative feasibility study of photolithography for development of all kinds of MEMS devices.
7. To apply top-down and bottom-up approach for MEMS devices.
8. Development of test rig for shape memory alloy (SMA).
9. To carry out experimental analysis for reliability aspects of microelectromechanical systems and SMA in different conditions.
10. Comparative analysis of shape memory alloy at different actuations like hot water actuations, laser actuation, electrical actuations, etc.
11. Detailed analysis of SMA-reinforced civil structures from all aspects.
12. Improvement in thermomechanical behavior, aging characteristics and number of working cycles of SMA in high loading conditions.
13. Vibrational analysis of SMA.
14. To develop physics-based model for SMA.
15. To apply bond graph theory for SMA analysis.

Table 1 Summary of the literature review

S. no.	Author	Investigation	Findings/observations/remark
1	Majeed and Ramesh [1]	Fabrication of capacitive MEMS pressure Sensor integrated with bossed diaphragm with advance techniques to improve sensor characteristics	Variations in developed device are low due to pressure exerted on rear end of diaphragm. Sensitivity is improved as backpressure plays a crucial role in the performance of developed sensor
2	Wang et al. [2]	Review on design optimization of MEMS has been done by evolutionary computation (EC) techniques	Global optimum solutions can be devised for multi-objective problems and performance enhancement by EC techniques. Future scope lies in determining way to decide priorities of multi-objective optimization
3	Dyuzhev et al. [3]	Comprehensive and parametric study of nanostructured membranes is carried out with the help of SEM/SIMS/AES. Development and fabrication of nanostructured membrane blanks for MEMS devices	Nanostructured membranes were successfully developed and used to determine mechanical stresses in silicon wafers
4	Fallah Ghanbari et al. [4]	Development of novel severe plastic deformation method titled as "multi-axial temperature forging (MATF)" for production of nanostructured Ti-6Al-4V	Results proved that the developed process can critically refine the grain size producing nanostructures
5	Bao et al. [5]	Reviewing synthesis, transfer and production of 2D hexagonal boron nitride (hBN) films, hBN-based composites and hBN-based van der Waals heterostructure	Future scope for research on union of 2D boron nitride films with variety of materials to optimize the performance of electronic devices
6	Natarajan et al. [6]	Utilization of tungsten electrode of 300 μm diameter for sake of analysis of consequences of micro-EDM drilling process parameters	Feed rate of 4.31 μm/s, capacitance of 0.1 μF and voltage of 120 V were the optimal machining conditions for which output values of MMR 0.016 mm^3/min, Ra of 0.49 μm and TWR of 1.08 were achieved

(continued)

Table 1 (continued)

S. no.	Author	Investigation	Findings/observations/remark
7	Cheng et al. [7]	Analysis of formation of the top burr during micro-end milling operation on Ti-6Al-4V	Least and best top burr size was obtained on the series of milling surfaces
8	Anand et al. [8]	Consideration of tool edge radius and minimum chip thickness for development and demonstration of mechanistic micro-drilling force model	The necessity of dynamic model for micro-drilling concerning consequences of drill run out, drill vibration, drill wandering, etc., has arisen due to several limitations
9	Smith et al [9]	Presentation of reasonable speculation to overpower obstacles that have cursed efficient modeling of additive manufacturing (AM) methods for metallic materials	Numerous problems are faced for implementation of AM for metallic components in mass production, and great work is to be done for overcoming computation and experimental problems
10	Hussein et al. [10]	Structural and surface morphological analysis of copper phthalocyanine thin film samples by the AFM, SEM, EDX, XRF, XRD to get optimum result at different annealing temperatures	Annealing temperature can control the crystal morphology and molecular orientation of CuPc film. Optimum results are obtained at the annealing temperature of 423 K
11	Xu and Zhang [11]	Simulational analysis and comparison of surface morphology of polished surface polished by new developed machine and the other traditionally polished surfaces by using MATLAB software	The simulation results and test results satisfy each other, proving superiority of new polished surface over traditionally polished surface, and simulation results are okay to select process parameters to polish actual surface
12	Lei et al. [12]	Discovery of influence of controllable diverse surface integrity parameters for optimizing performance during fabrication of components with collective properties through surface modification	Research concluded that excellent performance manufacturing was obtained due to improved surface integrity which was obtained from controllable multiple surface integrity parameters achieved through surface modification

(continued)

Table 1 (continued)

S. no.	Author	Investigation	Findings/observations/remark
13	Sahli et al. [13]	Spotlight of the work was aimed at prediction of shrinkages and densities by use of finite element analysis (FEA) method of a MIM micro-component	Research concluded that FEA and experimental data were in harmony with each other, and prediction of densities and shrinkages was thus validated
14	Takahashi et al. [14]	Microstructural investigation of weld zone, interface zones and base metals in dissimilar weld bead of Al and Ti after full penetration of dissimilar lap welding of Ti and Al done by single-mode fiber laser at ultrahigh welding speed	Ultrahigh welding speed and high solidification rate suppressed formation of brittle compounds; also, utilization of single-mode fiber laser with ultrahigh welding speed is efficient for welding dissimilar metallic components
15	Xiong et al. [15]	Exploration of advanced material processing methodology for production of nanomaterials and micro-/nanostructures in a wide range of dimensions	Laser-based techniques have upper hand in production of nano-/micro-functional structures and materials rather than conventional production techniques
16	Suganthi et al. [16]	Revision of present methodology for fabrication of micro-holes and also showing diverse strategies and hot research topics for micro-hole fabrication	Union of conventional individual techniques for fabrication of free-form geometries with time economy, flexibility and accuracy has given scope for R&D
17	Cecil et al. [17]	Reviewed technology and methodology related to micro-device assembly (MDA)	Scope for R&D in system and factory automation, framework design and development and utilizing IT system and technology for meeting increasing requirements
18	Park et al. [18]	Development of porous structure by combining polymer with CFA and then irradiating a 355-nm UV pulse laser on the sheet	Control of pore density, pore size and pore type can be achieved by varying process parameters consisting of CFA density, laser fluence and processing time
19	Chiou et al. [19]	Discovered the optimal parameters for micro-electro-discharge machines of SKH59 alloy utilizing WC electrodes	Resultantly, electrode coating is an excellent method exhibiting superior electrical and thermal aspects

(continued)

Table 1 (continued)

S. no.	Author	Investigation	Findings/observations/remark
20	Khanghah et al. [20]	Analyzing effects of cutting speed, feed rate and depth of cut on the burr thickness and burr height after up milling and down milling operation on 316 stainless steel	Results suggested that best configuration of 15,000 spindle speed of rpm, feed rate of 5 mm/tooth and depth of cut of 0.15 will diminish the burr sizes
21	Huo et al. [21]	Observation of aspects of micro-machined single crystal silicon with (100) orientation	Research concluded that micro-milling offers optional production route to clean room approaches for production of silicon components at a micron scale
22	Krestinin et al. [22]	Discovery of characteristics of single-wall carbon nanotubes products manufactured in Russia with novel technology and their scope for industrial use	By considering composition of nanotubes and cost-to-performance ratio, OCSiAl has released best synthesized SWCNTs under the TUBALL trademark
23	Aišman et al. [23]	Effective determination of effects of cooling rate, deformation in solid state and heat treatment on the evolution of microstructure in X210Cr12 tool steel	Solidification rate impacted carbide network quality without affecting austenite grain size; 1 K/s cooling rate causes coarsening of primary chromium carbide. Unconventional austenite martensite microstructure was obtained from heat treatment with a hardness of 800 HV30
24	Vella et al. [24]	Development of process chain for serial manufacturing of polymer-based devices possessing both micro- and nanoscale features	Developed process can be utilized for bulk production of polymer parts possessing diverse length scale characteristics
25	Kaushik et al. [25]	Review on the methodology for integration of silica nanowires into microscale devices within these recent years	Scope for using nanostructures in combination with antecedent microscale platforms as a chance to improve drug delivery, environmental monitoring, point of care testing and personalized diagnostics for health monitoring

(continued)

Table 1 (continued)

S. no.	Author	Investigation	Findings/observations/remark
26	Li et al. [26]	Examining mechanism of SnO_2 characterized by hierarchical micro-/nanostructure fabricated by hydrothermal, coordination, templating, and electrochemical deposition methods	Research concluded that controllable designs of produced hierarchical micro-/nanostructure facilitate diverse variety of industrial applications
27	Oborski [27]	Exhibition and execution of modern ideas regarding monitoring integration, single system supervision with machine control, company IT system, shop floor control and similar areas were spotlight of the review	Scope for uniting machinery, tools and workpiece condition monitoring, process, maintenance functions and product history documentation and integration of systems with modern IT system functioning at all levels by aid of cloud computing
28	Farshbaf Zinati and Razfar [28]	Utilization of polyamide 6/multi-walled CNTs composites for examination of surface roughness after end milling	Research concluded that the high spindle speed with low feed rate gives superior surface quality
29	Zhan et al. [29]	Development and utilizing micro-polycrystalline diamond (PCD) milling tool for the milling operation on the tungsten carbide	Superior surface quality can be obtained by the use of PCD tool for micro-machining of WC even in the condition of intense wear
30	Lou et al. [30]	Development of divide and conquer algorithm for superior performance over the alpha shape method	Consistent results and superior performance were obtained from divide and conquer algorithm over the alpha shape method
31	Pereira et al. [31]	Research on the wear and tear topography of mold of EN AW-6083 T4 material after interval of working cycles at marked points in the mold by optical interferometry	Wear rate was directly proportional to frontal obstacles in mold. Ssk decreased with increase of wear rate, while the Sku parameter increased with increase in wear rate
32	Nag Chaudhury [32]	Determination of quasi-static starting fracture toughness (J1c) of micro-alloyed steel (C = 0.19%) in rolled state lacking heat treatment through experimental means	Research on consequences of normalizing, shot peening and cyaniding discovered that it had good effects over commencing fracture toughness

(continued)

Table 1 (continued)

S. no.	Author	Investigation	Findings/observations/remark
33	Liu et al. [33]	Reviewing role of CNTs for advances in vital applications consisting of CNT-based thermal interface material, interconnection and cooling devices	Scope for advancement in growth technology is required for integration of CNT interconnection into the electronics system
34	Kumar and Rao [34]	Study the effect of post-annealing temperature treatment on surface morphology, topography and texture of ZAO films using AFM	Grain size rises with surface roughness with rise in annealing temperature, and thus significant change is seen in surface roughness with rise in temperature
35	Löchte et al. [35]	Proposing three compensation techniques which unite quality approach in a MPS for the sake of micro-fabrication	Dynamic view of the consequence of balancing errors throughout the complete fabrication process is possible
36	Gil et al. [36]	Investigation of the impact of the process variables on efficiency of ISEDM process	Economic production of micro-pins of aspect ratio of 0.2-mm diameter and 90:1 can be produced
37	Lee et al. [37]	Investigating a novel material composition and compatible process for hollow carbon nanofibers with best morphological quality with consistent hollow and wall thickness	Research concluded that SAN is an excellent choice due to its immiscibility with PAN satisfying the co-axial electrospinnability into core/shell nanofibers and thermal sustainability conditions throughout carbonization process
38	Kim et al. [38]	Development of technique named SRM utilizing sputtered particles obtained from offshoot of FIB for production of automatically sealed micro-/nanofluidic channel	Technique is effortless as elimination of sacrificial layer is pointless along with automatic sealing in a single process
39	Jimenez and Mavris [39]	Review on the progression of morphological analysis application in systems engineering	Morphological analysis plays a vital role in system engineering application, decomposition and synthesis in design phases, configuration selection and analysis and scenario generation

(continued)

Table 1 (continued)

S. no.	Author	Investigation	Findings/observations/remark
40	Liu et al. [40]	Demonstration of SEM as an excellent tool for analysis of microstructures, crystal growth evolution, grain size, chemical composition, etc.	SEM is a vital tool for analysis of microstructures of organic and inorganic matters along with its flexibility of being updated by the accessories which increase its wide area of application in morphology and nanofabrication
41	Zhovklyi et al. [41]	Studying the possibilities of using UTMP-6 and UTMP-7 for preparation of smooth surface specimens used in SEM and TEM	UTMP provides low temperature fixation of photo-film structure avoiding specimen deformation during section preparation of samples retaining water
42	Lee et al. [42]	Development of desktop micro-forming manufacturing system for producing multi-micro-patterns and optimizing performance	Herringbone grooves in FDB can be made using developed desktop press system by optimizing production time and cost
43	Kibria et al. [43]	Observation of micro-turned cylindrical-shaped Al_2O_3 prepared by pulsed Nd:YaG laser by changing laser turning parameters	Quality surface can be generated through proper controlling of the micro-turning process parameters
44	Saklakoglu and Kasman [44]	Inspection of influence of the diverse process parameters consisting of laser power, scan speed, frequency and fill spacing in the micro-milling operation on the AISI H13 material	Research concluded that superior surface finish is obtained by high scan speed and decreasing laser power and frequency. Decreasing scan speed and fill spacing will result in enhanced milling depth
45	Savanovic and Zeiler [45]	Analyzing performance of morphological charts in architectural and engineering meetings	Morphological charts are efficient as it offers complexity reduction for large data, time efficient, excellent research tool and good accuracy
46	Arenas-Alatorre et al. [46]	Use of OM, AFM, SEM, TEM and MFM in the study of archaeological samples	Diverse microscopy techniques help in analysis of archaeological samples for determining chemical composition, physical properties, raw materials, manufacturing process and other parameters

(continued)

Table 1 (continued)

S. no.	Author	Investigation	Findings/observations/remark
47	Cheng et al. [47]	Presentation of six-axis wire EDM machine with detailed analysis and design	Analysis of matrices concluded that there was necessity of three linear axes X, Y and Z as well as three rotational axes C, U and V
48	Brousseau et al. [48]	Reviewing the technologies corresponding to lithography-based process for the fabrication of components characterized by micro- and nano-scaled aspects	There is a future scope for evaluation of reach, reliability and development scope of current technologies
49	Shivareddy et al. [49]	Studying evolution of rough Cu surface during electropolishing on AFM and development of semiempirical function to describe the surface width evolution	Obtained results from the traditional fashion matched with developed function
50	Liou and Chen [50]	Investigation of polymer micro- and submicron structures from injection molding in micro- and submicron channels at micron/nanolevel	High mold temperature is vital for smooth functioning of injection molding, and main injection pressure, injection duration time and selection of polymer significantly influence molding characteristics
51	Desai and Haque [51])	Design and fabrication of MEMS tribometer with lateral (1 nN) and normal (100 nN) force resolutions along with bandwidth of 50 kHz	Developed tribometer permits qualitative and quantitative dynamic testing in electron microscope. Also, it is very stable for diverse environmental and operating conditions
Key literature on shape memory alloy			
52	Senf et al. [52]	Proposition of SMA-reinforced polymers and SMA pultrusion processing unit for its mass production was done	The novel SMA-reinforced polymer can be designed for actuation and high sensitivity sensing application

(continued)

Table 1 (continued)

S. no.	Author	Investigation	Findings/observations/remark
53	Alaneme and Okotete [53]	Discussed effects of process parameters on Cu- and Fe-based SMAs along with novel solutions with the aim to substitute NiTi SMA with Cu- and Fe-based SMA for low-cost applications	Cu- and Fe-based SMAs are not suitable to replace NiTi SMA because of low strain recovery, low speed of response, low sensitivity, low thermal response and phase transformation characteristics of Cu- and Fe-based SMAs, which gives scope for further R&D
54	Kang and Song [54]	Reviewed the fatigue failure analysis of NiTi SMAs by constant mechanical loading	Scope for research on morphological analysis and multi-axial fatigue failure analysis due to mechanical and thermal cycling of nitinols. Investigating relation between fatigue failure and martensitic transformation
55	Quan and Hai [55]	Investigated applications of SMA in aviation sector	SMA has promising applications for sensing, actuation, electromechanical energy conversion, auto-adjustable SMA-built aircraft wings for a wide range of altitude, temperature and pressure
56	Lobo et al. [56]	Review on constitutive behavior models of shape memory alloy for forecasting the behavior of material on stimulation	Superelasticity of NiTi can be stabilized by annealing and aging therapies in a cyclic order. Temperature aging and prestress cause transformation in SMA behavior. Dynamic test results of SMA along with 1D constitutive model equations were also conversed
57	Prashantha et al. [57]	Dissertation on thermal aging of Cu–Al–Be SMA's above austenite phase temperature	Increase in aging time results in increase in transformation temperature, formation of precipitates, hardness and decrease in the shape memory effect

(continued)

Table 1 (continued)

S. no.	Author	Investigation	Findings/observations/remark
58	Petrini and Migliavacca [58]	Reviewed and listed the biological applications of SMA in orthopedics, vascular, dental, surgical and neurological fields	Role of SMA in development of prosthetics has shown a great promise in the future
59	Dönmez and Özkan [59]	Proposed use of SMA actuators for flap mechanism of unmanned aviation units	Simulation ran on derived mathematical model displayed good results. Operation lag for re-actuation caused by prolonged cooling phase and false actuation of SMA actuators due to exposure to varying temperature conditions should be researched in the future
60	Huang et al. [60]	Overview of recent applications of Cu-, Fe- and NiTi-based SMAs, shape memory polymers and shape memory hybrid	Thermoplastic polyurethane shape memory polymer was suggested as economical alternative in terms of processing and material cost. Further research is must in optimizing performance and cost of SMAs
61	Song et al. [61]	Investigated civil engineering applications of SMA application of structural self-rehabilitation via implementation of reinforced SMA	Vibration suppression by active tuning of structural natural frequency and passive devices for structural damping are the applications of SMA

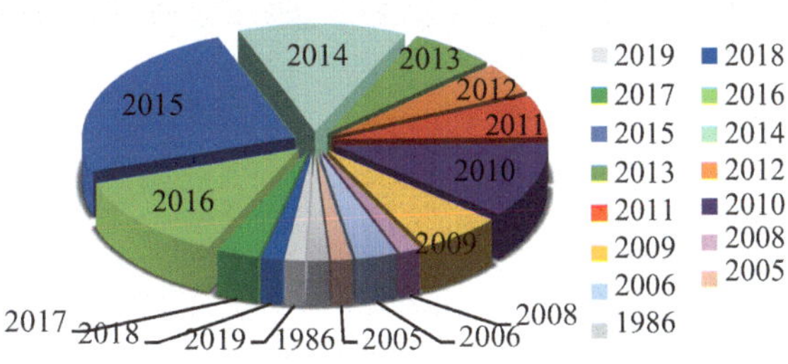

Fig. 1 Pie chart of publication year with percentage of papers

16. Cost optimization and investigation of alternative options for SMA.
17. Development of SMA-reinforced smart polymer and bio-prosthetics.
18. Feasibility study of SMA for surgical equipments.
19. Research on SMA behavior for different base material compositions and its morphological analysis in thermal conditions.

20. Case study of mass production industries of CNTs and MEMS.

3 Methodology

Taguchi's method of design of experiment [62] for improving product quality can be used with an aim to optimize the performance parameters of MEMS fabrication process. Figure 2 shows the ideal process flow chart for the methodology. Taguchi's main aim was to decide the optimal process parameters of the system so as to nullify the losses due to deviation in manufacturing, cumulative, environment and miscellaneous parameters, thus making the system robust.

3.1 Design of Experiment by Taguchi

Three-step design procedure was suggested by Taguchi to gain desired design product quality.
System Design. It is enhancing product/process quality by good innovation by means of analysis and visualization of product.
Parameter Design. It refers to Taguchi's approach of statistical DOE for finding optimal process parameter, making the process robust against uncontrollable parameters/noise and getting good functional performance.
Tolerance Design. This approach by Taguchi improves product quality without changing material/parts by stating the tightened tolerances in appropriate conditions.

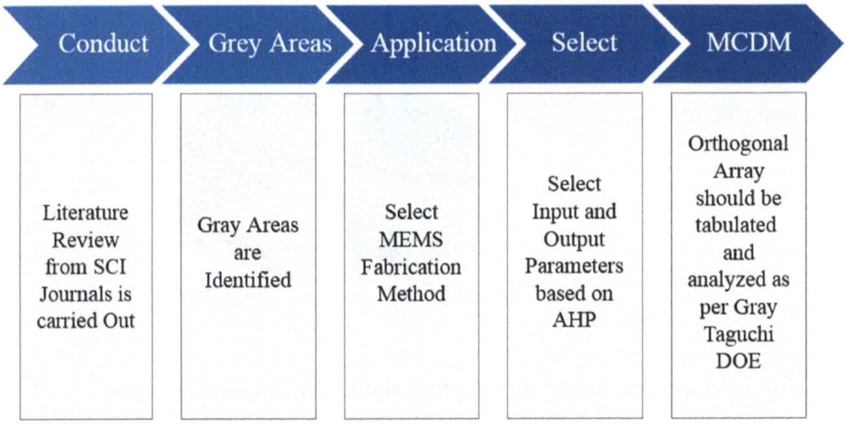

Fig. 2 Methodology process flowchart

3.2 Orthogonal Array

Taguchi invented methodology to conduct DOE which makes use of orthogonal array for investigating process parameters with optimal number of experiments. Varying and testing the influence of control factors are studied by implementing Taguchi's systematic tool of orthogonal array like L27, L18, L16, L12, L9, L8 and L4. Control factors and their levels are arranged in columns, whereas the experimental run executed in specified factors is arranged in rows. Also, the degree of freedom of process parameter under study should be equal to or less than degree of freedom of selected orthogonal array. Selection of orthogonal array is done by the following equation:

$$\text{NTaguchi} = 1 + \text{NV} \times (L - 1)$$

where

NTaguchi = number of experiments to be performed,
NV = number of variables,
L = number of levels (e.g., low, medium, high).

3.3 Stepwise Taguchi Methodology

Taguchi proposed an ordered approach for use of statistical DOE procedure to improve the process and product quality.

Define the objective output process parameter to be optimized. The ultimate output desired quality characteristic/parameter/variable to be enhanced is determined. The typical examples are surface roughness, metal removal rate, hole diameter, cost, electromagnetic radiation, weight, etc.

Recognize the uncontrollable noise factors, parameters and test conditions. The noise factors or the uncontrollable factors like environmental operating conditions, component depreciation, etc., which may have adverse effect on the process quality and performance are recognized.

Recognize the variable influential control parameters. Influential input control parameters must be identified, and appropriate levels/test values associated with them must be specified.

Plan the matrix experiment and data analysis methodology. Planning the matrix experiment by selection of appropriate orthogonal array for particular application/process must be done, and data analysis procedure should be stated in brief.

Actual experimentation. Actual planned series of matrix experiments should be performed, and output variables/response/process parameters/results should be noted. Experiments may be actual hardware experiment, mathematical model, algorithms or computer simulation models.

Investigate the data and conclude optimal levels. Calculation of optimal performance parameter/output variables must be done by analyzing test results. Mean-to-standard deviation or signal-to-noise ratio (SNR) should be calculated by statistical means to get an overview of performance.
Forecasting the performance at various levels. An experimental check will help us validating the forecasted optimal values for corresponding levels in various control parameters under consideration.

4 Conclusion and Future Scope

This paper presents the state-of-the-art literature review for various applications researched by the different researchers during the last decade. Gray areas for research have been identified in the advanced machining methods and nanotechnology. This paper deals with the state-of-the-art literature review, MEMS, NEMS and SMA. Thus, from the inferences of the literature review (Table 1), researcher can carry out substantial research work in cited gray areas to develop new technologies associated with MEMS and NEMS. With Taguchi design of experiment, analytical experimental orthogonal array for different MEMS fabrication processes can be formulated and process parameters can be optimized. The input and objective output parameters for such fabrication process can be selected by using multi-criteria decision making (MCDM) techniques like technique for the order of prioritization by similarity to ideal solution (TOPSIS), analytic hierarchy process (AHP), best worst method (BWM), gray relational analysis (GRA), simple multi-attribute rating technique (SMART), multi-attribute global inference of quality (MAGIQ) and weighted sum model (WSM).

Acknowledgements We are very thankful to various researchers whose work was of utmost importance for the formation of this review. Credit is given to them for their valuable share of the research in the field of micro-machining and MEMS. We are thankful to all.

References

1. Majeed A, Ramesh P (2019) Fabrication of MEMS Capacitive Pressure Sensor (MCPS) with Segmented Bossed Diaphragm. In: Sharma R, Rawal D (eds) The physics of semiconductor devices. IWPSD 2017. Springer Proceedings in Physics, vol 215. Springer, Cham
2. Wang P, Lu Q, Fan Z (2018) Cluster Comput. https://doi.org/10.1007/s10586-018-2085-3
3. Dyuzhev NA, Gusev EE, Gryazneva TA et al (2017) Nanotechnol Russ 12:426. https://doi.org/10.1134/S1995078017040073
4. Fallah Ghanbari B, Arabi H, Abbasi SM et al (2016) Int J Adv Manuf Technol 87:755. https://doi.org/10.1007/s00170-016-8343-8
5. Bao J, Jeppson K, Edwards M et al (2016) Electron Mater Lett 12:1. https://doi.org/10.1007/s13391-015-5308-2

6. Natarajan U, Suganthi XH, Periyanan PR (2016) Trans Indian Inst Met 69:1675. https://doi.org/10.1007/s12666-016-0828-5
7. Cheng J, Jin Y, Wu J et al (2016) Int J Adv Manuf Technol 86:2197. https://doi.org/10.1007/s00170-015-8312-7
8. Anand RS, Patra K, Steiner M et al (2017) Int J Adv Manuf Technol 88:241. https://doi.org/10.1007/s00170-016-8632-2
9. Smith J, Xiong W, Yan W et al (2016) Comput Mech 57:583. https://doi.org/10.1007/s00466-015-1240-4
10. Hussein M, Aadim K, Hassan E (2016) Structural and surface morphology analysis of copper phthalocyanine thin film prepared by pulsed laser deposition and thermal evaporation techniques. Adv Mater Phys Chem 6:85–97. https://doi.org/10.4236/ampc.2016.64009
11. Xu X, Zhang S (2015). https://doi.org/10.2991/icmra-15.2015.221
12. Lei MK, Zhu XP, Li YP et al (2016) Int J Adv Manuf Technol 82:1831. https://doi.org/10.1007/s00170-015-7467-6
13. Sahli M, Lebied A, Gelin JC et al (2015) Int J Adv Manuf Technol 79:2079. https://doi.org/10.1007/s00170-015-6983-8
14. Lee SJ, Takahashi M, Kawahito Y et al (2015) Int J Precis Eng Manuf 16:2121. https://doi.org/10.1007/s12541-015-0274-z
15. Xiong W, Zhou Y, Hou W et al (2015) Front Optoelectron 8:351. https://doi.org/10.1007/s12200-015-0481-3
16. Suganthi XH, Natarajan U, Ramasubbu N (2015) Int J Adv Manuf Technol 81:199. https://doi.org/10.1007/s00170-015-6900-1
17. Cecil J, Bharathi Raj Kumar MB, Lu Y et al (2016) Int J Adv Manuf Technol 83: 1569. https://doi.org/10.1007/s00170-015-7698-6
18. Park C, Shin BS, Kang MS et al (2015) Int J Precis Eng Manuf 16:1385. https://doi.org/10.1007/s12541-015-0182-2
19. Chiou AH, Tsao CC, Hsu CY (2015) Int J Adv Manuf Technol 78:1857. https://doi.org/10.1007/s00170-014-6778-3
20. Khanghah SP, Boozarpoor M, Lotfi M et al (2015) Trans Indian Inst Met 68:897. https://doi.org/10.1007/s12666-015-0525-9
21. Huo D, Lin C, Choong ZJ et al (2015) Int J Adv Manuf Technol 81:1319. https://doi.org/10.1007/s00170-015-7308-7
22. Krestinin AV, Dremova NN, Knerel'man EI et al (2015) Nanotechnol Russia 10:537. https://doi.org/10.1134/S1995078015040096
23. Aišman D, Mašek B, Jeníček Š (2014) Unconventional microstructures in tool steel obtained by semi-solid processing and subsequent heat treatment. Solid State Phenom 217–218:235–240 (2015)
24. Vella PC, Dimov SS, Brousseau E et al (2015) Int J Adv Manuf Technol 76:523. https://doi.org/10.1007/s00170-014-6148-1
25. Kaushik A, Kumar R, Huey E et al (2014) Microchim Acta 181:1759. https://doi.org/10.1007/s00604-014-1255-0
26. Hua L, Fang Z, Li M et al (2014) Wuhan Univ J Nat Sci 19:93. https://doi.org/10.1007/s11859-014-0984-6
27. Oborski P (2014) Int J Adv Manuf Technol 75:1613. https://doi.org/10.1007/s00170-014-6123-x
28. Farshbaf Zinati R, Razfar MR (2014) Int J Adv Manuf Technol 75:979. https://doi.org/10.1007/s00170-014-6178-8
29. Zhan Z, He N, Li L et al (2015) Int J Adv Manuf Technol 77:2095. https://doi.org/10.1007/s00170-014-6632-7
30. Lou S, Jiang X, Scott P (2013) An efficient divide-and-conquer algorithm for morphological filters. Procedia CIRP 10:142–147
31. Pereira A, Hernandez P, Martinez J, Perez J, Mathia T (2013) Study of morphology wear model of molds from alloys of aluminum EN AW-6082 in injection process. Key Eng Mater 554–557:844–849

32. Nag Chaudhury J (2014) J Mater Eng Perform 23:152. https://doi.org/10.1007/s11665-013-0709-6
33. Liu J, Jiang D, Fu Y et al (2013) Adv Manuf 1:13. https://doi.org/10.1007/s40436-013-0007-4
34. Kumar B, Rao T (2012) AFM studies on surface morphology, topography and texture of nanostructured zinc aluminum oxide thin films. Dig J Nanomater Biostruct 7(4):1881–1889
35. Löchte C, Kayasa J, Herrmann C, Raatz A (2012) Methods for implementing compensation strategies in micro production systems supported by a simulation approach. In: Ratchev S (eds) Precision assembly technologies and systems. IPAS 2012. IFIP advances in information and communication technology, vol 371. Springer, Berlin
36. Gil R, Sánchez JA, Ortega N et al (2013) Int J Adv Manuf Technol 65:1459. https://doi.org/10.1007/s00170-012-4270-5
37. Lee BS, Park KM, Yu WR et al (2012) Macromol Res 20:605. https://doi.org/10.1007/s13233-012-0087-1
38. Kim HB, Hobler G, Steiger A et al (2011) Int J Precis Eng Manuf 12:893. https://doi.org/10.1007/s12541-011-0119-3
39. Jimenez H, Mavris D (2010) An evolution of morphological analysis applications in systems engineering. In: American Institute of Aeronautics and Astronautics, 48th AIAA aerospace sciences meeting including the new horizons forum and aerospace exposition
40. Liu F, Wu J, Chen K, Xue D (2010) Morphology study by using scanning electron microscopy. In: Microscopy: science, technology, applications and education, pp 1781–1792
41. Zhovklyi VY, Chemeris AI, Filatova AG et al (2010) Bull Russ Acad Sci Phys 74:1034. https://doi.org/10.3103/S1062873810070312
42. Lee HJ et al (2010) Desktop micro forming system for micro pattern on the metal substrate. In: Ratchev S (eds) Precision assembly technologies and systems. IPAS 2010. IFIP advances in information and communication technology, vol 315. Springer, Berlin
43. Kibria G, Doloi B, Bhattacharyya B (2010) Int J Adv Manuf Technol 50:643. https://doi.org/10.1007/s00170-010-2527-4
44. Saklakoglu IE, Kasman S (2011) Int J Adv Manuf Technol 54:567. https://doi.org/10.1007/s00170-010-2953-3
45. Savanovic P, Zeiler W (2009) Morphological analysis of design concepts emergence in design meetings. In: Stanford, international conference on engineering design, ICED'09, vol 6, pp 179–188
46. Arenas-Alatorre J, Silva-Velazquez Y, Alva Medina A et al (2010) Appl Phys A 98:617. https://doi.org/10.1007/s00339-009-5451-4
47. Cheng X, Wang ZG, Kobayashi S et al (2010) Int J Adv Manuf Technol 46:179. https://doi.org/10.1007/s00170-009-2094-8
48. Brousseau EB, Dimov SS, Pham DT (2010) Int J Adv Manuf Technol 47:161. https://doi.org/10.1007/s00170-009-2214-5
49. Shivareddy S, Bae S-E, Brankovic SR (2008) Cu surface morphology evolution during electropolishing. Electrochem Solid State Lett 11:D13–D17
50. Liou AC, Chen RH (2006) Int J Adv Manuf Technol 28:1097. https://doi.org/10.1007/s00170-004-2455-2
51. Desai AV, Haque MA (2005) Tribol Lett 18:13. https://doi.org/10.1007/s11249-004-1700-z
52. Senf B, Mäder T, y de Sosa IN, Bucht A, Knobloch M, Löpitz D, Drossel WG (2017) Sensing and actuating functions by shape memory alloy wires integrated into fiber reinforced plastics. Procedia CIRP 66:249–253
53. Alaneme K, Okotete E (2016) Reconciling viability and cost-effective shape memory alloy options—a review of copper and iron based shape memory metallic systems. Eng Sci Technol Int J 19:1582–1592
54. Kang G, Song D (2015) Review on structural fatigue of NiTi shape memory alloys: pure mechanical and thermo-mechanical ones. Theor Appl Mech Lett 5:245–254
55. Quan D, Hai X (2015) Shape memory alloy in various aviation field. Procedia Eng 99:1241–1246

56. Lobo PS, Almeida J, Guerreiro L (2015) Shape memory alloys behaviour: a review. Procedia Eng 114:776–783. https://doi.org/10.1016/j.proeng.2015.08.025
57. Prashantha S, Mallikarjun US, Shashidhara SM (2014) Procedia Mater Sci 5:567–574. https://doi.org/10.1016/j.mspro.2014.07.301
58. Petrini L, Migliavacca F (2011) Biomedical applications of shape memory alloys. J Metall 2011, Article ID 501483, 15 p. https://doi.org/10.1155/2011/501483
59. Dönmez B, Özkan B (2011) Design and control of a shape memory alloy actuator for flap type aerodynamic surfaces. In: Proceedings of 18th world congress IFAC, Milano, Italy, Sept 2011
60. Huang WM, Ding Z, Wang CC, Wei J, Zhao Y, Purnawali H (2010) Mater Today 13:54–61
61. Song G, Ma N, Li H-N (2006) Applications of shape memory alloys in civil structures. Eng Struct 28:1266–1274
62. Taguchi G (1986) Introduction to quality engineering: designing quality into products and processes (No. 658.562 T3)

Chapter 18
Shunt Active Power Filter (SAPF) Design and Analysis of Harmonics Mitigation in Three-Phase Three-Wire Distribution System

Shivangi Upadhyay and Sachin Singh

1 Introduction

Earlier, instruments were configured in a way to resist the disruptions such as flash, sudden overloading, etc. without additional expenditures. Nowadays, the electronic instruments would cost much higher if it is configured with the same strength. Nonlinear loads like transformer, saturated coils, etc. introduced the pollution into the power system. The fast switching and nonlinearity of the power electronics (PE) devices are creating most of the pollution issues and this keeps growing mainly due to their fast-growing capability. A competition is currently going on between increasing issues due to PE and new-found corrective devices based on PE for the attenuation of these issues.

Power filters play a major role when it comes to the attenuation of power quality issues. Active filters are more feasible than passive filters, as passive filters have fixed compensation, large in size and dismiss the resonance conditions. The main purpose of active filter is to resolve these listed problems by combining to a required passive element of reduced rating. Different network topologies for active power filters have been formulated [1–12]. Even during highly nonlinear load, the current control voltage source PWM converter-based shunt active filter has been effective [1, 4, 11]. To sense the harmonics active filters are developed [7, 10, 11] and to compensate the reactive power of nonlinear loads [1, 3, 12, 13]. A new topology is given in [10], where to determine the compensating current load currents are detected which is further improved by only sensing the line current [8, 14, 15]. To generate a template for reference current PI controller is used.

S. Upadhyay (✉) · S. Singh
Department of Electrical Engineering, Institute of Engineering and Technology, Lucknow, India

V. K. Giri et al. (eds.), *Computing Algorithms with Applications in Engineering*, Algorithms for Intelligent Systems, https://doi.org/10.1007/978-981-15-2369-4_18

This paper represents both PI control and hysteresis current control based; three-phase SAPF in a three-wire three-phase system is implemented to even out the reactive power, harmonics and improving the total harmonic distortion for the nonlinear loads to improve power quality. This compensation process is a different approach only to sense the line currents and not the sensing of harmonics or load components of reactive power used in conventional methods. The voltage and current sensors are used to measure voltage and current. To calculate the reference current template the DC capacitor voltage is regulated. A MATLAB/Simulink environment is used to carry out the simulations. The obtained simulation results are the demonstration of improvement in power quality.

2 PI Control Scheme

The input of the PI controller is the error signal and the obtained output is the peak value of reference current which gets multiplied by unit vectors (U_a, U_b, U_c) to determine the reference currents (I_a*, I_b*, I_c*). By comparing the actual and reference current in the PWM converter based on hysteresis controller, the switching signals are generated.

The upper switch is turned ON for the reduction in current; similarly, the lower switch will be turned ON for the current increment. The switching of devices takes place after proper amplification and the current will flow through inductor filter L_f, to compensate the reactive load and load current harmonics.

2.1 Reference Current Signal Generation

To maintain the constant level DC-link voltage and to compensate load current harmonics, SAF is used, and to achieve this, the DC-link voltage sensing is done, and then with the help of low pass filter of the first order, the present ripples in it are removed. It further gets compared with the reference V_{dc} and provides the V_{dcerr}. The nth sampling of V_{dcerr} is given as:

$$V_{dcerr(n)} = V^*_{dc(n)} - V_{dc(n)} \tag{1}$$

This V_{dcerr} is given as input to the PI controller and its output obtained as the peak (I_m) value of reference current and it further gets multiplied by unit vectors which are in the same phase with source voltage to determine the reference currents for SAPF as shown in Fig. 1.

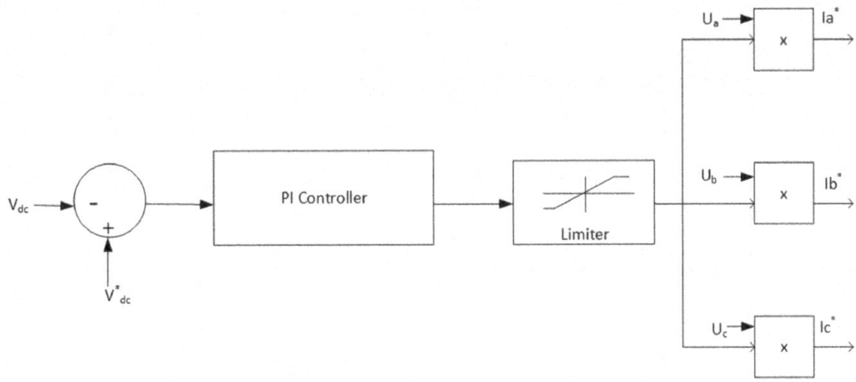

Fig. 1 Reference current signal generation

The instantaneous values of three-phase reference current are given as:

$$I_a^* = I_m U_a \tag{2}$$

$$I_b^* = I_m U_b \tag{3}$$

$$I_c^* = I_m U_c \tag{4}$$

2.2 *Hysteresis Current Controller*

After obtaining the signals for SAPF reference current, the next step is the switching of IGBTs which will allow the inverter to follow the reference current signals. The necessary gate signal is obtained with the help of hysteresis current controller (HCC) by controlling the voltage source inverter, to generate the output current waveform. It controls the switching asynchronously and provides ramp current through inverter so that it follows the reference current. In this method, the sensed current is compared within the hysteresis controller limits and the current feedback is required as well to control the switching of IGBTs. The advantages of this method are low switching losses and high speed. The actual currents (I_a, I_b, I_c) and reference currents (I_a*, I_b*, I_c*) being compared to estimate the current error signals as given below:

$$I_{aerr} = I_a^* - I_a \tag{5}$$

$$I_{berr} = I_b^* - I_b \tag{6}$$

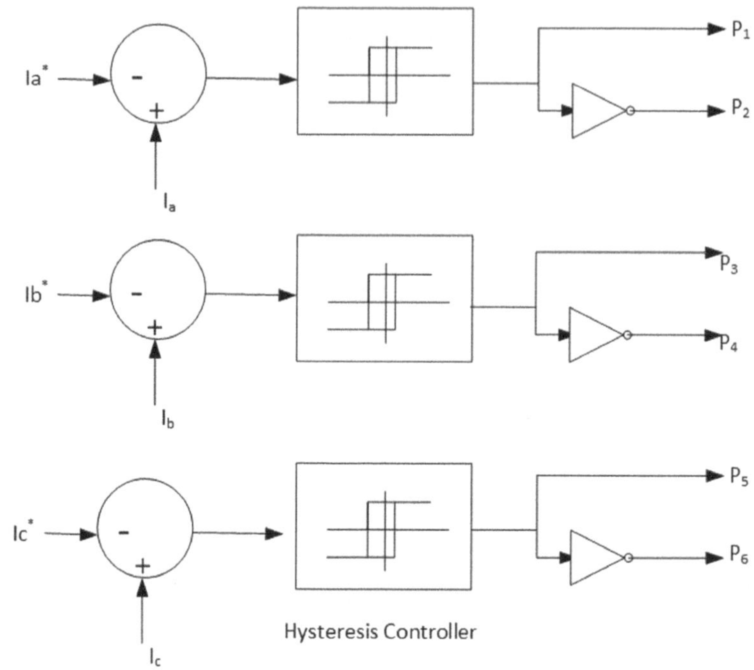

Fig. 2 Hysteresis current controller

$$I_{\text{cerr}} = I_{\text{c}}^{*} - I_{\text{c}} \tag{7}$$

These current errors are given to the HCC which will provide the switching signals (P1–P6) as the gate signals for the inverter as shown in Fig. 2.

The switching sequence is explained as:

If $i_{sa} < (i_{sa}^{*} - h_{b})$, the lower switch is switched ON and upper one is OFF in leg 'a' of the SAPF;
If $i_{sa} > (i_{sa}^{*} - h_{b})$, the lower switch is switched OFF and the upper one is ON in leg 'a' of the SAPF.

This will be same for the legs 'b' and 'c' and h_{b} is the hysteresis bandwidth and the supply current is regulated within the hysteresis limits.

3 Description of the Test System Model

3.1 Topology

The system description of investigated SAPF and its connection with the grid is shown in Fig. 3. The proposed system consists of DC-link and PWM inverter. The topology consists of current control voltage source operating voltage source inverter (VSI). The VSI ports the DC source to the grid and provides the generated power so it is a key element in the system. Table 1 shows the selected parameters for the simulation.

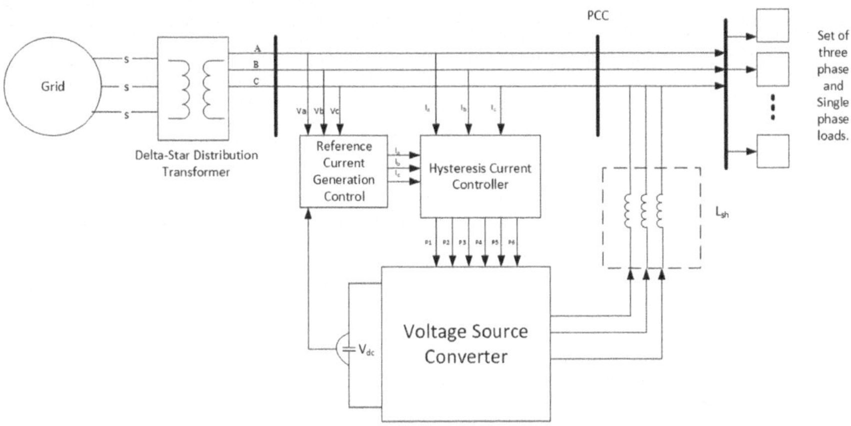

Fig. 3 Proposed system

Table 1 Parameters for the simulation

Selected parameters	Values
Voltage source (V_s), frequency (f)	100 V (rms), 50 Hz
Source impedance (R, L)	0.1 Ω, 4 mH
Filter impedance (R_c, L_c)	0.01 Ω, 9 mH
Load impedance (R_1, C_1)	15.16 Ω, 1 μF
Load impedance (R_1, L_1)	15.16 Ω, 10 mH
Reference DC-link voltage (V_{dcref})	283 V
DC-link capacitance	1100 μF

3.2 *Nonlinear Load Modeling*

A nonlinear load is a combination of three-phase diode rectifier's input impedance with a RC- or RL-load [16, 17]. A cycle with six-overlapping and six-non-overlapping intervals occurs as per the presence of source inductance. Three devices of the bridge conduct for an overlapping interval while two devices will conduct during a non-overlapping interval.

3.3 *Inverter Design*

The inverter output voltages in the form of DC voltage instantaneous values and switching pulse can be written as:

$$V_{\text{inva}} = \frac{(P1 - P2)}{2} V_{\text{dc}} \tag{8}$$

$$V_{\text{invb}} = \frac{(P3 - P4)}{2} V_{\text{dc}} \tag{9}$$

$$V_{\text{invc}} = \frac{(P5 - P6)}{2} V_{\text{dc}} \tag{10}$$

where V_{inva}, V_{invb} and V_{invc} are inverter's generated output switching voltages. Each IGBT's switching pattern inside the inverter will be decided on the basis of difference between the inverter's reference and actual current.

4 Design of Shunt Active Power Filter (SAPF)

The design procedure depends on the following system specifications [18]:
$S = 20$ KVA, V_s (LL) $= 240$ V, $f = 50$ Hz, $f_s = 10$ kHz, RAF $= 5\%$, $m_f = 200$, $m_a = 1$.

There are three main parameters for the design of SAPF power circuit:

a. Selection of V_{dc}
b. Selection of L_c
c. Selection of C_{dc}.

And, the component's design depends on the assumptions given below [18]:

a. Sinusoidal voltage source.
b. Line current distortion is 5% assumed for the designing of L_c.
c. Active filter fixed capability of reactive power compensation.
d. Operation of PWM converter in the linear modulation ($0 \le m_a \le 1$).

Reference Capacitor Voltage Estimation V_{dc}. As assumed the linear modulation operation of PWM converter, i.e., $(0 \leq m_a \leq 1)$, then [18]

$$m_a = \frac{2\sqrt{2}V_{f1}}{V_{dc}} \tag{11}$$

for $m_a = 1$, $V_{dc} = 2\sqrt{2}V_{f1}$; where V_{f1} is the AC side fundamental component. According to the principle of reactive power compensation:

$$V_{f1} = V_s + j\omega L_f I_{f1} \tag{12}$$

In this case, V_{f1} must vary as the function of system's capacity requirement $V_s < V_{f1} \leq 2V_s$, one can set the value of V_{dc} (tolerable ripples in voltage $V_{dc(p\text{-}p)max}$ also found) after knowing the values of L_f and I_{f1}. But to a particular nonlinear load, the equation below shows the voltage as the load rated power function and to be compensated maximum order of harmonic as [18],

$$V_{dc} = 2\sqrt{2}V_{(fh)max} \tag{13}$$

$$V_{f(t)} = V_s + \frac{6}{\pi}\omega\,.\,L_f I_{f1}(-\cos 5\omega t - \cos 7\omega t + \cos 11\omega t + \cdots) \tag{14}$$

where $V_{(fh)max}$ is the maximum voltage value of $V_{f(t)}$ including corresponding harmonic terms to be compensated. Due to the high switching frequency choice, if output of the inductance filter is small, then V_{f1} will be equal to the source voltage approximately;

$$V_{dc} = 2\sqrt{2}\,V_{f1} \tag{15}$$

(a) Output Filter Inductor Estimation L_c. The standard chosen for designing is the inductor peak ripple current, and to estimate the ripple current, the inverter reference voltage and the supply voltage will be equal if the effect of inductor resistance is negligible and no-load condition is considered. The required filter inductance is calculated as [18]:

$$L_c = \frac{V_s}{2\sqrt{6}f_s \Delta I_{f(p-p)\max}} \tag{16}$$

Design of the DC capacitor C_{dc}. The determination of the energy storage capacitor value is to support the step-up/down in the power exhausted by the load, using the concept of energy balance. The value of the capacitor is determined by [18]:

$$C_{dc} = \frac{2\,.\,E_{max}}{V_{dc}^2 - V_{dcmin}^2} \tag{17}$$

where E_{max} is the maximum energy that will be supplied by the capacitor in the worst case of a transient.

5 Simulation Results

To simulate the PI control-based SAPF in MATLAB, a simulation model is formulated. The SAPF system is the combined set of a source, a PWM converter, a nonlinear load and a PI controller. Each component will be collected and designed individually and then integrated and solved for the simulation of the system. The simulated results for the proposed RL-load and RC-load model are shown below:

5.1 For RL-Load

Without Shunt Active Power Filter.

Source Voltage (V_{abc}). The assumed balanced and sinusoidal three-phase voltage source. The source voltage waveform is shown in Fig. 4.

Source Current ($\mathbf{I_{abc}}$). The three-phase source current with the presence of harmonic component is obtained without using shunt active filter as shown in Fig. 5.

THD of I_a, I_b, I_c. Total harmonic distortion (THD) of $I_a = 20.11\%, I_b = 20.11\%$ and $I_c = 20.10\%$ is shown in Fig. 6a–c at the fundamental frequency of 50 Hz and start time of 0.7 s (Table 2).

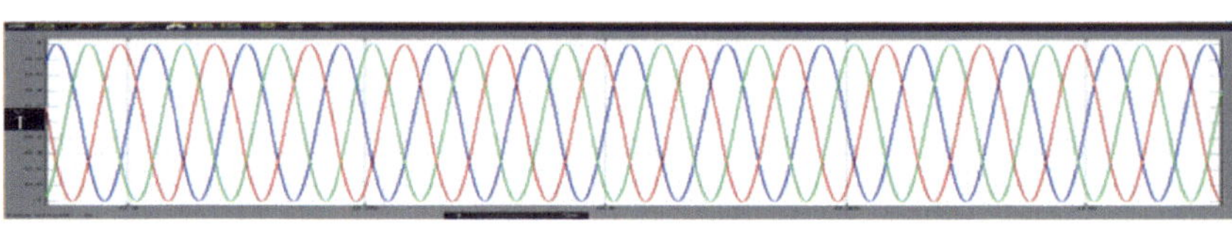

Fig. 4 Source voltage waveform

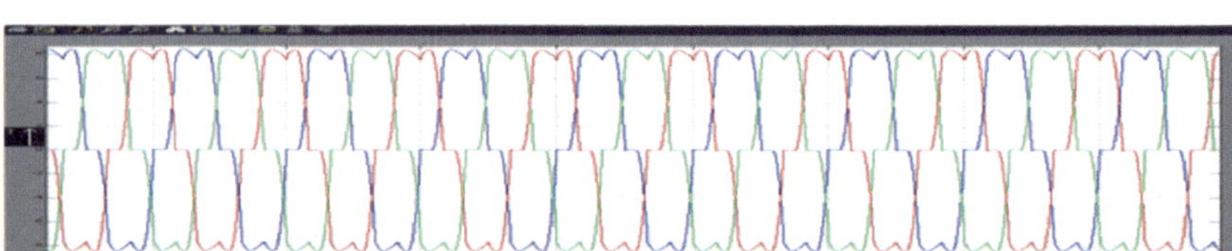

Fig. 5 Source current waveform without APF

(a)

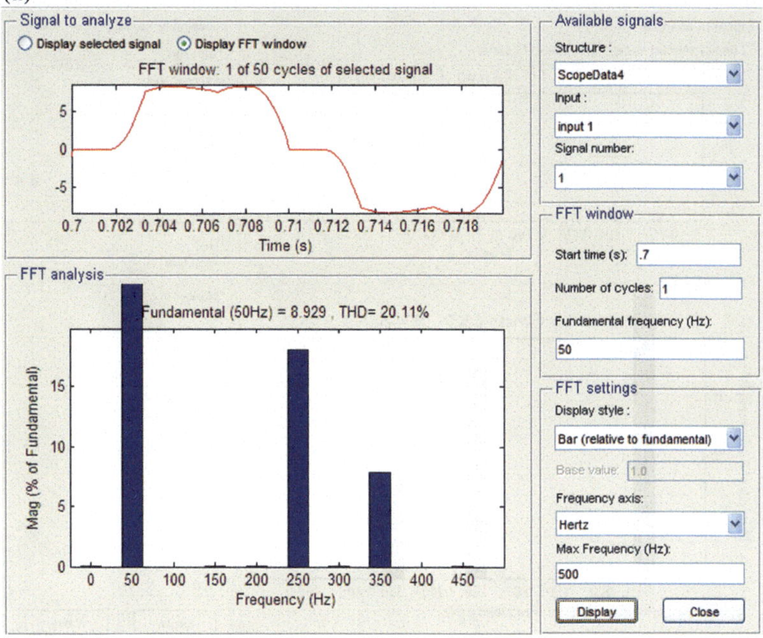

(b)

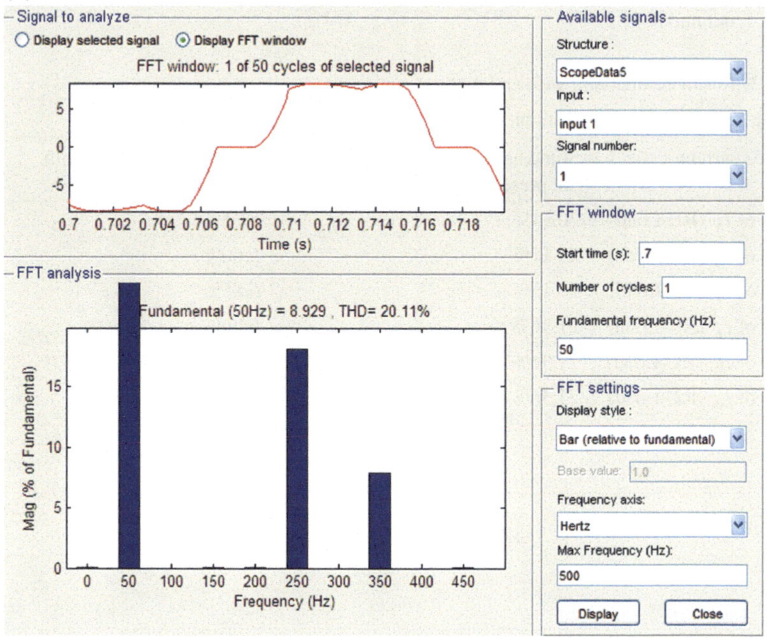

Fig. 6 THD of I_a, I_b, I_c without APF

(c)

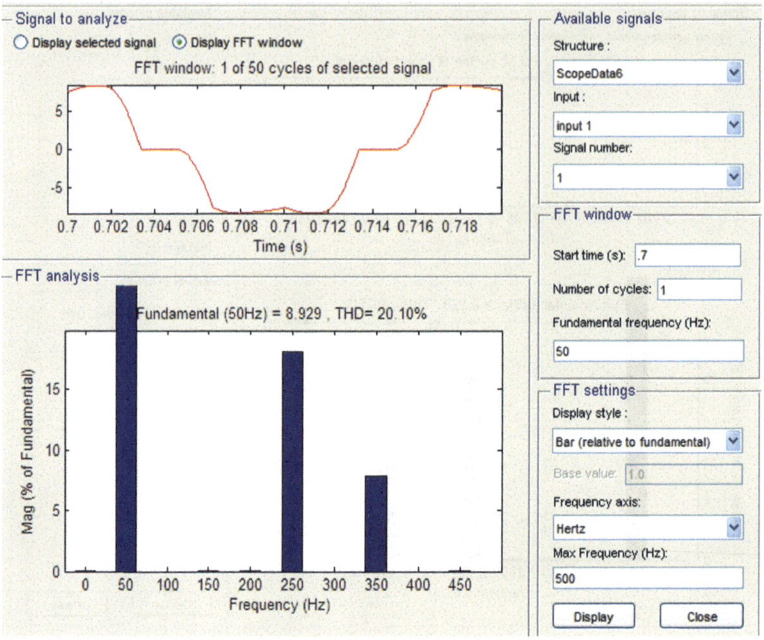

Fig. 6 (continued)

Table 2 Result comparison with and without controller

Result	Without controller	With controller
Source current	Some harmonics and distortion are present	Mitigate harmonics with minimum distortion
THD of I_a (RL-Load)	20.11%	5.35%
THD of I_b (RL-Load)	20.11%	5.60%
THD of I_c (RL-Load)	20.11%	5.85%
THD of I_a (RC-Load)	21.36%	6.76%
THD of I_b (RC-Load)	21.36%	6.50%
THD of I_c (RC-Load)	21.36%	6.29%

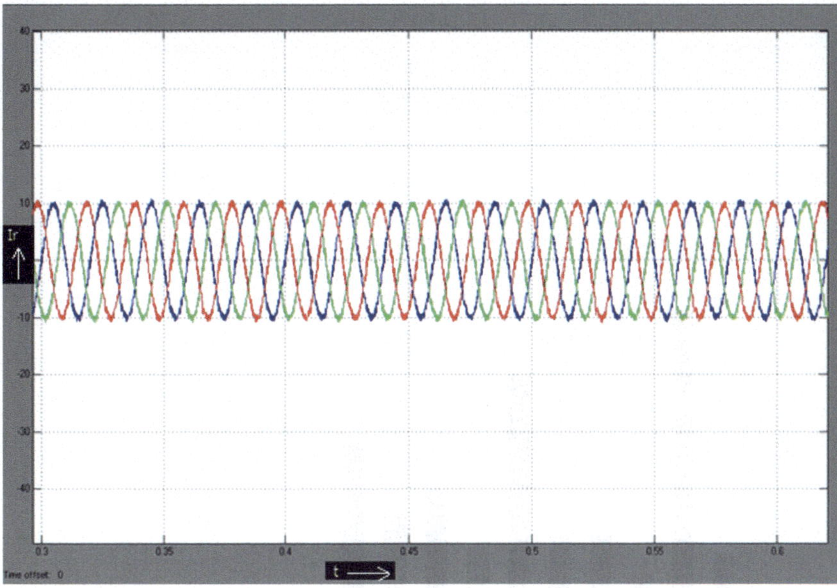

Fig. 7 Source current waveform with APF

With Shunt Active Power Filter.

Source Current (I_{abc}). Figure 7 shows the waveform of three-phase source current in the presence of SAPF.

THD of I_a, I_b, I_c. Total harmonic distortion (THD) of $I_a = 5.35\%$, $I_b = 5.60\%$ and $I_c = 5.85\%$ is shown in Fig. 8a–c at the fundamental frequency of 50 Hz and start time of 0.7 s.

5.2 *For Parallel RC Load*

Without Shunt Active Power Filter.

THD of I_a, I_b, I_c. Total harmonic distortion (THD) of $I_a = 21.35\%$, $I_b = 21.36\%$ and $I_c = 21.36\%$ is shown in Fig. 9a–c at the fundamental frequency of 50 Hz and start time of 0.7 s.

With Shunt Active Power Filter.

THD of I_a, I_b, I_c. Total harmonic distortion (THD) of $I_a = 6.76\%$, $I_b = 6.50\%$ and $I_c = 6.29\%$ is shown in Fig. 10a–c at the fundamental frequency of 50 Hz and start time of 0.7 s.

(a)

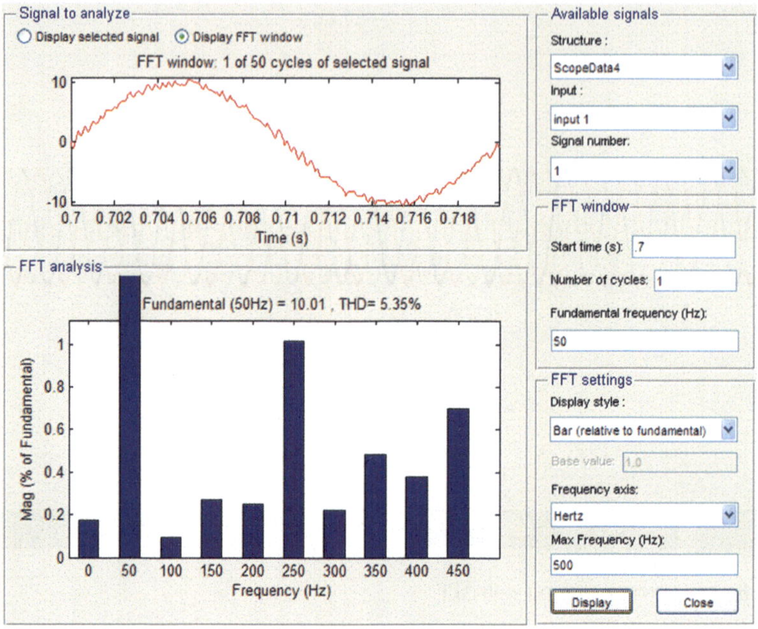

(b)

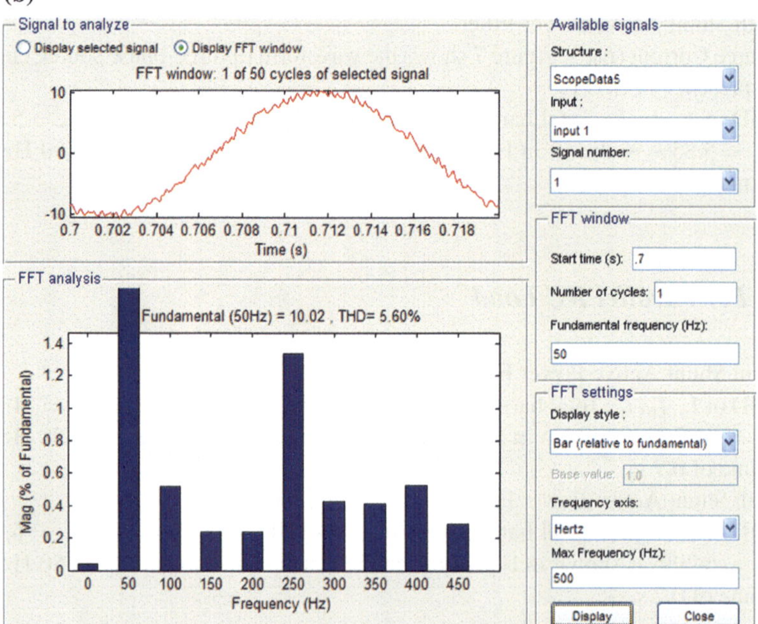

Fig. 8 THD of I_a, I_b, I_c with APF

(c)

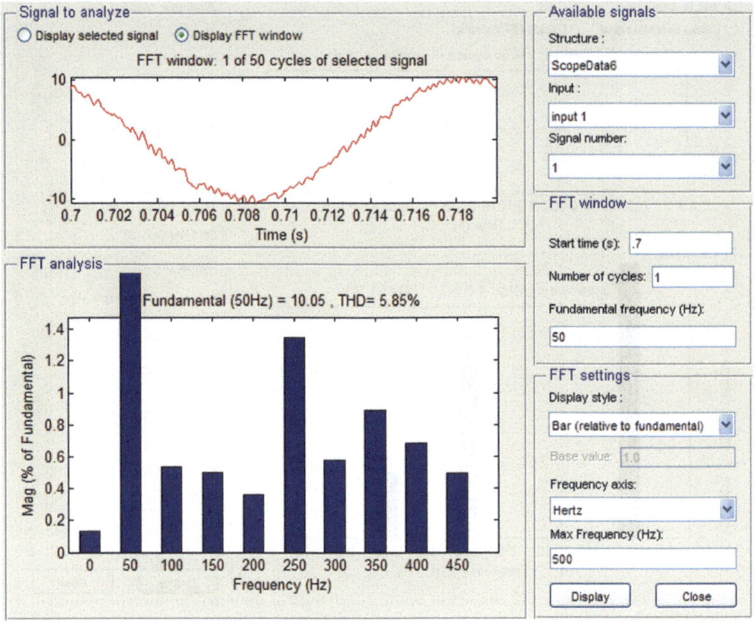

Fig. 8 (continued)

(a)

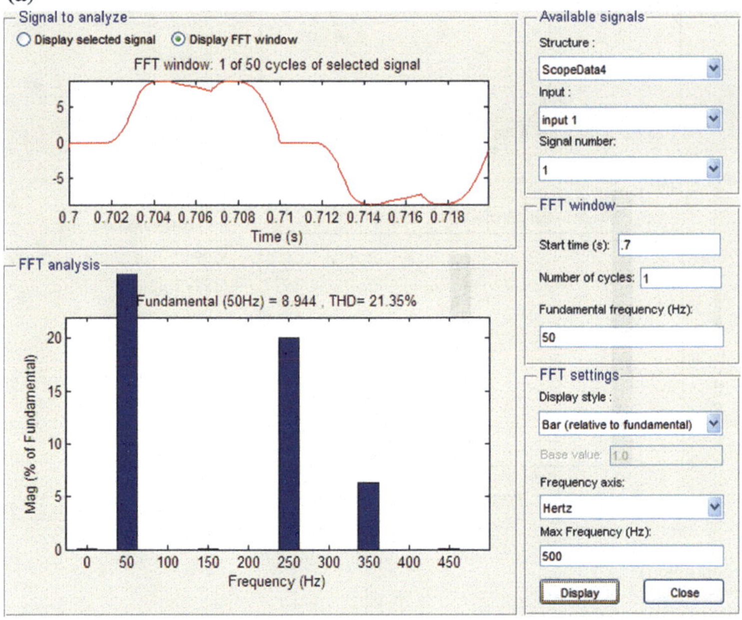

Fig. 9 THD of I_a, I_b, I_c without APF

(b)

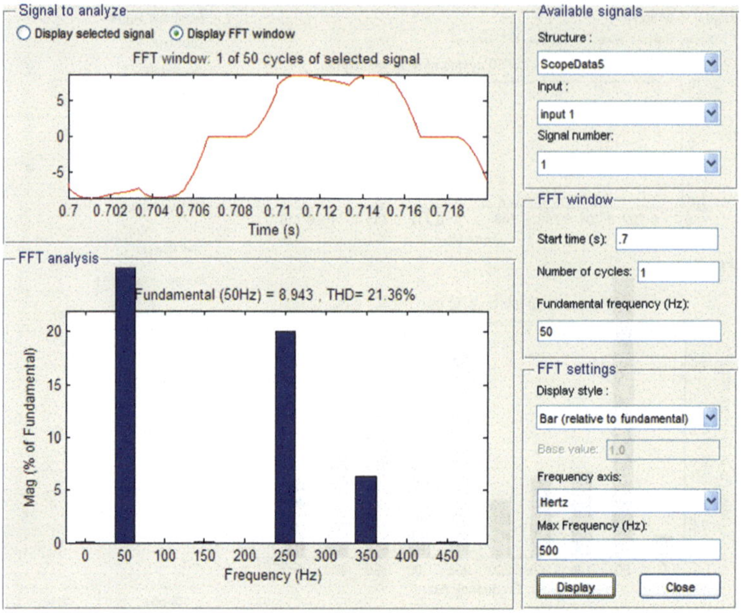

(c)

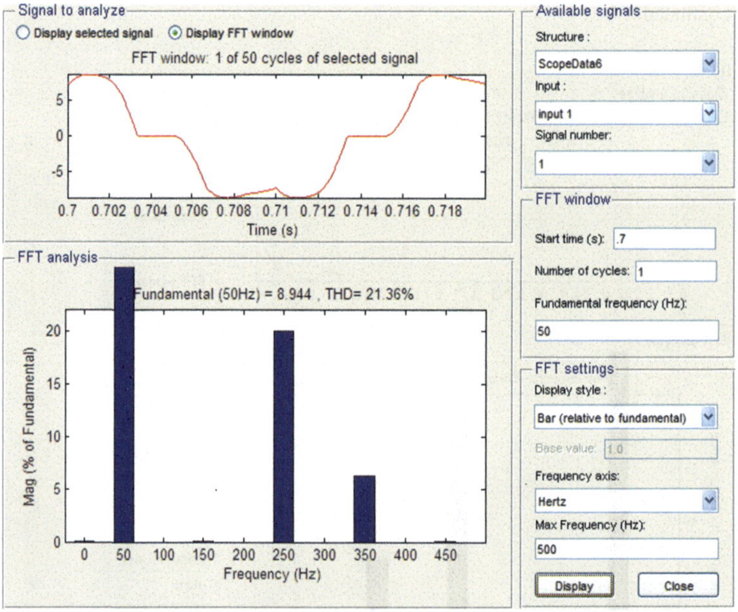

Fig. 9 (continued)

6 Conclusions

(a) For the purpose of power quality improvement, various simulations and the investigation of SAPF have been performed for system analysis. SAPF based on PI control has been exerted to minimize the harmonics and to compensate the reactive power of nonlinear loads. To implement the same, a model on MATLAB/Simulink platform has been developed.

(b) SAPF improves the power quality by eliminating the harmonics and making the load current sinusoidal; also, the controller performance has been studied. The THD of the current as found from the simulation results is 5–6% which is the harmonics limit imposed by IEEE.

(c) The presented validation can also be applied in different operations such as linear/nonlinear load balancing, PF correction etc. The results obtained from simulations help in defining the overall possibilities for the control strategy of shunt active power filter in a bigger scenario.

(a)

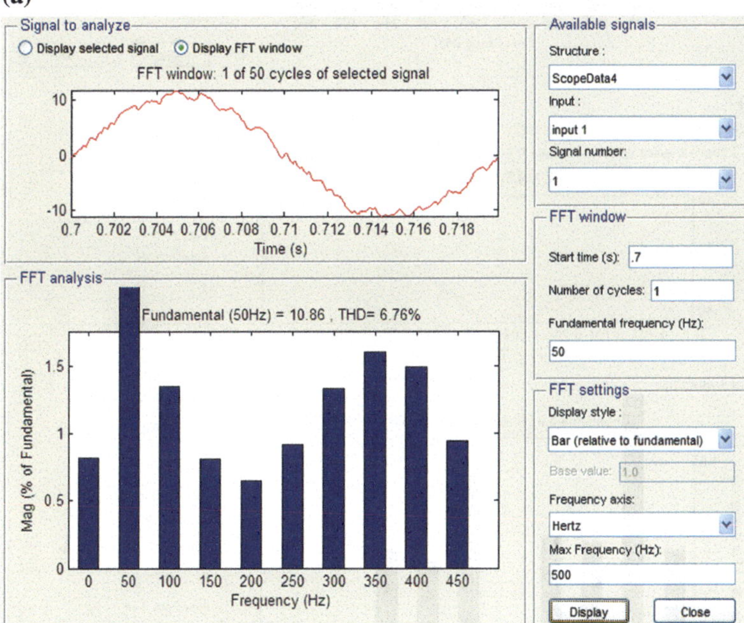

Fig. 10 THD of I_a, I_b, I_c with APF

(b)

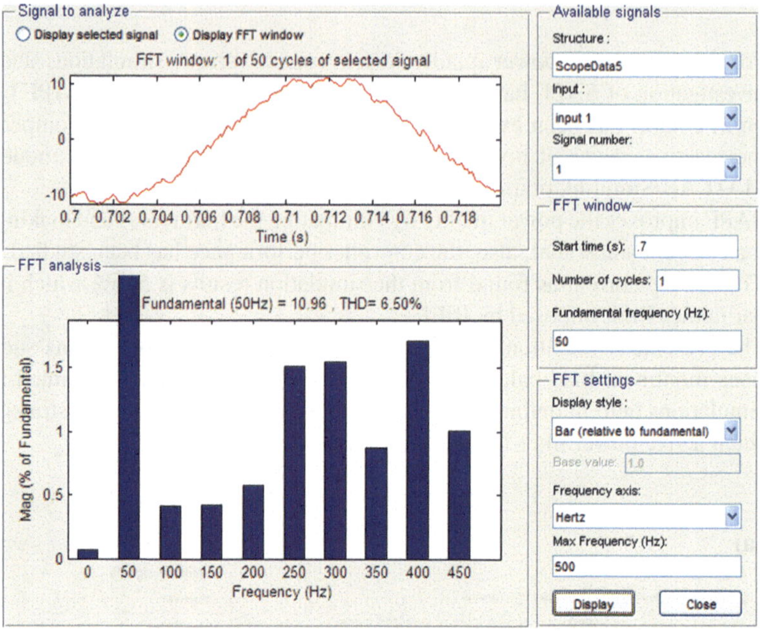

(c)

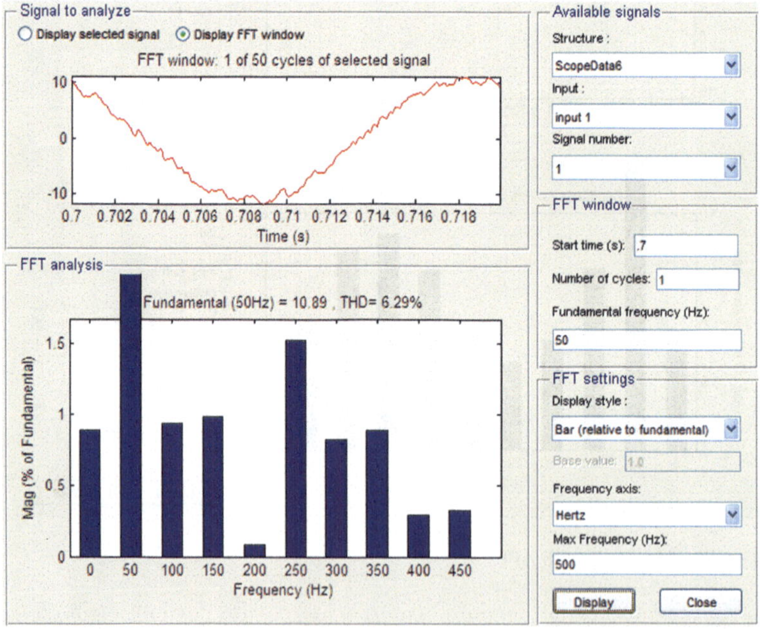

Fig. 10 (continued)

References

1. Grady WM, Samotyj MJ, Noyola AH (1990) Survey of active power line conditioning methodologies. IEEE Trans Power Delivery 5(3):1536–1542
2. Akagi H, Kanazawa Y, Nabae A (1984) Instantaneous reactive power compensators comprising switching devices without energy storage components. IEEE Trans Ind Appl IA-20(3):625–630
3. Jain S, Agrawal P, Gupta HO (2003) Design simulation and experimental investigations on a shunt active power filter for harmonics and reactive power compensation. Electr Power Compon Syst 32(7):671–692
4. Peng FZ, Akagi H, Nabae A (1990) Study of active power filters using quad series voltage source PWM converters for harmonic compensation. IEEE Trans Power Electron 5(1):9–15
5. Akagi H (1994) Trends in active power line conditioners. IEEE Trans Power Electron 9(3):263–268
6. Jain SK, Agrawal P, Gupta HO (2002) Fuzzy logic controlled shunt active power filter for power quality improvement. Proc Inst Electr Eng Electr Power Appl 149(5):317–328
7. Morgan LA, Dixon JW, Wallace RR (1995) A three phase active power filter operating with fixed switching frequency for reactive power and current harmonics compensation. IEEE Trans Ind Electron 42(4):402–408
8. Singh B, Chandra A, Al-Haddad K (1999) Computer-aided modeling and simulation of active power filters. Electr Mach Power Syst 27:1227–1241
9. Singh B, Chandra A, Al-Haddad K (1999) A review of active filters for power quality improvement. IEEE Trans Ind Electron 46(5):1–12
10. Duke RM, Round SD (1993) The steady state performance of a controlled current active filter. IEEE Trans Power Electron 8:140–146
11. Dixon JW, Garcia JJ, Morgan L (1995) Control system for three phase active power filter which simultaneously compensates power factor and unbalanced loads. IEEE Trans Ind Electron 42(6):636–641
12. Watanbe EH, Stephan RM, Aredes M (1993) New concepts of instantaneous active and reactive powers in electrical systems with generic loads. IEEE Trans Power Delivery 8(2):697–703
13. Soares V, Verdelho P, Marques GD (2000) An instantaneous active and reactive current component method of active filter. IEEE Trans Power Electron 15(4):660–669
14. Chatterjee K, Fernandes BG, Dubey GK (1999) An instantaneous reactive volt- ampere compensator and harmonic suppressor system. IEEE Trans Power Electron 14(2):381–392
15. Singh B, Chandra A, Al-Haddad K (1998) Performance comparison of two current control techniques applied to an active filter. In: 8th International conference on harmonics and power quality ICHQP, pp 133–138
16. Huang S-J, Wu J-C (1999) A control algorithm for three-phase three-wired active power filters under non-ideal mains voltages. IEEE Trans Power Electron 14(4):753–760
17. Torey DA, Al-Zamel AM (1995) A single phase active filter for multiple nonlinear load. IEEE Trans Power Electron 10:263–272
18. Chaoui A, Gaubert J-P, Krim F, Rambault L (2008) On the design of shunt active filter for improving power quality. In: 2008 IEEE symposium on industrial electronics

Chapter 19
All-Optical Combinational Logic Design Based on Optical Amplifier

Devendra Kumar Tripathi

1 Introduction

Communication is the primary need of human being since primitive days. People have used numerous techniques to establish communication in between distantly located parties. Altogether with growth in technology day-by-day need of Internet-based services uses mounting, which demands huge bandwidth. So, the need for the communication bandwidth is growing instantaneously in every era. Accordingly, numerous multiplexing techniques have been formulated along with their pros and cons. Altogether with traditionally used back bone electronic network have reached to their optimum capacity, exhibits sluggish processing of signal and incompetent to fulfill modern day broad band spectrum needs. Accordingly, growing requisite for all-optical networks due to its inherent established attributes of optical domain endows with huge bandwidth need for numerous modern-day executions. Although communication networks usually are hybrid type, well consisting of both electronic and optical arrangements, in which there are so many optical–electronic–optical (O/E/O) translations, by miscellaneous optoelectronic conversions power consumption is more. So, all-optical computing is one of the emergent technologies which employ signal processing in all-optical modes. Accordingly, all-optical digital operation requires numerous digital logic devices. With numerous applications as buffering of signals, packet forwarding, switching, signal regeneration, and for the bit length processing optically, etc, several investigations have been explored with the progress of the modern advanced semiconductor device-based optical switching devices [1–8]. In recent years altogether with the development of semiconductor technology, the semiconductor optical amplifiers are implemented for the realization of optical logic gates. All-optical inverter binary logic gates executed with Mach–Zehnder interferometer and semiconductor optical amplifier (SOA) (SOA-MZI) with aptly picked control and continuous clock pulses. Digital logic optical gates' performance

D. K. Tripathi (✉)
Department of Electronics Engineering, REC Sonbhadra, Churk, U.P., India

V. K. Giri et al. (eds.), *Computing Algorithms with Applications in Engineering*, Algorithms for Intelligent Systems, https://doi.org/10.1007/978-981-15-2369-4_19

is evaluated for the output performance metrics as contrast ratio (CR) and extinction ratio (ER), its outcome with good optical performance [9, 10].

Other investigations for an all-optical inverter logic at data rate of 100 Gbps for the return-to-zero (RZ) and non-return-to-zero (NRZ) formats were resulted with good extinction ratio (ER) performance [11]. Whereas for the all-optical universal logic gate (NOR), two designs were explored, and the designs relying upon the nonlinear act of semiconductor optical amplifier, with Mach–Zehnder interferometer (SOA-MZI) configurations. Evaluations for the optical NOR logic gate showed very good extinction factor performance [12]. Further using non-return-to-zero format OR/NOR/Buffer binary, logic optical network is designed and realized with successfully at 100 Gbps data rate [13]. An all-optical half adder link was realized using three optical gates to function up to 1 Gb/s, and later, it was efficiently tested for the bit error rate and quality factor performance up to a 80 Gbps of data rate [14, 15]. Another combinational optical logic network for the half subtractor was explored with microring and nanoring devices in which the concurrent function of half adder/subtractor arithmetic was exercised by the applications of dark-bright soliton conversion control inside the schematic [16]. Accordingly, several outstanding exertions have been done designing different all-optical logic links with each design having their respective pros and cons; in view of that a variety of investigations have been explored infer that all-optical signal processing (ASOP) is one of the emergent fields of research.

In this vision, an all-optical link is proposed, which is based on the semiconductor optical amplifier that can concomitantly generate logic AND operation and exclusive operation simultaneously. In the subsequent sections, the proposed design, results and discussion, conclusion parts are illustrated.

2 The Design Presentation

High-speed communication networks are always desired for the signal processing. For that as carrier of information and energy optical fibers have been an established mode globally. Accordingly, an all-optical schematic design is as demonstrated here in Fig. 1. To investigate the design, the input bit pattern generation is with the help of the pseudo-random pattern binary signal (PRBS) generator. The incoming pulsed pattern is passed through the non-return-to-zero pattern modulator, which is driven by the raised cosine driver. It is further allowed to go through the Mach–Zehnder modulator (MZM) which accompanied by a continuous wave laser source with high power and operating at a wavelength (λ) of 1550 nm. Now, the modulated input data pattern with an operating data rate of 20 Gbps is allowed propagating all the way through optical coupler. It is also united with a continuous wave (CW) laser operating wavelength (λ) of 1555 nm. The next input data pattern is produced by another pseudo-random pattern binary signal generator source at the wavelength (λ) of 1550 nm. This generated signal is further modulated and combined to another optical coupler. This coupler data is further allowed to go through branch one that is

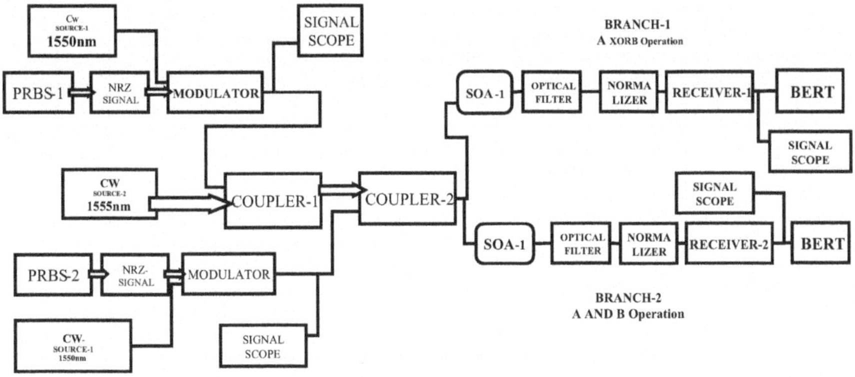

Fig. 1 Schematic for optical combinational logic network

designed to obtain the two-bit exclusive OR operation over two applied data inputs. It is also joined with lower power probe continuous wave (CW) laser source well operating at wavelength (λ) of 1550 nm. The designs of two branches consist of the semiconductor optical amplifiers in the two branches whose nonlinearity is exploited.

Accordingly, united pulse pattern propagates through the semiconductor optical amplifier. Because of swift carrier dynamics that is taking place inside at a picoseconds timescale. The gain responds in step with the deviations in the arriving input power on a bit-by-bit basis. It creates an alteration in the gain for the signals at dissimilar wavelengths and is amplified by the semiconductor optical amplifiers. On the output side when output is passed through Fabry–Perot optical filter. It filters out the wavelength of 1555 nm. The filtered signal goes through the normalization process for the signal shaping purpose. Finally, for the essential measurement is to be done with the help of bit error rate tool (BERT). This is used to observe the consequential extinction ratio (ER) and bit error rate (BER) parameters.

3 Results and Discussion

The explored all-optical high-speed combinational network performance is successfully estimated for the applied data inputs at 20 and 50 Gbps of data rate. The design's performance is investigated using the nonreturn to zero (NRZ) pulse pattern along with continuous wave (CW) lasers operating as probe and clock pulse, raised cosine drive and Mach–Zehnder modulation (MZM) pattern is used with suitably selected necessary operating parameters. Different investigations successfully carried out and the investigated results are demonstrated in Fig. 2, 3, 4, 5, 6, 7, 8, 9 and 10. Figure 2 illustrates applied input data pattern A [0 0 0 0 0 1 1 1 1 0 0 0 0 1 1 1 1 0 0 0 0 1 1 1 1 0 0 0 0 1 1 1 1 0 0 0 0 1 1 1 1 0 0 0 0 0 0 0 0], and Fig. 3 illustrates the applied input data pattern B [0 0 0 0 0 0 0 0 0 0 0 0 0 1 1 1 1 0 0 0 0 1 1 1 1 0 0 0 0 0 0 0 0 0 0 0 0 0 1 1 1 1 0 0 0 0 0 0 0], which are used to observe digital logic operation. Extra bits are added

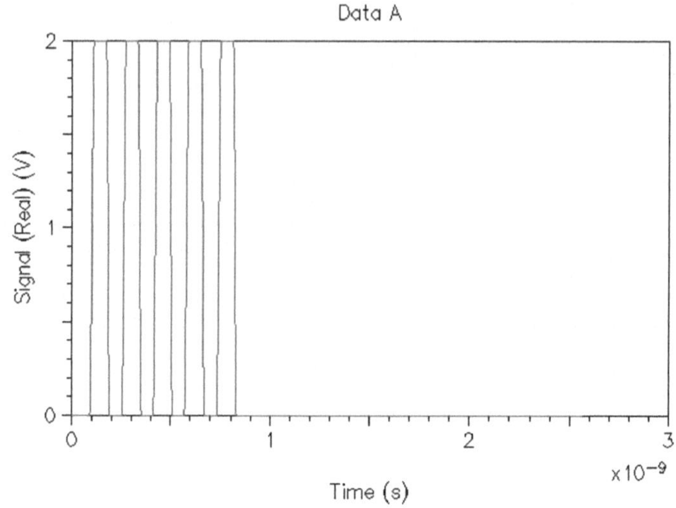

Fig. 2 Input data A [0 0 0 0 0 1 1 1 1 0 0 0 0 1 1 1 1 0 0 0 0 1 1 1 1 0 0 0 0 1 1 1 1 0 0 0 0 1 1 1 1 0 0 0 0 0 0 0]

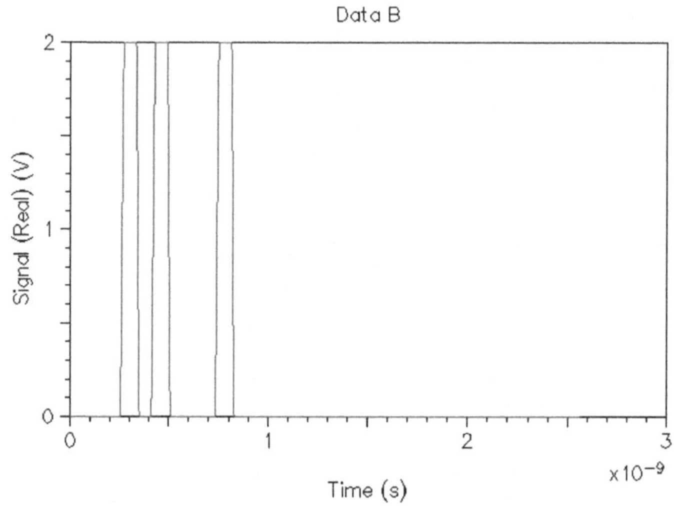

Fig. 3 Input data B [0 0 0 0 0 0 0 0 0 0 0 0 0 1 1 1 1 0 0 0 0 1 1 1 1 0 0 0 0 0 0 0 0 0 0 0 0 0 1 1 1 1 0 0 0 0 0 0 0]

in begin and end of the data inputs for god accuracy. Correspondingly, Fig. 4a, b illustrates output patterns subsequent to the optical couplers. A coupler is an optical device competent of joining one or more optical fiber ends so as to let the spread of light signal in the manifold paths. The simulated outcome is shown in Fig. 5. It depicts the exclusive OR operation of the two applied data inputs. The output pulse

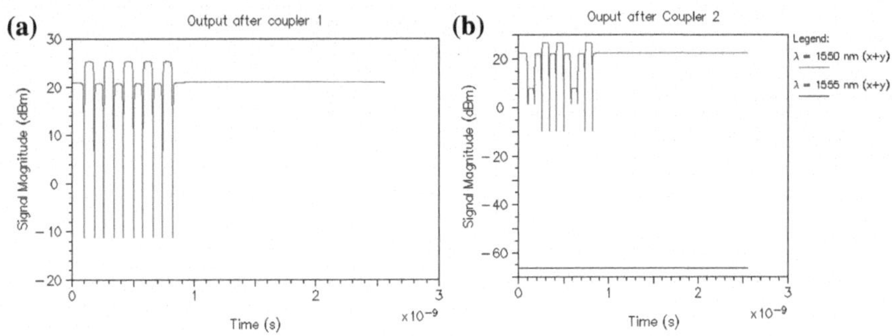

Fig. 4 Output pattern (**a**) after first coupler (**b**) and second coupler

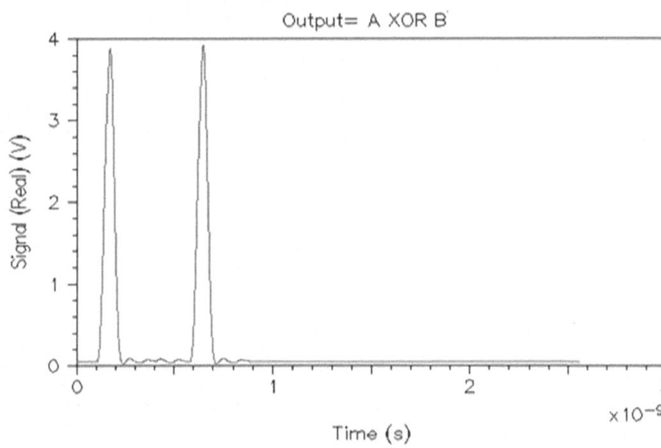

Fig. 5 Output XOR logic over data A and data B

pattern which is binary logic XOR operations of the data input A and data input B will be (0 0 0 0 0 1 1 1 1 0 0 0 0 0 0 0 0 0 0 0 0 0 0 0 0 0 0 0 0 0 1 1 1 1 0 0 0 0 0 0 0 0 0 0 0 0 0 0 0). As for the exclusive OR logic operation over the applied digital data inputs, logic high (1) output is obtained when only one of the data input signals is high, whereas low logic output pulses when both input signals are high (1) or low (0). Figure 5 demonstrates output pattern for the exclusive OR logic and validates the output identical to theoretical logic operation. Figure 6 shows the logical AND operation with applied data inputs. Theoretical logical AND operation over the two applied data inputs A and data B will result as output pattern is [0 0 0 0 0 0 0 0 0 0 0 0 0 0 1 1 1 1 0 0 0 0 1 1 1 1 0 0 0 0 0 0 0 0 0 0 0 0 0 1 1 1 1 0 0 0 0 0 0 0 0 0]. It shows that there will be three HIGH pulses in the output pattern. The output shown in Fig. 6 verifies theoretical results, and simulated results for the logical AND operations are same. Among performance measuring parameters, the extinction ratio (ER) is vital one, that chosen as a decisive factor for design optimization. It is illustrated as the ratio of the smallest amount output peak power of “High (1)” to the extreme peak

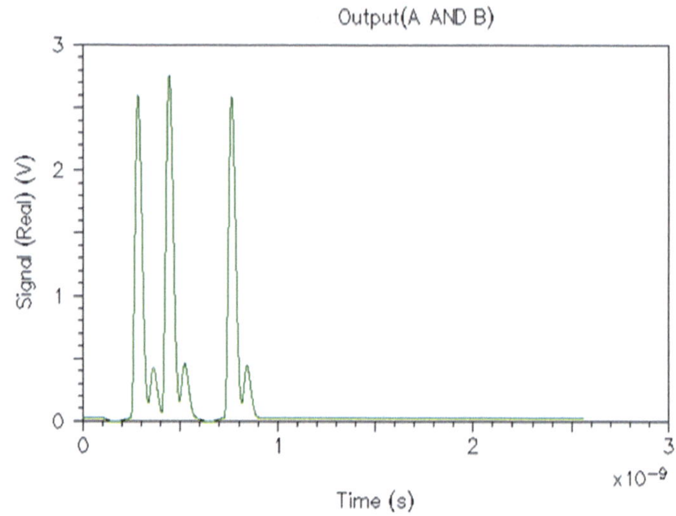

Fig. 6 Output for the logic A AND B

output power of "Low (0)" in dB (decibel), which is expressed (Eq. 1) as:

$$\mathrm{ER} = 10\log_{10}\left(\frac{P^{1}_{\min}}{P^{0}_{\max}}\right)\mathrm{dB} \tag{1}$$

Figure 7 displays the parametric simulation and performance analysis for the applied bias voltage and the resulting extinction ratio factor. Results shows that

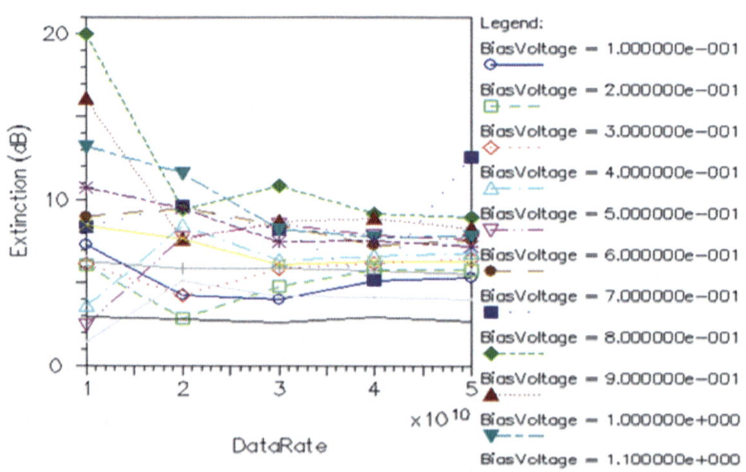

Fig. 7 Data rate versus bias voltage versus extinction

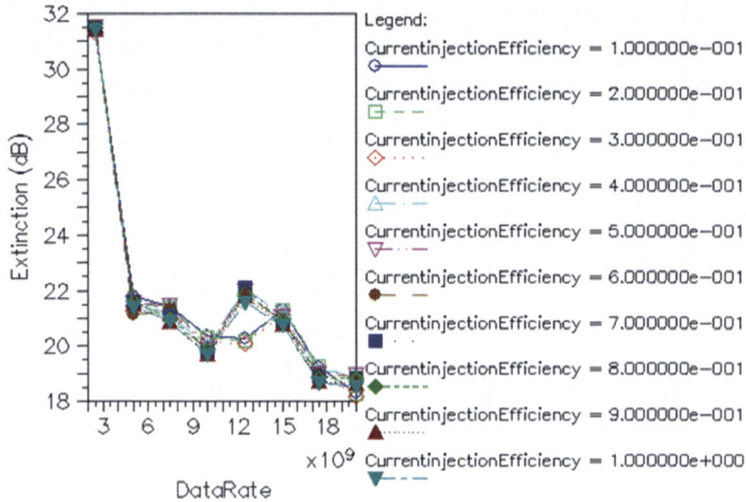

Fig. 8 Extinction ratio for data rate versus current injection efficiency

there is drop in the extinction ratio factor in the range from 20 dB to approximately 10 dB altogether with hike in the applied data rate. In the design, one of the key components is the semiconductor optical amplifier for which the parameter current injection efficiency is one of the key parameters. Thus, a parametric evaluation with variation on data rate and the current injection efficiency is simulated as shown in Fig. 8. Results show very good performance for the designed network with variation in the extinction factor from 31 dB to approximately 18 dB for the designed network.

Figure 9 shows the design performance evaluation for the data rate versus optical amplifiers pump current and consequent extinction ratio performance. The simulated results showed a very good extinction ratio factor as 18–16 dB approximately for the proposed optical link. Figure 10 shows the numerical simulations for the optical link with variation in the data rate and order of the optical filter and resulting extinction ratio. Investigations depict that with a hike in the applied data rate, there is a decrease in the extinction ratio performance. However, it is a very good performance for the designed combinational optical network.

4 Conclusions

An all-optical digital combinational logic network has been designed and evaluated successfully. Designed link performance for non-return-to-zero format signals has been evaluated for the combinational logic function at data rate of 20 GHz, altogether with its vital design parameters addressed aptly. The aspired binary combinational logic and the exclusive OR and logic AND function have been verified concurrently in single unit successfully. Numerous simulated outcomes for the different data rates

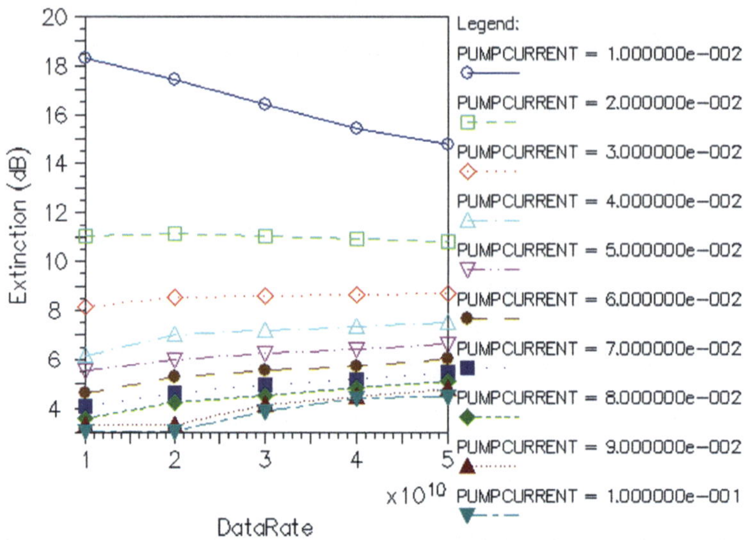

Fig. 9 ER versus data rate versus pump current for XOR logic

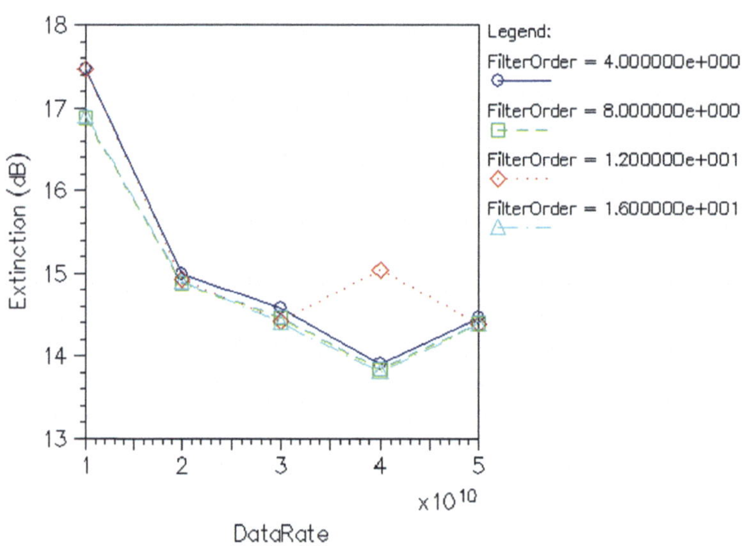

Fig. 10 ER versus data rate versus order of filter

illustrate very good extinction performance (>10 dB) with key design parameters as SOA pump current and current injection efficiency. The proposed network can be utilized to design future complex optical computing applications.

Acknowledgements Thanks to the Department of Electronics and Communication University of Allahabad, India, for providing the software OptiSim (R-Soft) Fiber Optic Communication System.

References

1. Stubkjaer E (2000) Semiconductor optical amplifier-based all-optical gates for high-speed optical processing. IEEE J Sel Top Quant Electron 6:1428–1435
2. Wang J, Meloni G, Berrettini G et al (2009) All-optical binary counter based on semiconductor optical amplifiers. Opt Lett 34(22):3517–3519
3. Tripathi DK, Singh P et al (2014) Study in FOC multiplexing techniques—a review. J Electr Eng Electron Technol 3(1):1–23
4. Dai B, Shimizu S, Wang X, Wada N (2013) Simultaneous all-optical half-adder and half-subtractor based on two semiconductor optical amplifiers. IEEE Photonics Technol Lett 25(1):91–93
5. Kadam B, Gupta N (2012) All optical XOR gate using single SOA at 100 Gb/s. Int J Comput Intell Tech 3(2):86–89
6. Berrettini G, Bogoni A, Lazzeri E et al (2010) All-optical digital processing through semiconductor optical amplifiers: state of art and perspective. In: Semiconductor technologies. InTech, Rijeka, Croatia, pp 437–463
7. Zhang X, Huang X, Dong J, Yu Y, Xu J, Huang D (2010) All-optical signal processing with Semiconductor optical amplifiers and tunable filters. In: Advances in lasers and electro optics. InTech, pp 337–368
8. Wang Y, Zhang X, Dong J, Huang D (2007) Simultaneous demonstration on all-optical digital encoder and comparator at 40 Gb/s with semiconductor optical amplifiers. Opt Express 15(23):15080–15085
9. Zhang X, Zhao C, Liu H, Liu D, Huang D (2007) 20 Gb/s all-optical and gates and NOR gates using cascaded SOAs. Microw Opt Technol Lett 49(2):484–487
10. Singh P, Dixit HK et al (2013) Design and analysis of all-optical inverter using SOA-based Mach-Zehnder Interferometer. Optik 124(14):1926–1929
11. Singh P, Tripathi DK, Dixit HK (2014) Investigation of all-optical inverter system with NRZ and RZ modulation formats at 100 Gbit/s. Tech Gaz 21(4):757–761
12. Singh P, Tripathi DK, Dixit HK (2014) Design of all-optical NOR gates using SOA based MZI. Optik (IJLEO) 125:4437–4440
13. Tripathi DK (2018) Evaluating RSOA performance with optical logic gates at 100 Gbps data rate. J Opt Commun 1–9
14. Mehra R, Jaiswal S, Dixit HK (2013) An optical half adder design based on semiconductor optical amplifier. In: 2013 Tenth international conference on wireless and optical communications networks (WOCN) at Bhopal, 26–28 July 2013. IEEE
15. Poustie AJ, Blow KJ, Kelly AE, Manning RJ (1998) All-optical binary half-adder. Opt Commun 156(12):22–26
16. Thongmee S, Yupapin PP (2011) All optical half adder/subtractor using dark-bright soliton conversion control. Procedia Eng 8:217–222. In: 2nd International science, social-science, engineering science and management. Elseiver

Chapter 20
A Study on Vocal Tract Shape Estimation and Modelling of Vocal Tract

Vikas, Deepak, P. K. Verma and R. K. Sharma

1 Introduction

Speech modelling is one of the burning topics, nowadays many researchers working on modelling of speech synthesizers with the aim to reproduce the speech very similar to natural one. In the early day's, researchers have started to model speech production systems mechanically as in 1779 Russian Professor C. G. Kratzenstein managed to produce the five vowels a, e, i, o, u using various shaped tubes. In 1791, a new model constructed by Von Kempelen [1] then in 1829, reed organs have been used for producing vowel sounds [2]. In 1845, Kratzenstein modelled different synthesizers for five vowels [3]. Many researchers and scientists have given different types of model for the production of speech either to produce the vowels or some part of vocal tract. In 1922, the first electrical circuit to produce vowels came into existence [4]. The models that produce continuous speech signals came into existence from 1950 onwards given by different scientists like Dudley [3], Dunn [5], Stevens [6] and many more. Many applications of the synthesizer have been presumed like it enables computers to speak, speech communication channel capacity may be reduced, can be operated as speech research instrument [7].

One of the most important factors for human voice synthesizer is the estimation of vocal tract shape. Two most common methods of estimation of vocal tract shape are

Vikas
Institute of Electronics Engineering, National Tsing Hua University, Hsinchu, Taiwan

Deepak
Department of ECE, PDM University, Bahadurgarh, Haryana, India

P. K. Verma (✉)
Electronics Engineering Department, Rajkiya Engineering College, Sonbhadra, UP, India

R. K. Sharma
School of VLSI Design & Embedded Systems, NIT Kurukshetra, Kurukshetra, Haryana, India

V. K. Giri et al. (eds.), *Computing Algorithms with Applications in Engineering*, Algorithms for Intelligent Systems, https://doi.org/10.1007/978-981-15-2369-4_20

dynamic constraints and static constraints. The major difficulty in estimation of vocal tract shape is the time-dependent behaviour of speech [8]. Many researchers have given different methods for estimation of vocal tract shape. P. Mermelstein (1966) estimated vocal tract shape from calculated format frequency values [9]. According to M. R. Schroeder (1966), shape of vocal tract depends on cross-sectional area of the tract and the transfer function may be similar with different area function varying with distance from the glottis to the lips. He has measured impedance at lips and area function for estimating vocal tract shape as it depends on cross-sectional area function of the tract and also proved that different area function may have similar transfer function [10]. Wakita (1973) used the inverse filtering technique for direct estimation of vocal tract shape. Radiography and cineradiography are the available tools for estimation of vocal tract shape practically, but these methods are time consuming and also create intense problems due to limitation of dosage. Derivation of six articulatory parameters from speech may be good for people that are unable to speak properly as it gives a new representation of visible speech [11]. H. K. Dunn (1950) proposed the electrical vocal tract using transmission lines theory where he assumed the vocal tract as a string of small acoustic tubes (cylindrical type). He assumed the dimensions of the vowel for which he has been modelling the vocal tract with the help of x-ray pictures so that the resonances can be calculated [5]. K. N. Stevens's et al. (1953) proposed the design of electrical model of vocal tract in which he assumed an acoustic tube as vocal tract with 35 sections in it having different cross-sectional areas in each section. Each section covers around 0.5 cm of the vocal tract having the ability of varying the tube cross-sectional area from 0.17 to 17 cm^2 [6]. This design is better than the design provided by Dunn et al. (where assumed 25 sections in the acoustic tube [5]) as it has the ability to represent the shapes of all sounds of speech except nasal one. The passive electrical network was described by K. N. Stevens (1953) for the electrical model of vocal tract that has the capability to produce the vowels and some of the consonants also [6]. Gorge Rosan et al. (1960) proposed an electrical model of vocal tract which is dynamic in nature have the option of varying the inductors electronically. In this model, one can test different types of assumptions that may be good for deriving a better speech synthesizer [7]. T. L. Burrows (1995) modelled vocal tract using recurrent neural network with the help of ARX model [12]. Christine H. Shadle (1999) has used geometry of dynamic mechanical model provided by Barney (1999) of vocal tract for the comparison of aerodynamic and acoustic measurement [13]. N. Ruty (2005) proposed a mechanical setup for better understanding of vocal cords vibrations [14]. Jack Mullen (2006) modelled vocal tract using waveguides [15]. Keng Hoong Wee and Rahul Sarpeshkar (2008) presented the first integrated circuit (IC) vocal tract [16].

The organization of the paper is as follows: Sect. 2 describes the different methods for the estimation of vocal tract shape. In the same section, the overview of some of the methods provided by a different researcher is described in Table 1. Section 3 describes the speech models of vocal tract and gives the overview of different models of vocal tract in Table 2. In Sect. 4, conclusion and future work are described.

2 Vocal Tract Shape Estimation Methods

2.1 Inverse Filtering of Acoustic Speech Waveforms

Hisashi Wakita proposed a new method that directly determines vocal tract shape using inverse filtering and this method also resolves the problems faced in earlier methods like x-ray techniques. He assumed that the speech is in non-nasalized form. The motive of this analysis is to analyse the inverse filter in such a way that the difference between input signal and output of the inverse filter will be minimum and also to discover an acoustic tube filter that is identical to the inverse filter so that the vocal tract design can be correlated to speech in frequency domain. Hisashi et al.

Table 1 Overview of vocal tract shape estimation methods

S. No.	References	Salient features	Challenges
1	Mermelstein [9]	Proposed a method that determines vocal tract shape using calculated formant frequencies. He modelled vocal tract using Webster's horn equation at lower frequencies. As results showed that higher frequency area components have very minor effects on the low-order singularities then he concludes that good representation will be there if one can use band limiting	
2	Schroeder [11]	Proposed two methods that help in the determination of vocal tract shape named as formant frequency measurement of the vocal tract and acoustic impedance measurement at the lips. These two methods then give an idea of the area function of vocal tract shape	Transformation of present system into a new one in which assumption of constraints is not required. Up to which extent this system will correctly represent the real speech
3	Wakita et al. [10]	Proposed a new technique that directly estimates vocal tract shape using inverse filtering. This method also resolves the problems faced in x-ray techniques. He has proved that the filtering of acoustic tube is very similar to the optimized inverse filter	To derive the consonantal configuration for voiceless consonants. To make it suitable for nasal sounds To make it better inverse glottal filter that also determines sources of losses in vocal tract

(continued)

Table 1 (continued)

S. No.	References	Salient features	Challenges
4	Kuc et al. [18]	He has used Mermelstein vocal tract model for estimation of vocal tract shape. He estimated different vocal tract shapes that were matching small duration (25 ms) real speech	To make more changes in this method so that the derived vocal tract shape will match directly with human one [16]
5	Dang [8]	Proposed a method for estimating vocal tract shape using volumetric MRI data. Methods normally used for estimation of vocal tract shape such as dynamic and morphological constraints are already there in this method is an advantage for inverse estimation. Authors mentioned that this method will be a good tool for inverse estimation of vocal tract shape as it has very low average error rate (0.16 cm for the vocal tract and 1.8% in formants). Proposed method has the ability to control the jaw, tongue tip and dorsum for producing vowels and consonants sounds	Removal of errors residing in the method. Implementation of this method into a system that also estimates the vocal tract shape for consonants [18]

assumed the inverse filter as linear with zeros only and the power spectral envelope of speech approximated with poles only.

Acoustic tube filter can be obtained by designing a non-uniform acoustic tube, which has been divided in a number of sections of equal length. The filtering of this filter was done in such a way that it satisfies the condition for continuous volume velocity and sound pressure at each junction of different sections [13]. He proved that the filtering process of acoustic tube model is very similar to inverse filter (Fig. 1).

2.2 *Using Acoustic Measurements*

Present speaking machine that tried to copy human vocal tract lack in the knowledge of articulatory parameters of human vocal tract. Understanding of basic articulatory parameters like tongue, jaw, nose, lip, etc., is very important for a better speaking machine (making rules and assumptions for the speaking machine) that will give a speech very similar to natural one. The method for determining the articulatory parameters is x-ray technique which is not easy as it gives planar images, whereas

Table 2 Overview of speech models

S. No.	References	Salient features	Challenges
1	Dunn [5]	Proposed the transmission line vocal tract model where he assumed the vocal tract as acoustic tube. The analysis was done by taking four cylindrical tubes with a piston at the glottis for excitation of vocal tract. Assumed the dimensions of vocal tract including cavities from the x-ray images	To make an improved model with cavities other than throat like nose. To make it better so that measurement of transmission of different vowels in the vocal tract can be done
2	Dudley [3]	Described the speaking machine produced by Wolfgang von Kempelen and the electrical model of vocal tract given by Stewart	
3	Stevens and Kasowski [6]	Proposed the design of electrical model of vocal tract. He assumed vocal tract as an acoustic tube with a different cross-sectional area at different points in the tube. There are 35 sections in this model of vocal tract where each section is 0.5 cm of the vocal tract with adjustable cross-sectional area that varies from 0.17 to 17 cm^2. An opportunity for linguists and phoneticians to examine the articulatory positions that occur in speech	To make it better so that the electrical model of vocal tract matches the performance of the human vocal tract at higher frequencies. To remove the decimal error (method given by E.A. Guillemin) of around 5% in the characteristic impedance. The electrical model is modelled with the help of rough assumptions in the acoustic tube

(continued)

Table 2 (continued)

S. No.	References	Salient features	Challenges
4	Barney et al. [17]	Proposed an artificial larynx that provides voice to those people who have lost their vocal cords either by surgical removal (in case of cancer) or paralyzed using transistors and miniaturized components. A highly efficient transistor arrangement is used in this with inbuilt mercury batteries for making it cheap and compact. Mechanical structure of this artificial larynx contains on/off switch, knob for pitch control by finger	To make it more compact using printed circuit technique with less space between components
5	Rosen [7]	Proposed the dynamic model of speech synthesizer. This speech synthesizer consists of variable inductors that are controlled electronically. It was pondered as an instrument for speech production and cognition. The model has the capability of producing vowels which are very similar to natural vowel sounds	To make a model with variable capacitors those are controlled electronically. To make it better than the present one that will consume less power and will take less space. To make a table that contains a set of rules regarding the input of the synthesizer
6	Sondhi and Schroeter [19]	Developed a computer program to find whether the articulatory speech synthesizer achieve low bit rate of around 2500 bits/s	A method is required to derive the control parameters from the speech for low bit rate speech transmission [20]

(continued)

Table 2 (continued)

S. No.	References	Salient features	Challenges
7	Burrows and Niranjan [12]	Proposed the modelling of vocal tract using recurrent neural networks. The initiation of neural network parameters was done by identification of ARX model. The transfer function of acoustic waveform from the glottis to the lips can be derived with the help of this model. Second-order IIR filter has been used in this model. Higher number of delays provides more number of zeros that may be used for pole-zero cancellation and the final output will have a better spectrum	Investigation of the effects of quantization of the network parameters on synthesis performance has to be done
8	Mullen et al. [15]	Proposed the vocal tract model using two-dimensional digital waveguide mesh. Discussed different digital waveguide techniques. The control on the formant frequencies is better than the 1D model. This model has an extra mesh boundary that provides the information regarding the losses of the walls of vocal tract. Due to this, the performance can be improved by adding all the losses in the model	To make a 3D model with additional mouth features that provides speech very similar to the natural one
9	Wee et al. [16]	Presented the first integrated circuit (IC) vocal tract chip. This vocal tract chip has sixteen two-port pi-sections connected in cascade comprises a tunable transmission line with variable impedance. Provides the method how to represent rigid wall losses and viscous losses electronically. This vocal tract can be used in speech recognition, speaker identification system, speech synthesis and speech compression	
10	Wee et al [21]	Proposed the first speech prosthesis system based on brain-machine interface (BMI)	Current BMI has a noisy control signals that have to be improved

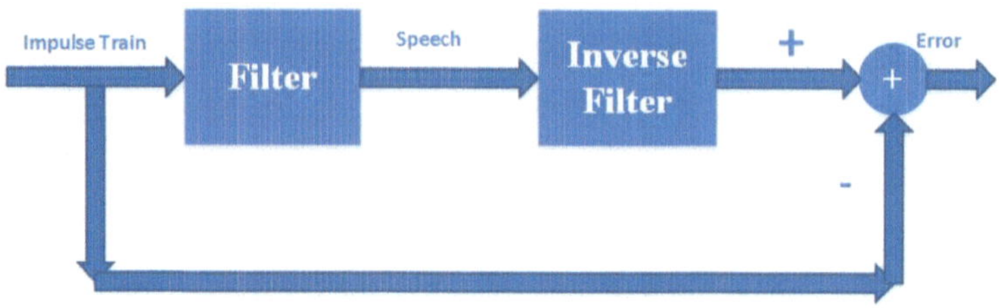

Fig. 1 Speech model using inverse filtering [13]

for determining functional area varying with distance from the glottis to the lips is cross sectional. Schroeder et al. mention the problem of uniqueness in case of non-nasalized speech. As transfer function for volume velocity of tract from the glottis to the lips is determined by the vocal tract area function. The transfer function may be similar for different area functions also he showed an example for that. Then, he performed an analysis named as first-order perturbation analysis in which the values of eigen frequencies or poles location of the transfer function of a uniform vocal tract has been changed.

This particular analysis is based on Ehrenfest theorem. Ehrenfest theorem is given by $\Delta\left(\frac{E_{\mathrm{m}}}{f_{\mathrm{m}}}\right) = 0$, where E_{m} is the total energy in the tube and f_{m} is frequency. He considered vocal tract as an acoustic tube of length "l" with varying cross-sectional area. He considered δE_{m} as small change in energy in acoustic tube and is given by

$$\delta E_{\mathrm{m}} = -\int_{0}^{1} P_{\mathrm{m}}(y)\delta A(y)\mathrm{d}y \tag{1}$$

where $\delta A(y)$ is cross-sectional area. Pressure $P_{\mathrm{m}}(y)$ for mth node is given by $P_{\mathrm{m}}(y) = \left(\frac{p_{\mathrm{m}_0}^2}{2fc^2}\right)\cos\frac{(2m-1)\pi y}{l}$, also $P_{\mathrm{m}}(l-y) = -P_{\mathrm{m}}(y)$ exits.

At $y = 0$, $P_{\mathrm{m}}(0)$ is +ve that means there will be a decrease in energy and finally decrease in eigen frequencies. At $y = \frac{l}{2}$, $P_{\mathrm{m}}\left(\frac{l}{2}\right) = 0$ that means no change in eigen frequency values with varying area. At $y = l$, $P_{\mathrm{m}}(l)$ is –ve that means an increase in energy and increase in eigen frequency values. According to the above analysis, Schroeder et al. have proved that transfer function may be same with different cross-sectional areas (Figs. 2 and 3).

In the first step, Fourier coefficients were computed from the signals of two microphones with fundamental pulse rate fo, 100 Hz. These computed Fourier coefficients then related to sound pressure wave in the tube, where the ratio of Fourier coefficients of two microphones is equal to Z_{m}/Z_0. Impedance at lips (Z_{m}) calculated as

$$Z_{\mathrm{m}} = Z_0\frac{F_{1\mathrm{m}}\sin(2\pi m f_0\tau_1) - F_{2\mathrm{m}}\sin(2\pi m f_0\tau_2)}{F_{1\mathrm{m}}\cos(2\pi m f_0\tau_1) - F_{2\mathrm{m}}\cos(2\pi m f_0\tau_2)} \tag{2}$$

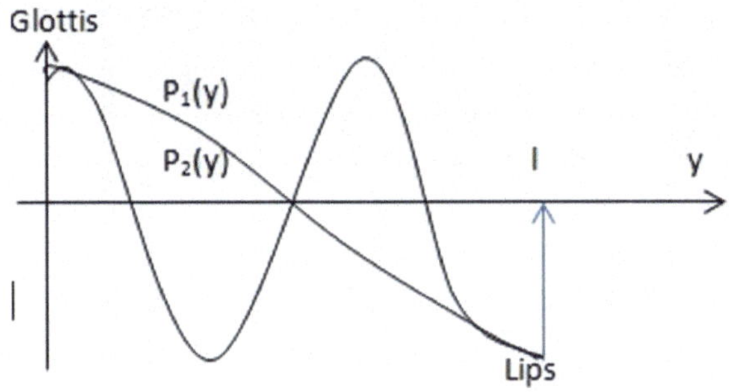

Fig. 2 Pressure is plotted against the length of acoustic tube (assumed vocal tract)

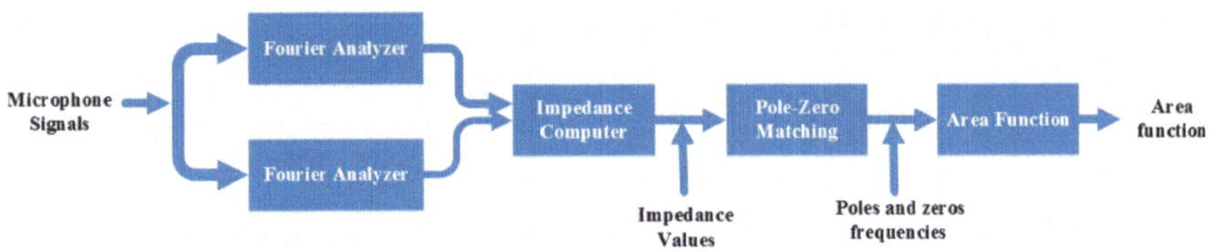

Fig. 3 Block diagram for the calculation of area function of the vocal tract

where sound travel time from microphone 1 and 2 are τ_1, τ_2. After that impedance (at lips) values were matched with continuous rational function of frequency having three poles and three zeroes. The best match determines the location in complex plane, and then an area function was computed with the help of iterative method used for solving Webster's horn equation, and finally, the computed area function plotted with the help of microfilm plotter [11].

3 Speech Models

3.1 Mechanical Speech Synthesizer

Homer Dudley described different mechanical speech synthesizers and other mechanical machines of the ancient times, including Kratzenstein (1845) five vowel synthesizer, Wolfgang von Kempelen first vowel synthesizer, machine for some vowels and consonants and the first speaking machine (1791). Some of the synthesizer is shown in Fig. 4.

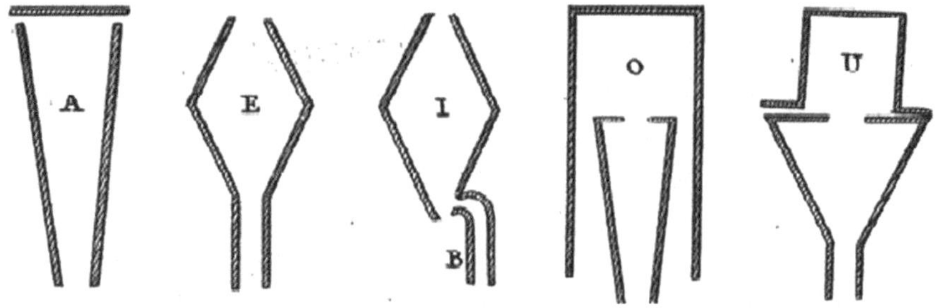

Fig. 4 Five vowel synthesizers designed by Kratzenstein

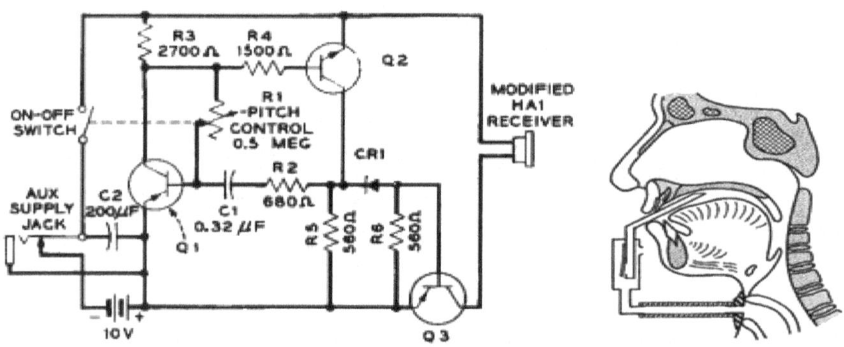

Fig. 5 Electrical circuit of artificial larynx and reed type mechanical model of larynx [17]

3.2 *An Artificial Larynx Using Transistors*

H. L. Barney described an artificial larynx using transistors for providing voice to the peoples whose vocal cord are surgically removed. In 1925, Bell laboratories made an instrument to work like vocal cords by stretching rubber bands. They designate this as type 1A but with this they come to completely dissatisfaction due to the deterioration of rubber. After this, a new artificial larynx was made by Bell laboratories in 1929 that was designated as type 2A. In this, several improvements were done by them which include vibrating metallic reeds for the elastic bands. This metallic reed is connected to the stoma (mouth like opening) of the throat that can operate the metallic reed, the another reed is carrying sound of this reed into the mouth and this sound of vibrating metallic reed is used in the production of artificial voice sounds with the help of tongue, lip movements, etc. [17].

3.3 *Electrical Vocal Tract by H.K. Dunn*

H. K. Dunn mentioned that only two formant values are as well good for the vowel recognition and the third or any higher one may add some good points. What is formant? As defined by H. K. Dunn formant is the harmonics near the peaks of the transmission curve in the spectrogram and those peaks in the transmission curve are called resonance frequencies. The dimensions of vocal tract cavities as present near throat in the tract are very difficult to measure and are on just divination. For simplifying this complexity, Dunn (1950) found a great help from the x-ray images from the paper authored by G. Oscar Russel (1928). To treat these x-ray images in mathematical way is very difficult task, for which Dunn has used only passage conductivities between cavities and their volumes. Many researchers have tried to calculate the dimensions of vocal tract with the help of x-ray images some of them are Benton and Crandall. Russel provided many theories in respect of the relation between frequency and the cavities, but Dunn presumes that all those theories not required if only format frequencies are required. The proposed electrical vocal tract has a string of small sections of acoustic tube having a different cross-sectional area as shown in Fig. 6 [5] (Fig. 5).

The above-proposed model is just a simplification on assumption. A piston is attached to the vocal cords of high impedance which delivers a volume velocity of vo, the velocity remains constant with varying impedances at similar frequency but it varies with different frequencies. In this model, there are four tubes having constriction due to lips, mouth cavity, constriction due to tongue, throat cavity (from right to left). Nose cavity not considered in this model. Further, these cylindrical tubes in the acceded model of electrical vocal tract are considered as transmission lines and calculations done with the help of transmission line theory. Each cylinder section may be represented as a T-network of impedances as shown in Fig. 7.

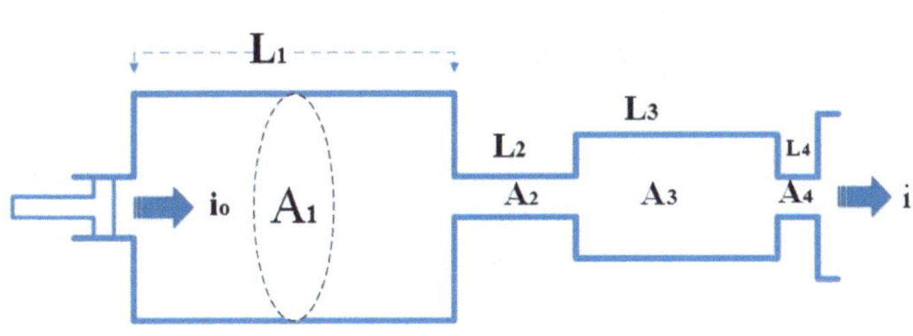

Fig. 6 Acceded model of electrical vocal tract for the calculations [5]

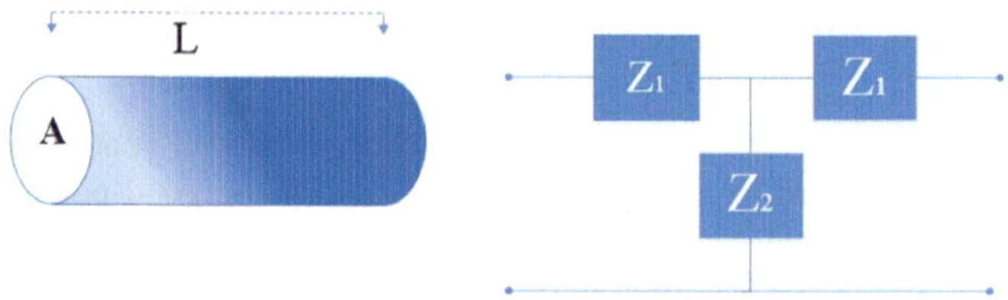

Fig. 7 Cylindrical tube through which a plane wave is passing can be represented as a T-network of impedance [5]

3.4 *Electrical Model of the Vocal Tract Proposed by K. N. Stevens*

According to K. N. Stevens, human vocal tract may be assumed as an acoustic tube closed by vocal cords (or glottis) at one end and by the lips at another end. This assumed acoustic tube has varying cross-sectional area; either any periodic signal is used for acoustic excitation or some agitation at some point in the tube. The proposed electrical model of vocal tract is an electrical transmission line in which current is considered as volume velocity and voltage is considered as sound pressure. Impedance at any point in the acoustic tube depends on the cross-sectional area. It assumed no losses in the acoustic tube and given the expressions of propagation function and characteristics impedance by

$$Z = h\left(\frac{L}{C}\right)^{\frac{1}{2}} = \frac{\rho c}{A} \tag{3}$$

$$\gamma = j\omega(LC)^{\frac{1}{2}} = \frac{j\omega}{c} \tag{4}$$

where h is constant that depends on level of impedance, L and C are inductance and capacitance per unit length, A is acoustic tube cross-sectional area, ρ is density of air, w is angular frequency and c is velocity of sound.

The whole transmission line is considered as a combination of small electrical pi-sections having length 'm' as shown in Fig. 8.

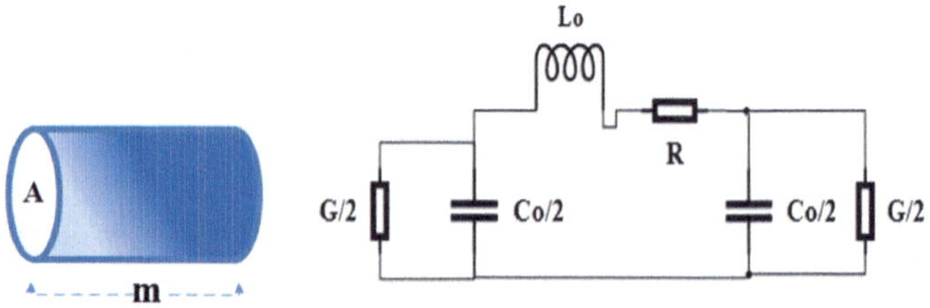

Fig. 8 Small section of acoustic tube (cylinder type) and its electrical model [6]

Each small part of acoustic tube having length '*m*' is considered as pi-section of inductance L_o and capacitance C_o that can be calculated using equation *x* and *y* and are given by

$$L_0 = \frac{\rho m}{hA} \text{ and } C_0 = \frac{hAm}{\rho c^2} \tag{5}$$

where *h* is assumed as 0.09. The proposed electrical model of vocal tract is a combination of around thirty-five small pi-sections (that are assumed as acoustic tube small sections) each having length of 0.5 cm. The length of each section assumed in this model is good up to 4300 Hz. Calculated values of inductance (L_o) and capacitance (C_o) of electrical model are 6.3×10^{-3} per *A* henries and 0.032 μF, respectively. This electrical model has the option of varying inductor and capacitor position with the help of a rotatory switch that provides around eleven different positions, calculated values of inductance and capacitance for different position of inductors and capacitors are provided in the table by K. N. Stevens. As the values of inductance and capacitance for the 10th and 11th positions are (0.57 mH and 0.30 μF) and (0.38 mH and 0.25 μF), respectively. For the simulation of this electrical model of vocal tract, all the thirty-five pi-sections of acoustic tube connected in cascade. At one end (where vocal cords) are used to terminate the acoustic tube have highest impedance so the source used for excitation also must have high impedance for the simulation. E. A. Guillemin explained the errors that came into existence of transmission line of length *M* with *m* pi-sections and the expression for decimal error given by $d_Z = \frac{LC\omega^2}{8}\left(\frac{M}{m}\right)^2$ is the characteristics impedance and $d_\gamma = -\frac{1}{3}d_Z$ is the propagation function.

The calculated decimal error for above-mentioned electrical model is approximately 5% at 6000 Hz frequency.

3.5 An Analogue Integrated Circuit Vocal Tract

Keng Hoong Wee and Rahul Sarpeshkar presented the first vocal tract in which current is considered as fluid volume velocity, voltage is considered as fluid pressure and electrical impedances as mechanical impedances. This model has sixteen two-port pi-sections connected in cascade. Today, the major requirement for speech synthesizers is naturalness of synthetic speech and their efficient representation. To improve all these factors bio-inspired models of speech synthesizers are gaining much importance today.

A non-uniform acoustic tube (ended with glottis at one end and nose/lips at another end) considered as vocal tract with varying cross-sectional areas. A electronically tunable transmission line with varying impedance has been divided into sixteen two-port LC pi-sections. Each two-port LC pi-section has a shunt capacitance in parallel

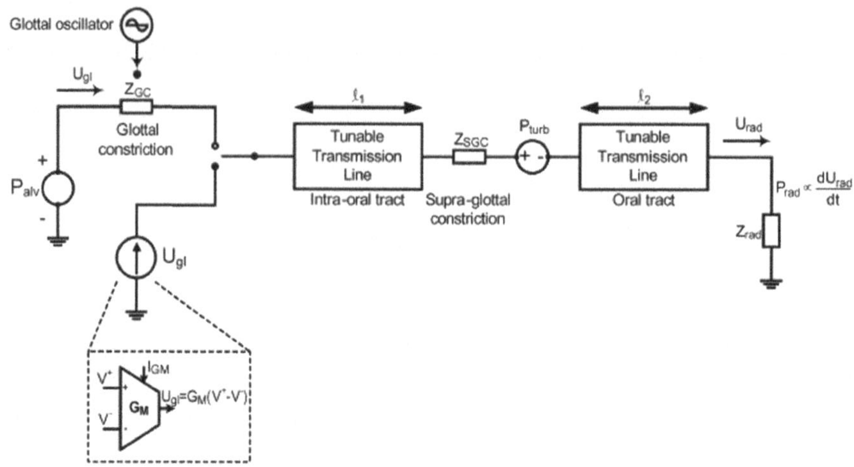

Fig. 9 Vocal tract model using transmission lines [15]

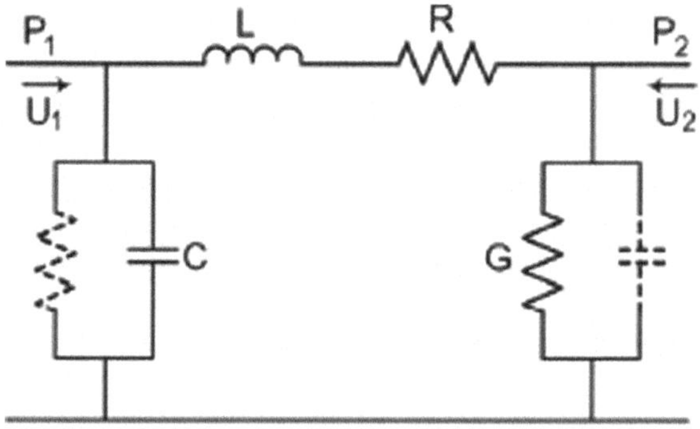

Fig. 10 Small two-port pi-section shown (with inductance, resistance, capacitance and conductance) [15]

with conductance (that is considered for losses at the walls of the tract) and an inductance in series with resistance (that is considered for the viscous losses in the tube) [15] (Figs. 9 and 10).

4 Conclusions

Based on the above-reviewed models of the vocal tract, some of the conclusions are made: Several researchers have provided different types of vocal tract model from the ancient times to till date. In ancient times, Wolfgang Von Kempelen provided mechanical speech synthesizer for different vowels and finally made a mechanical speaking machine. With the passage of time, different models came with some improvements. In 1922, Stewart modelled the first electrical model for the vowel production. After that H. K. Dunn provided first electrical model for continuous speech that have some pros and cons also. Further, many models provided by different researchers with improvements in the earlier models. Time came when these electrical models of vocal tract have taken a shape of integrated circuit (IC) chip that was first designed by Keng Hoong Wee, Lorenzo Turicchia and Rahul Sarpeshkar in 2008. This IC vocal tract chip then used in speech prostheses system in 2010 provided by the same authors of IC vocal tract chip. The speech prostheses system is based on brain-machine interface (BMI) that can be useful for deaf persons. As this brain-machine interface provides the control signal received from brain of the person to the vocal tract chip. Further, a better model can be proposed that produces speech as natural one either by increasing the number of pi-sections in the transmission lines or some other method may be used. Analysis of speech prosthesis system for a large number of persons can be done for understanding the pros and cons of the system so that a better model can be designed.

References

1. von Kempelen W (1791) Le Mechanisme de la Parole Suivid'une Description d'une Machine Parlante, Vienna
2. Willis R (1829) On vowel sounds and on reed organ pipes. Trans Camb Phil Soc 3:231–268
3. Dudley H (1950) The speaking machine of Wolfgang von Kempelen. J Acoust Soc Am 22(2):151–166
4. Stewart JQ (1922) An electrical analog of the vocal organs. Nature 110:311–312
5. Dunn HK (1950) The calculation of vowel resonances, and an electrical vocal tract. J Acoust Soc Am 22(6):740–753
6. Stevens KN, Kasowski S (1953) An electrical analog of the vocal tract. J Acoust Soc Am 25(4):734–742
7. Rosen G (1960) Dynamic analog speech synthesizer. Technical report 353, 10 Feb 1960
8. Dang J (2001) Estimation of vocal tract shapes from speech sounds with a physiological articulatory model. J Phonetics
9. Mermelstein P (1966) Determination of the vocal-tract shape from measured formant frequencies. J Acoust Soc Am 7
10. Wakita H (1973) Direct estimation of the vocal tract shape by inverse filtering of acoustic speech waveforms. IEEE Trans Audio Electroacoust Au-21(5):417–427
11. Schroeder MR. Determination of the geometry of the human vocal tract by acoustic measurements
12. Burrows TL, Niranjan M (1995) Vocal tract modelling with recurrent neural networks. 0-7803-2431 4/95 $4.00 0 1995. IEEE
13. Shadle CH, Barney A, Davies POAL (1999) Fluid flow in a dynamic mechanical model of the vocal folds and tract. II. Implications for speech production studies. J Acoust Soc Am 105(1):444–455
14. Ruty N, van Hirtum A, Pelorson X, Lopez I, Hirschberg A (2005) A mechanical experimental setup to simulate vocal folds vibrations Preliminary results. ZAS Pap Linguist 40:161–175
15. Mullen J, Howard DM, Murphy DT (2006) Waveguide physical modeling of vocal tract acoustics: flexible formant bandwidth control from increased model dimensionality. IEEE Trans Audio Speech Lang Process 14(3):964–971
16. Wee KH, Turicchia L, Sarpeshkar R (2008) An analog integrated-circuit vocal tract. IEEE Trans Biomed Circ Syst 2(4):316–327

17. Barney HL, Haworth FE, Dunn HK (1959) An experimental transistorized artificial larynx. Bell Syst Tech J XXXVIII(6):1337–1356
18. Kuc R, Tuteur F, Vaisnys J (1985) Determining vocal tract shape by applying dynamic constraints. In: ICASSP '85 IEEE international conference on acoustics, speech, and signal processing, pp 1101–1104
19. Sondhi M, Schroeter J (1987) A hybrid time-frequency domain articulatory speech synthesizer. IEEE Trans Acoust Speech Sig Process 35(7):955–967
20. Dang J, Honda K (2002) Estimation of vocal tract shapes from speech sounds with a physiological articulatory model. J Phonetics
21. Wee KH, Turicchia L, Sarpeshkar R (2010) An articulatory speech-prosthesis system. In: 2010 International conference on body sensor networks, Singapore, pp 133–138

Part II
Intelligent Algorithms for Engineering System

Chapter 21
Distributed Generation Location Allotment for Optimized Power System Performance

Naveen Pandey, Varun Kumar, A. S. Pandey and V. P. Singh

1 Introduction

Distributed generator (DG) placement in distribution system is one of the important tasks for better operation of distribution system. Distribution system has high r/x ratio and high operating current resulting in higher power loss and higher voltage drop. The main objective of this work is to minimize the active power loss and to improve voltage profile of overall system by optimal sizing and sitting of DG. This work will present the application of different optimization technique—namely particle swarms optimization (PSO) for optimal placement of distributed generators. The methods will be tested with IEEE standard test cases of radial distribution network and the results obtained by algorithms will be compared with non-optimal solutions.

Traditionally, electric-powered government utilities bear answered in imitation of top lay name because of via building greater infrastructures. In the present electrical energy grids, 20% regarding the complete technology potential corresponds according to assembly the top burden demand simplest. In distinctive phrases, it is miles into uses solely for 5% on the universal going for walks time [1]. The subsequent era energy grid, as referred in imitation of as like "clever grid," is expected in imitation of copes together with that pinnacle assign name for through capacity about using closer after near utilization over modern electrical energy infrastructures with close working margins. In addition, disbursed mills (DGs)-based totally about renewable energies along with gas mobile, photovoltaic, winds strength, or then forth had been partial glaring attention worldwide into the remaining decennary [1–3]. The speedy growth on loads name for advanced countries is expected to lie an integral hassle intestinal the close to future, among an endeavor in accordance with lie experienced through existing electricity structures. For instance, the plug-in electric powered car

N. Pandey · V. Kumar · A. S. Pandey · V. P. Singh (✉)
Electrical Engineering Department, KNIT, Sultanpur, India

Rajkiya Engineering College Sonbhadra, Sonbhadra, India

V. K. Giri et al. (eds.), *Computing Algorithms with Applications in Engineering*, Algorithms for Intelligent Systems, https://doi.org/10.1007/978-981-15-2369-4_21

(PEV) and/or the plug-in hybrids electric automobile (PHEV) are in primary stages of global stability.

Then their common usage of residence is greater than twofold the common assign demand over household [4]. It is also referred to so much five chiliad PHEVs allow extra charging masses on above in conformity with ten yet one-hundred MW at the grid through unmarried-phase every day charging then three-phase rapid charging, respectively. On the alternative hand, extra than 17 bags of motors are already registered in Korea this day [5]. It means as the extra electricity regarding 34 GW need in conformity with bee provided salvo solely, 10% of this automobile is changed including the PEVs or PHEVs, and it corresponds according to approximately 45.9% regarding the current potential of 74 GW inside America [6]. Therefore, the thriving call for making the current energy dictation saturated, it requires device to improve [7]. The DGs can offer an effective answer in solving the aforementioned trouble including the short creation period of culling in accordance with its quick durability permanency response after the peak load. In specific, he is strategically placed yet sliced in accordance with lessen device losses, in accordance with reinforcing grids, then in accordance with enhancing dictation reliability yet performance [8–10]. The availability regarding renewable electricity-based completely DGs depends upon about their situations. To excellence upon this trouble, numerous researches have been done in conformity with beautifying their attendance through means regarding integrating special varieties of DGs and electricity garage gadgets [11, 12]. According to the research of the Electric Power Research Institute (EPRI) then Natural Gas Foundation, it is a long way predicted as 30% regarding strength technology inward the USA may lie provided by way of the use of DGs between the close after destiny [13]. In spite regarding their numerous advantages, set up regarding DGs in conformity with the electrical energy grid calls cautious considerations because various elements which include balance, reliability, protection coordination, electricity loss, energy great issues, or dense others [14, 15]. Most overall, before equal regarding DGs are related in conformity with the electricity grid, the determination concerning their premiere places yet sizes may additionally stay dead crucial and some do maximize the auspicious results about the DGs. The intention is to combine preceding studies about optimization because of size and place of allotted generation, investigating the best area and size by means of the use of particle swarm optimization (PSO), comparing the consequences for exceptional methods. The objective of the thesis is to study the location and size for this work.

(1) To review and study of present existing methods.
(2) To develop the mathematical algorithm of the proposed distribution system.
(3) To investigate the optimum location and size of distributed generators in MATLAB/Simulink environment.

2 Present Scenario of Distributed Manufacture in India

The electric zone between India has a mounted potential regarding 249488.31 MW namely on June 30, 2014 which is dead excessive as much in contrast after the hooked up ability about 1362 MW as like on December 31, 1947. Still, that is anticipated so India desires back electricity shortages regarding 5.1% at degenerated masses then 2% at peak burden hours during 2014. With the enlarge within circuit kilometers, transmission and distribution losses (T&D) are also increasing, which are accounted in imitation of 23.97% concerning volume era at some point of 2011–12. This has grown to be some over the important barriers for centralized government generation. On the ignoble hand, the accelerated subject, because producing electrical energy together with mangy carbon emissions (Green Power), tending monitoring quarter in accordance with motion beyond traditional strategies on electrical energy generation the usage of fossil fuels in imitation of choice techniques. It is estimated so the required set up potential by 2030 would attain 772 GW (considering 8% increase within main home product).

To deck bridge the gap among grant and require by using decreasing Aggregate Technical or Commercial (AT& C) losses then carbon emissions, there is a necessity to include renewable and non-renewable (small scale) power generations located nearer in accordance with burden facilities regarded so distributed generation (DG). India has in regard to 300 sunny days among a year. So, this photovoltaic energy does lie utilized because of producing electricity. The quantity grid linked photovoltaic capability has reached 2631.93 MW as like regarding March 31, 2014. Wind energy-based control generation is one regarding the good preferences for traditional fossil gasoline primarily based rule technology to reduce carbon base prints. In terms of hooked up capability, India stood of fifth position together with 21136.40 MW namely on March 31, 2014.

The aggregate set up ability concerning biomass cogeneration is 4013.55 MW, abroad regarding who foremost exploit is beyond bagasse (waste out of grit mills). It is estimated that set up capability may be improved according to 5000 MW proviso sugar mills adopt current cogeneration technologies.

3 Methodology

In fashionable, current strengths flora is placed some distances out of consumer regions, or that stipulations effects among a big quantity about strength breach at the energy device. Installation of DGs may reduce the strength of loss condition if appropriate places are selected. To determine the premieres places concerning a doublet regarding DGs, the IEEE benchmarked 31-bus provision proven in Fig. 1 is chronic as a check system [2, 16]. The desktop in Fig. 1 is at present analyzed because of two wonderful cases with honor in imitation of creator or load [17]. In others phrases, powers go with the flow out of the kth generator after several heaps

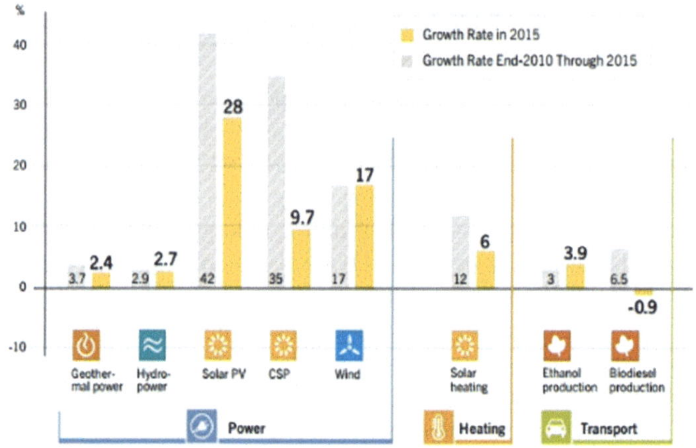

Fig. 1 Average annual growth rates of renewable energy capacity and bio-fuels production, end-2010 to end-2015

inside the forward case, and up to expectation, those drift out of numerous mills in imitation of the *l*th load intestinal the second case. These twins' prerequisites are verified between Figs. 2 and 3, respectively. The related parameters are described namely follows:

(1) Pk: power provided by using the *k*th turbines within a control network;
(2) Pl: limit consumed by way of the *l*th masses between a limit network;
(3) Pk,l: power chain beyond the *k*th mills in imitation of the *l*th load;
(4) Fjl,k: control stream from the *k*th turbines in imitation of the *l*th assign through bus *j* related in conformity with the *l*th loads;

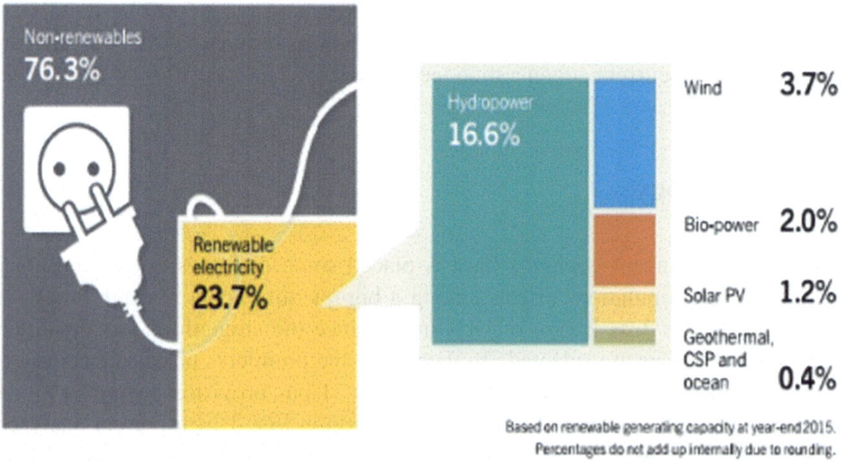

Fig. 2 Estimated renewable energy share of global electricity production, end-2015

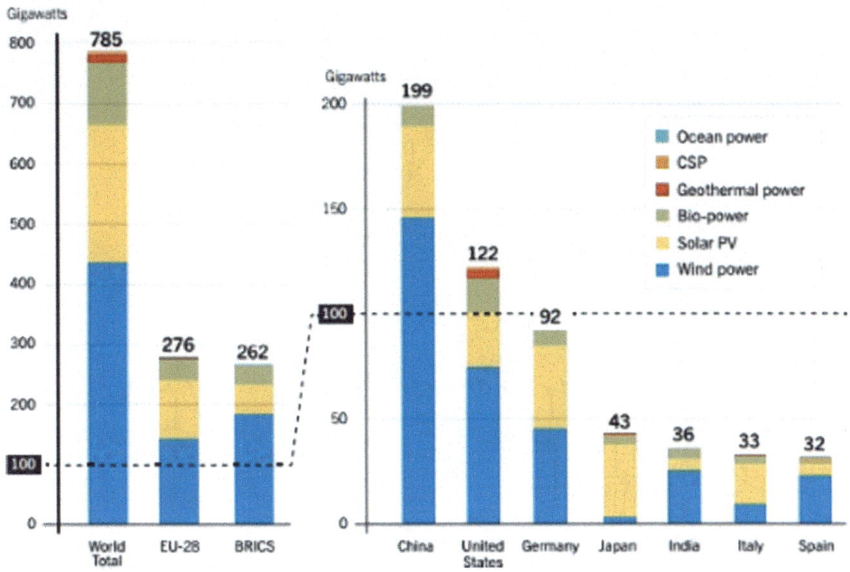

Fig. 3 Renewable power capacities in world, EU-28, BRICS and top seven countries end-2015

(5) Fkj,l: power stream out of the *k*th factor in imitation of the *l*th burden thru bus *j* linked in conformity with the *k*th generator.

In an aggregate concerning joining cases described previously, the law of Fig. 1 does keep expressed by way of the simplified tour shown in Fig. 4 together with the concerns over only power generations and assign consumptions. The branch in buses *i* then *j* of Fig. 4 may stay represented together with the simplified soloist circuits as much shown in Fig. 5. Then the amount of government loss on the entire systems may lie calculated through summing the losses about all branches whenever the DG is linked after other bus [2]. In others words, the system among Fig. 8 could

Fig. 4 Hydropower global capacity shares of top six countries and rest world, 2015

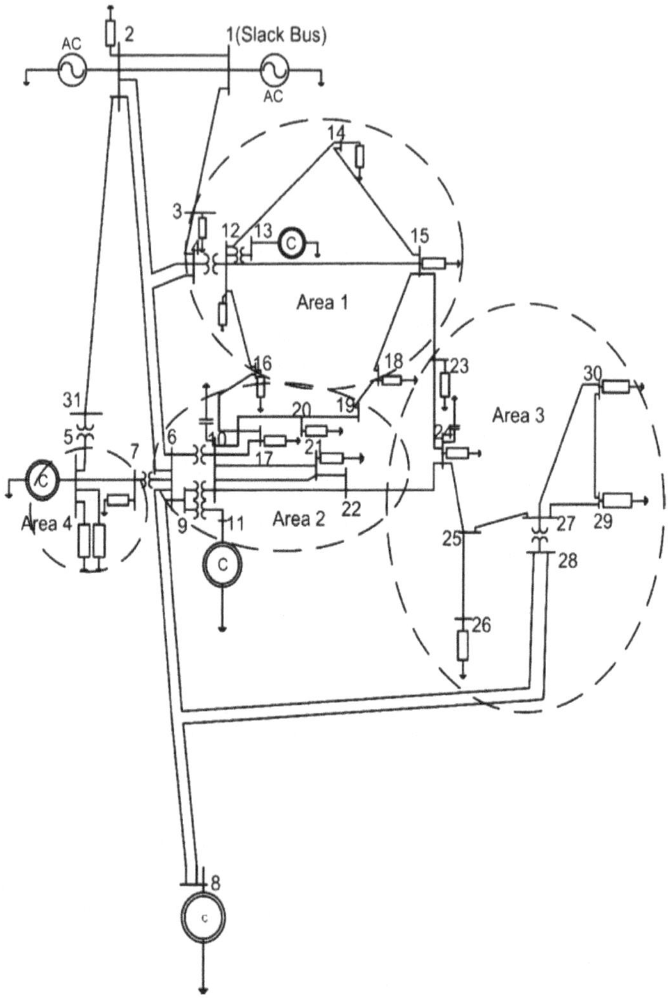

Fig. 5 IEEE benchmarked 31-bus system

be simplified as a circuit of Fig. 9 by focusing on the relationship among the DG and loss of power. In the equal manner, the whole powers system (Fig. 5) do also stand simplified.

In DAPSO2, if there were many particles far away from the global best position, then the velocities should be given a larger value. If there were many particles near the global best position, then the velocities should be given a smaller value. DAPSO1 only adjusts the velocity of the certain particle, but in DAPSO2, the velocities of all particles are adjusted together.

The general flow of DAPSOs and the flowchart of DAPSO are shown as follows.

Step 1. Initialization on a populace over particles together with random positions then velocities.
Step 2. Evaluation of particles.
Step 3. Calculate the distance from each particle to the global best position and save the farthest distance in the memory.
Step 4. Adjust particle's velocity according to its distance from itself to the global best position.
Step 5. Update particle's position by the adjusted velocity.
Step 6. Repeat Steps 2–5 until termination criteria are met.

3.1 Particle Swarm Optimization

Evolutionary computation offers realistic benefits to the researcher facing hard optimizations problems. These advantages are multifold, inclusive of the simplicity of the method, its robust responses to converting situation, its flexibility, and many different facets. The evolutionary set of rules can be implemented to issues in which heuristic answers are not available or commonly results in unsatisfactory consequences. As an end result, evolutionary algorithms have recently acquired improved interest, especially in regards to the manners wherein they will be implemented for practical hassle solving. Usually grouped beneath the time periods evolutionary computations or evolutionary algorithms, they locate the domain names of genetic algorithms [1], evolution strategies [6, 7], evolutionary programming [2], and genetic programming [3]. They all share common conceptual bases of simulating the evolution of persons systems via procedures of choice, mutation, and replica. The procedures rely upon the perceived performance of the individual structures as defined with the aid of the hassle. Compared to other worldwide optimization strategies, evolutionary algorithms (EA) are clean to implement and very regularly they provide ok answers. A population of candidate answers (for the optimization venture to be solved) is initialized. New answers are created via applying replica operators (mutation and/or crossover). The health (how true the answers are) of the resulting solutions is evaluated and appropriate choice approach is then carried out to decide which solutions are to be maintained into the following generation. The technique is then iterated.

For numerous troubles, a simple evolutionary algorithm might be accurate enough to locate the favored answer. As said in the literature, there are numerous styles of issues where an instantaneous evolutionary set of rules could fail to attain a convenient (most beneficial) solution [4, 5, 9, 10].

In Fig. 6, $x1$, $x2$, and $x3$ all have difference distance from itself to global best position. $X1$ drops within the radius of (0.5− ac)* FDd. The distance from $x2$ to global best position is between (0.5− ac)* FDd and (0.5 + ac)* FDd. $x3$ drops beyond the radius of (0.5 + ac)* FDd in DAPSO1. In DAPSOs, we define the "long distance" as the distance from the particle to the global best beyond (0.5 + ac)* FDd and the "short distance" as the distance from the particle to the global best is smaller

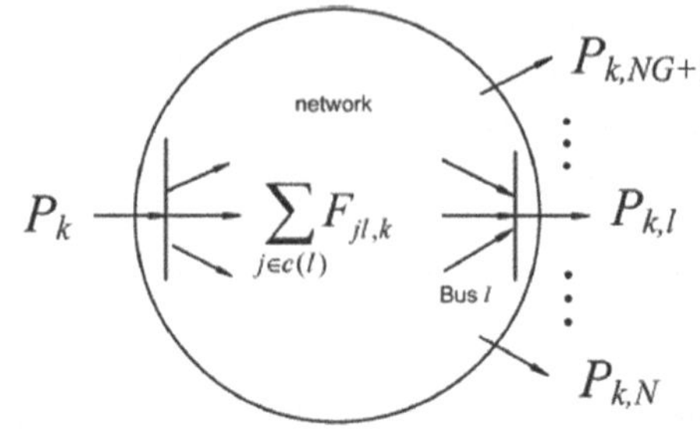

Fig. 6 Power flows from the kth generator to the several loads

than (0.5− ac)* FDd. In DAPSO1, the particles far away from the global best should be given larger value of velocity so it may explore an unknown region, whereas those close to the global best should be given smaller value of velocity so that it may exploit the neighborhood of the global best. In DAPSO2, if there were many particles far away from the global best position, then the velocities should be given a larger value. If there were many particles near from the global best position, then the velocities should be given a smaller value. DAPSO1 only adjusts the velocity of the certain particle, but in DAPSO2, the velocities of all particles are adjusted together.

The general flow of DAPSOs and the flowchart of DAPSO are shown as follows.

Step 1. Initializations of populations of particles with random positions and velocities
Step 2. Evaluation of particles.
Step 3. Calculate the distance from each particle to the global best position and save the farthest distance in the memory.
Step 4. Adjust particle's velocity according to its distance from itself to the global best position.
Step 5. Update particle's position by the adjusted velocity.
Step 6. Repeat Step.2 – Step.5 until termination criteria are met (Figs. 7 and 8).

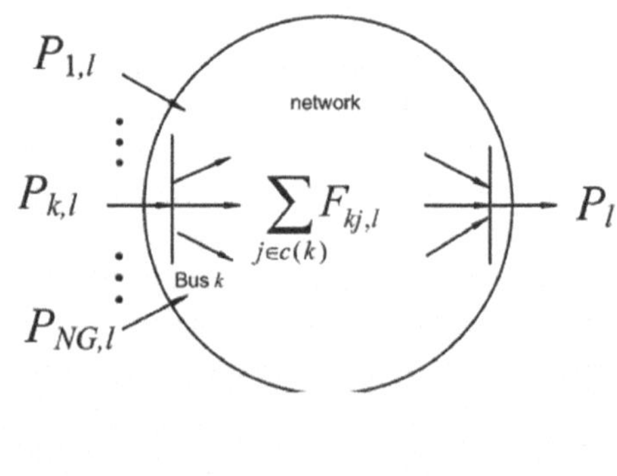

Fig. 7 Power drifts from several turbine to lth load

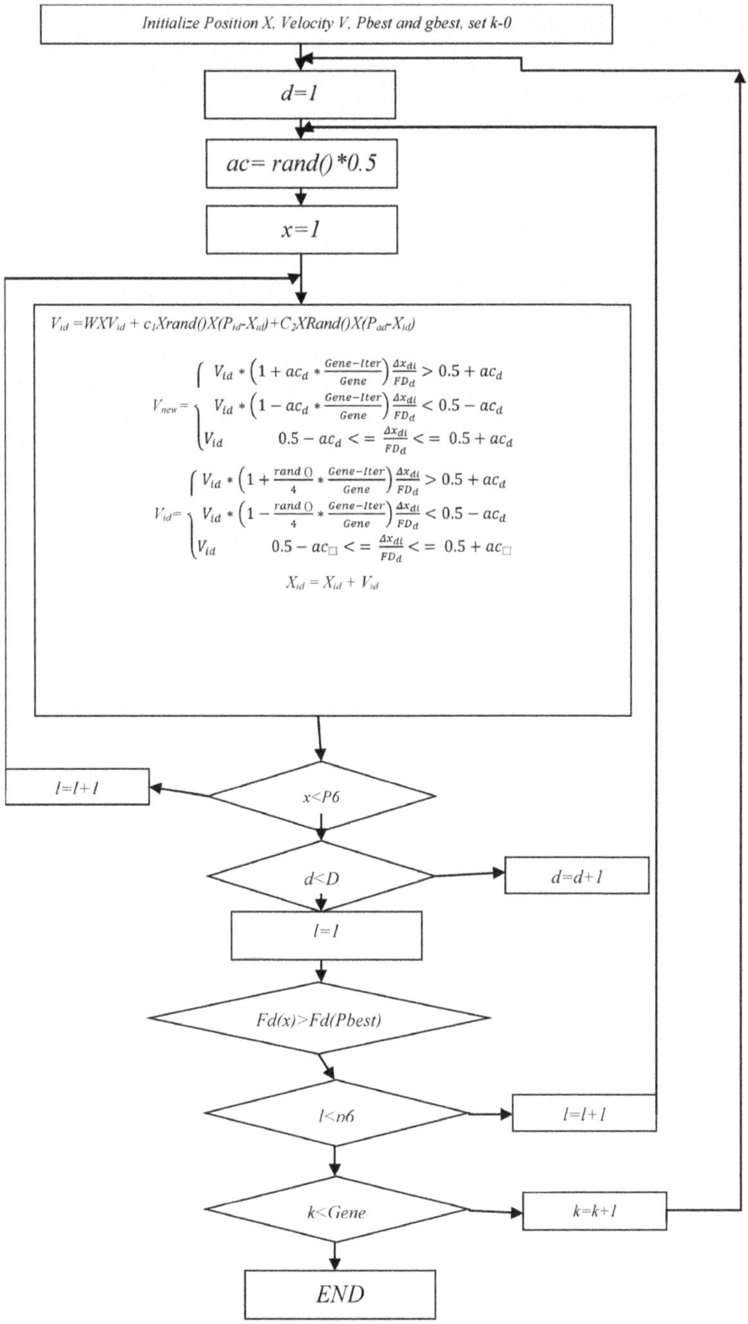

Fig. 8 Flowchart of DAPSO

4 Result and Discussion

Optimized parameter value:

Line no.:	66.00
Distance from line:	0.25
Voltage:	1.28 pu
Power:	15.00 MW

Minimum cost vs iteration number is presented in Fig. 9. Further, convergence plot of proposed algorithm is given in Fig. 10. Figure 11 shows various cost finally obtained by 10 particle-based optimized solution at the end of optimum search.

Figure 12a shows the parametric value of all the optimized 10 particle at respective id obtained on reaching the final round of optimization search in terms of DG location as per line id, its distance from to bus for line id, its voltage rating in pu, and its rated power in MW. Similarly, at different iteration and number of particles several times, the PSO optimization algorithm is run for several times and best optimized solutions with better cost and lower losses are considered. The comparative values with respect to 57 bus and optimized 58 bus with DG placement are shown in Fig. 13.

Fig. 9 Minimum cost versus iterations

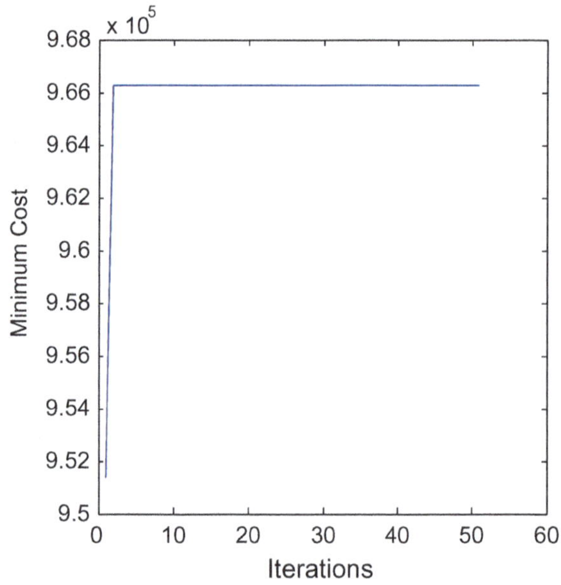

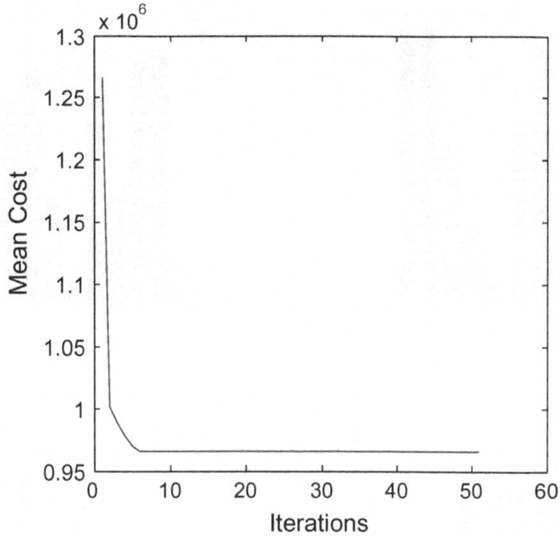

Fig. 10 Convergence plot

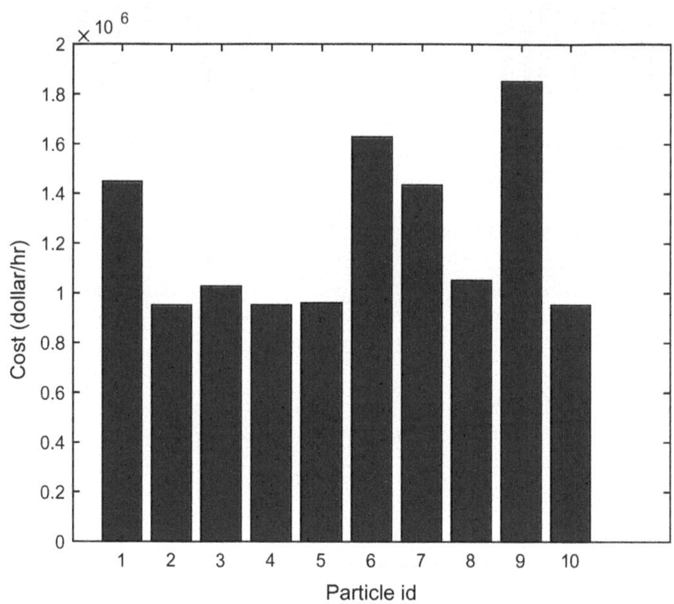

Fig. 11 Cost obtained at different particle id solution by PSO

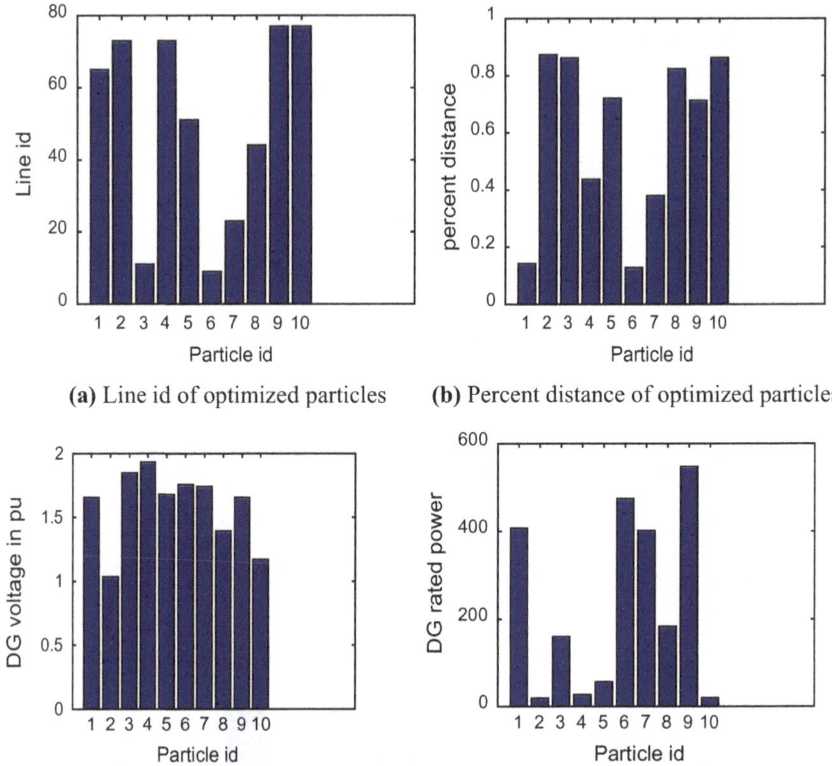

(a) Line id of optimized particles **(b)** Percent distance of optimized particles

(c) DG voltage of optimized particles **(d)** DG rated power of optimized particles

Fig. 12 Parametric value of all particles

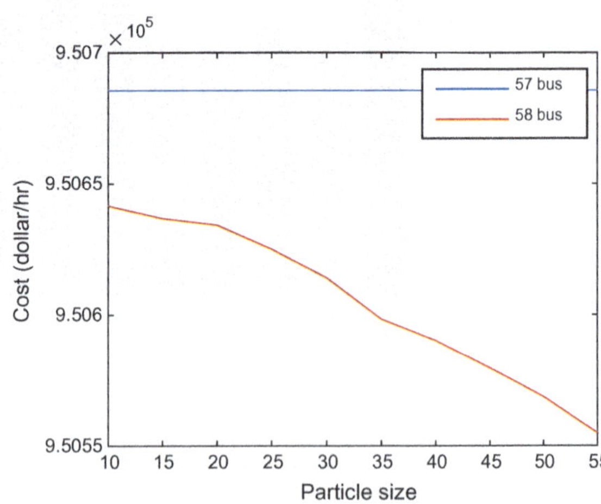

Fig. 13 Selected optimized result with better performance cost with respect to 57 (blue) and optimized 58 bus systems

5 Conclusions

The PSO based totally most advantageous DG parameters estimator is designed the uses of MATLAB programming algorithm. It has been proved that this optimization algorithm is quite effectives for swiftly finding the modifications required in insertion of extra DG's area, voltages, and electricity scores in both the simulated and real-global power machine eventualities considered. The new optimization method has validated a potential to no longer handiest rapidly find big changes to electricity machines modes, however has additionally been able to perceives the mode which has modified. Multisite measurements may be extensively utilized to provide greater self-belief insides the detection alarming. This has substantial implications for energy application intervention techniques. Importantly, the method is computationally greens and might easily be carried out in real time

References

1. Farhangi H (2010) The path of the smart grid. IEEE Power Energy Mag 8(1):18–28
2. Lee SH, Park JW (2009) Selection of optimal location and size of multiple distributed generations by using Kalman filter algorithm. IEEE Trans Power Syst 24(3):1393–1400
3. Chowdhury AA, Agarwal SK, Koval DO (2003) Reliability modeling of distributed generation in conventional distribution systems planning and analysis. IEEE Trans Ind Appl 39(5):1493–1498
4. Ipakchi A, Albuyeh F (2009) Grid of the future. IEEE Power Energy Mag 7(2):52–62
5. Statistics Korea (2010) Car Registration Status, Daejeon, South Korea (Online). Available: http://www.index.go.kr/egams/stts/jsp/potal/stts/PO_STTS_IdxMain.jsp?idx_cd=1257&bbs=INDX_001&clas_div=A
6. Korea Power Exchange (2010) Demand and Supply Result, Seoul, South Korea (Online). Available: http://www.kpx.or.kr/KOREAN/htdocs/main/sub/bidYesterdayList.jsp?cmd=&curpage=103
7. Hegazy YG, Salama MMA, Chikhani AY (2003) Adequacy assessment of distributed generation systems using Monte Carlo simulation. IEEE Trans Power Syst 18(1):48–52
8. Singh D, Misra RK, Singh D (2007) Effect of load models in distributed generation planning. IEEE Trans Power Syst 22(4):2204–2212
9. El-Khattam W, Hegazy YG, Salama MMA (2005) An integrated distributed generation optimization model for distribution system planning. IEEE Trans Power Syst 20(2):1158–1165
10. Wang C, Hehrir MH (2004) Analytical approaches for optimal placement of distributed generation sources in power systems. IEEE Trans Power Syst 19(4):2068–2076
11. Karki R, Billinton R (2001) Reliability/cost implications of PV and wind energy utilization in small isolated power systems. IEEE Trans Energy Convers 16(4):368–373
12. Teleke S, Baran ME, Bhattacharya S, Huang AQ (2010) Rule-based control of battery energy storage for dispatching intermittent renewable sources. IEEE Trans Sustain Energy 1(3):117–124
13. Kumar A, Gao W (2010) Optimal distributed generation location using mixed integer non-linear programming in hybrid electricity markets. IET Gen Transmiss Distrib 4(2):281–298
14. Carpinelli G, Celli G, Mocci S, Pilo F, Russo A (2005, July) Optimization of embedded generation sizing and siting by using a double trade-off method. Proc Inst Elect Eng—Gen Transm Distrib 152(4):503–513

15. Senjyu T, Miyazato Y, Yona A, Urasaki N, Funabashi T (2008, April) Optimal distribution voltage control and coordination with distributed generation. IEEE Trans Power Del 23(2):1236–1242
16. Saadat H (2004) Power system analysis. In: 2nd ed. McGraw-Hill, Singapore, pp 227–234
17. Chang YC, Lu CN (2002, July) Bus-oriented transmission loss allocation. Proc Inst Elect Eng—Gen Transmiss Distrib 149(4):402–406

Chapter 22
Application of Firefly Algorithm Optimized Fuzzy 2DOFPID Controller for Diverse-Sourced Multi-area LFC

More Raju, Upasana Sarma and Lalit Chandra Saikia

1 Introduction

It is known fact that the frequency is correlated with active power, and therefore, a mismatch between the power output of the generators and the load can lead to frequency deviations (Δf), and deviations from scheduled power interchange (ΔP_{tie}) between the interconnected areas in power system. The main objective of load frequency control (LFC), more commonly known as automatic generation control (AGC), is to diminish the deviations in Δf and ΔP_{tie}, therefore to ensure the operation of the system within suitable limits [1]. Also, today, as the complexity of power system has increased owing to the increased number of generating units, therefore such large power system requires enhanced notion of fault/load tolerance with improved LFC for each area. The design of secondary controllers (SCs) in LFC plays a major role to minimize Δf and ΔP_{tie} in an interconnected environment. Classical controllers (I, PI and PID), controllers with double derivative (DD) terms [2], fractional-order controllers, two or dual degree of freedom (2-DOF) [3], etc., have been applied in LFC of various types of electrical systems. Fuzzy logic control (FLC) forms another important field of research in this discipline. Authors have shown that implementation of such FLC greatly improves the operational performance of the controller without negatively affecting consumer's quality of supply [2]. Fuzzy logic-based classical controllers and integral-double derivative (IDD) controllers have already been implemented in the LFC systems [2, 4, 5]. But, the literature survey shows that combination of fuzzy logic and 2DOF controllers is not found in LFC so far. Hence, it requires further study.

M. Raju (✉)
Maturi Venkata Subba Rao (MVSR) Engineering College, Nadergul, Hyderabad 501510, India

U. Sarma
Indian Institute of Technology Guwahati, Guwahati 781039, India

L. C. Saikia
National Institute of Technology Silchar, Silchar 788010, India

V. K. Giri et al. (eds.), *Computing Algorithms with Applications in Engineering*, Algorithms for Intelligent Systems, https://doi.org/10.1007/978-981-15-2369-4_22

However, the main problem of LFC does not lie in the selection of secondary controller alone. Proper selection of the controller tuning parameters is vital, as improper optimum controller gains may lead to suboptimal solutions. Heuristic search techniques prove immensely useful in this regard. Firefly algorithm (FA) is one such optimization technique proposed by Yang [6]. Recent studies show that FA could outperform most of the other meta-heuristic search techniques, in dealing with stochastic test functions and noisy optimization problems. However, the FA optimized fuzzy logic controller is not evaluated in LFC so far. Thus, it needs further study.

The small hydropower plant (SHPP) produces hydropower in a limited scale that serves a smaller community or industry which is becoming more popular not only as a clean energy but also as an economic power plant. The power output of a SHPP varies due to reasons like climatic conditions, shortage of water etc., Therefore, the SHPP can be operated in island mode or in an interconnected mode [7]. In the field of SHPP application, LFC of isolated SHPP has been carried out by Dolla et al. [8], where the authors have projected two/three pipe systems in order to control the flow of water to turbine for dump load management. In [7], the authors have developed the equivalent transfer function for SHPP which is used in multi-area LFC studies. The diverse-sourced multi-area LFC along with SHPP is yet to be investigated.

The key contributions of the present manuscript are: to development of diverse-sourced multi-area LFC model considering SHPP; to apply FA-based F2DOFPID controller as SCs and to prove its supremacy over PID and 2DOFPID controllers for various system conditions.

2 Proposed System

The interconnected diverse-sourced multi-source-thermal-thermal system is considered for analysis (Fig. 1). The controllers F2DOFPID, 2DOFPID and PID are assumed as SCs. Area1 is a diverse-sourced control area in which SHPP, diesel, gas thermal and thermal generations are incorporated. The ratio 1: 3: 5 is assumed as the power generation ratio for area1, area2 and area3, respectively. Area participation factor of 15% (0.15) is taken for the combination of SHPP, gas and diesel generations. For thermal generations, the constraints for generation are assumed as 3%/min with reheating feature. From [7, 9, 10], the system nominal values are taken. Using FA, integral squared error (ISE) technique (1), the SC's parameters are optimized.

$$J = \int_0^T (\Delta f_1^2 + \Delta f_2^2 + \Delta P_{\text{tie}12}^2)\mathrm{d}t \tag{1}$$

For this study, FA parameters are tuned as: fireflies (n) = 100, iterations (S) = 200, randomization parameter (α) = 0.5, attractiveness (β) = 0.2 and absorption coefficient (γ) = 0.5. These values are obtained by a number of trial runs for different values of these parameters and finally obtained corresponding to the minimum values of cost function.

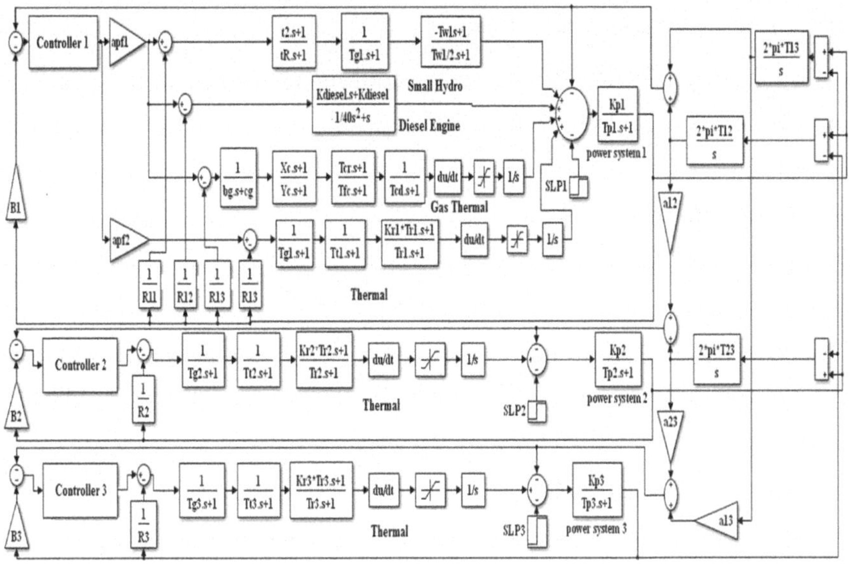

Fig. 1 Three-area diverse-sourced multi-area system with secondary controller

3 Fuzzy Two-Degree-of-Freedom PID (F2DOFPID) Controller

Researchers have already shown the advantage of 2DOF controllers over 1DOF ones, in terms of its flexibility in set-point tracking as well as regulation in case of disturbed inputs [3]. The combination of fuzzy and 2DOF controllers gives an edge over simple 2DOF structure, as it improves the reliability further in dealing large power system disturbances and nonlinearity. In the present manuscript, fuzzy-based 2DOFPID is considered for studies.

The fuzzy 2DOFPID controller consists of the fuzzy logic controller (FLC) along with the conventional 2DOFPID controller. The FLC characterizes by its four components, (a) fuzzifier, (b) the inference engine, (c) the rule base and (d) the defuzzifier. The schematic block diagram to describe the operation of FLC scheme is given in Fig. 2. The area control error (ACE) (2) and its rate of change ($\dot{ACE}$) define the inputs to FLC.

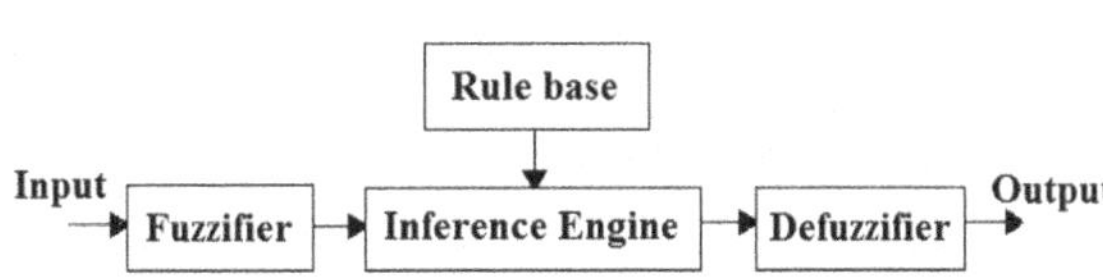

Fig. 2 Block diagram of the fuzzy control system

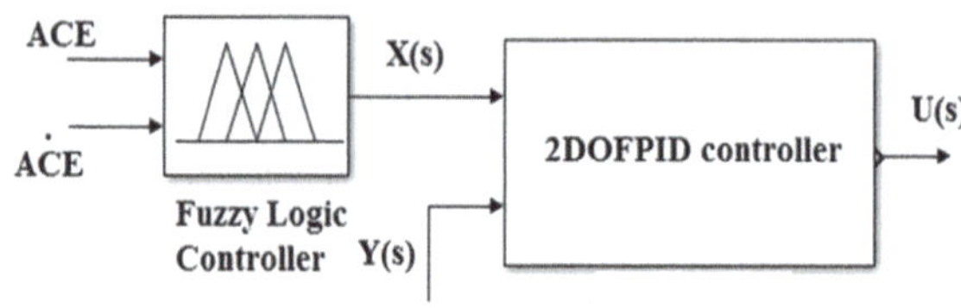

Fig. 3 Structure of the F2DOFPID controller

$$\mathrm{ACE} = \Delta P_{\mathrm{tie}} + B\Delta f \tag{2}$$

The power system output $Y(s)$ and FLC output $X(s)$ act as inputs for 2DOFPID controller. The output of F2DOFPID controller (Fig. 3) acts as input to the plant (4).

$$U(s) = [K_P\{b\,.\,X(s) - Y(s)\} + K_I\frac{1}{s}\{X(s) - Y(s)\} + K_{\mathrm{DD}}\frac{Ns}{s+N}\{c\,.\,X(s) - Y(s)\}] \tag{3}$$

For FLC, seven sets of Gaussian membership functions (MFs) are taken for input and output. The FA is used to decide MFs centre, width, set-point weights gains and filter coefficients of controllers. For instant, one of MFs (input) is shown in Fig. 4.

The control rules are adopted from [5] and are shown in Table 1. The terms N-L, N-M, N-S, Z-E, P-S, P-M and P-L stands for negative-large, negative-medium, negative-small, zero-error, positive-small, positive-medium and positive-large, respectively. Mamdani fuzzy inference system (FIS), min as AND method, max as OR method and centroid method for defuzzification [2] are utilized.

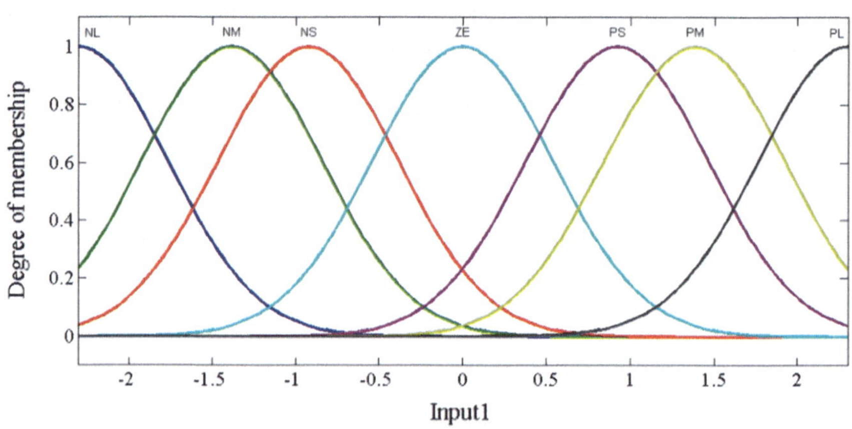

Fig. 4 Membership functions for input of F2DOFIDD controller

Table 1 Control rule for F2DOFPID controller

$\dot{\text{ACE}}$	ACE						
	N-L	N-L	N-L	Z-E	P-S	P-M	P-L
N-L	P-L	P-L	P-L	P-M	P-M	P-S	Z-E
N-L	P-L	P-L	P-L	P-L	P-M	P-S	N-S
N-L	P-L	P-L	P-L	P-M	Z-E	N-S	N-M
Z-E	P-L	P-L	P-L	Z-E	N-S	N-M	N-L
P-S	P-M	P-S	Z-E	N-S	N-S	N-M	N-L
P-M	P-S	Z-E	N-S	N-M	N-M	N-M	N-L
P-L	Z-E	N-S	N-M	N-M	N-L	N-L	N-L

4 Results and Analysis

4.1 System Dynamics When F2DOFPID, 2DOFPID and PID Controllers Acted as SCs

The system dynamics using PID, 2DOFPID and F2DOFPID controllers as SCs are investigated at 50% loading condition, inertia constant $H = 5$s and 1% SLP applied to area1. The controller gains, filter coefficients, and centre of Gaussian membership function (in case of F2DOFPID) are optimized using FA. The retrieved optimum values of the controller and other parameters are as follows: for PID controller, $K_{P1}^{+} = 0.27578$, $K_{P2}^{+} = 0.20655$, $K_{P3}^{+} = 0.126426$, $K_{I1}^{+} = 0.379$, $K_{I2}^{+} = 0.27888$, $K_{I3}^{+} = 0.19101$, $K_{D1}^{+} = 0.19925$, $K_{D2}^{+} = 0.26088$, $K_{D3}^{+} = 0.46625$, $N_{1}^{+} = 0.000017$, $N_{2}^{+} = 0.00011$, $N_{3}^{+} = 0.00001$, for 2DOFPID the values are $K_{P1}^{+} = 0.87578$, $K_{P2}^{+} = 0.20655$, $K_{P3}^{+} = 0.226426$, $K_{I1}^{+} = 0.329$, $K_{I2}^{+} = 0.228888$, $K_{I3}^{+} = 0.239101$, $K_{D1}^{+} = 0.19925$, $K_{D2}^{+} = 0.26088$, $K_{D3}^{+} = 0.46625$, $N_{1}^{+} = 0.000017$, $N_{2}^{+} = 0.00011$, $N_{3}^{+} = 0.00001$, $b_{1}^{+} = 0.9977$, $b_{2}^{+} = 0.7800$, $b_{3}^{+} = 0.9089$, $c_{1}^{+} = 0.4985$, $c_{2}^{+} = 0.4230$, $c_{3}^{+} = 0.4328$, and for F2DOFPID controller, the values are $K_{P1}^{*} = 2.2299$, $K_{P2}^{*} = 0.2001$, $K_{P3}^{+} = 0.1920$, $K_{I1}^{+} = 2.1999$, $K_{I2}^{+} = 0.0107$, $K_{I3}^{+} = 0.3011$, $K_{D1}^{+} = 0.0032$, $K_{D2}^{+} = 0.9414$, $K_{D3}^{+} = 0.6802$, $N_{1}^{+} = 0.9990$, $N_{2}^{+} = 0.5160$, $N_{3}^{+} = 0.4023$, $b_{1}^{+} = 0.2239$, $b_{2}^{+} = 0.5448$, $b_{3}^{+} = 0.9973$, $c_{1}^{+} = 0.3931$, $c_{2}^{+} = 0.6941$, $c_{3}^{+} = 0.2590$. Corresponding to these optimum values, the system responses Δf and ΔP_{tie} are obtained for each controller and shown in Fig. 5. Critical observation of the responses proves the superiority of F2DOFPID controller over the PID and 2DOFPID controllers in view of settling time and minimum deviations.

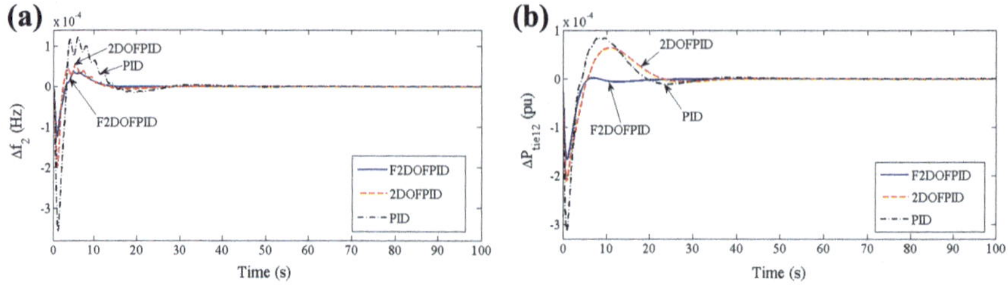

Fig. 5 System dynamic responses for PID, 2DOFPID and F2DOFPID controllers **a** Δf and **b** ΔP_{tie} versus time

Table 2 Cost function for 2DOFPID (J_X) and F2DOFPID (J_Y) controllers in case of various step load perturbations

SLP (%)	$J_X(\times 10^{-7})$ (F2DOFPID)	$J_Y(\times 10^{-7})$ (2DOFPID)	$\frac{J_X - J_Y}{J_y} * 100$ (%)
1	3.6	8.3	56
2	15.1	33.85	53
3	34.3	76.7	55
4	61.3	13.7	55
5	95.9	21.4	55

4.2 *Performance of 2DOFPID and F2DOFPID Controllers for Wide Changes in System Conditions*

In this subsection, the system oscillations for conventional 2DOFPID and F2DOFPID controllers are assesses for changes in SLP up to 5% in steps of 1%. In each of the changed condition, the controller gains as well as other parameters are fixed to that of obtained at nominal system condition. The cost values for two controllers are shown in Table 2. Observation of cost functions reveals that the F2DOFPID provides lesser cost function than 2DOFPID, which shows the superiority of F2DOFPID controller. The proposed F2DOFPID performs well even in the event of larger perturbations.

5 Conclusion

A successful attempt has been made to apply FA-based F2DOFPID controller for diverse-sourced multi-area LFC study. The SHPP is considered in area1 of the system. Simulation results carried out at various cases of SLP show that the F2DOFPID controller exhibits enhanced performance as compared to the others, in view of minimized settling time, reduced deviations and oscillations. The F2DOFPID controller also provides smaller cost function values. The performance of various algorithms such as symbiotic organisms search (SOS) [11, 12], sine cosine algorithm (SCA)

[13], teaching–learning-based optimization (TLBO) [14, 15], grey wolf optimization (GWO) [16] and Jaya optimization [17] can be implemented in the proposed system in future studies.

References

1. Singh SP, Prakash T, Singh VP, Babu MG (2017) Analytic hierarchy process based automatic generation control of multi-area interconnected power system using Jaya algorithm. Eng Appl Artif Intell 60(3):35–44
2. Saikia LC, Sinha N (2010) Maiden application of fuzzy logic based IDD controller for automatic generation control of multi-area hydrothermal system: a preliminary study. In: Universities power engineering conference (AUPEC), pp 1–6
3. Dash P, Saikia LC, Sinha N (2014) Comparison of performances of several Cuckoo search algorithm based 2DOF controllers in AGC of multi-area thermal system. Int J Electr Power Energy Syst 55:429–436
4. Chown GA, Hartman RC (1998) Design and experience of Fuzzy logic controller for automatic generation control. IEEE Trans Power Syst 13(3):965–970
5. Pothiya S, Ngamroo I, Runggeratigul S, Tantaswadi P (2006) Design of optimal fuzzy logic based PI controller using multiple tabu search algorithm for load frequency control. Int J Control Autom Syst 4(2):155–164
6. Yang XS (2008) Nature-inspired meta-heuristic algorithms. Luniver Press, Beckington, UK
7. Fan RX, Zhao J, Pan B, Chen N, Wang T, Ma H ((2014)) Automatic generation control of three-area small hydro system based on fuzzy PID control. In: International conference on power system technology (POWERCON), pp 2522–2528
8. Doolla S, Bhatti TS (2006) Load frequency control of an isolated small hydro power plant with reduced dump load. IEEE Trans Power Syst 21(4):1912–1919
9. Saikia LC, Chowdhury A, Shakya N, Shukla S, Soni PK (2013) AGC of a multi area gas-thermal system using firefly optimized IDF controller. In: Annual IEEE India conference (INDICON), 1–6
10. Bhatt P, Roy R, Ghoshal SP (2010) GA/particle swarm intelligence based optimization of two specific varieties of controller devices applied to two-area multi-units automatic generation control. Int J Electr Power Energy Syst 32(4):299–310
11. Rathore NS, Singh VP, Phuc BDH (2019) A modified controller design based on symbiotic organisms search optimization for desalination system. J Water Supply Res Technol AQUA 68(5):337–345
12. Singh SP, Prakash T, Singh VP (2019) Coordinated tuning of controller-parameters using symbiotic organisms search algorithm for frequency regulation of multi-area wind integrated power system. Eng Sci Technol Int J (in press)
13. Prakash T, Singh SP, Singh VP (2019) Analytic hierarchy process based model reduction of higher order continuous systems using sine cosine algorithm. Int J Syst Control Commun (in press)
14. Singh SP, Singh V, Singh VP (2019) Analytic hierarchy process based approximation of high-order continuous systems using TLBO algorithm. Int J Dynamics Control 7(1):53–60
15. Prakash T, Singh VP, Singh SP, Mohanty SR (2018) Economic load dispatch problem: quasi-oppositional self-learning TLBO algorithm. Energy Syst 9(2):415–438
16. Rathore NS, Singh VP, Kumar B (2018) Controller design for DOHA water treatment plant using grey wolf optimization. J Intell Fuzzy Syst 35(5):5329–5336
17. Prakash T, Singh VP, Mohanty SR, Singh SP (2017) Binary Jaya algorithm based optimal placement of phasor measurement units for power system observabilty. Int J Control Theory and Appl 10(5):515–523

Chapter 23
Induction Motor Bearing Fault Classification Using PCA and ANN

R. K. Patel, S. Agrawal and V. K. Giri

1 Introduction

The recording of vibration signal is an indispensable element for bearing fault diagnosis in an induction motor whose analysis further leads to identify the fault existence along with its location, fault progressive nature and its severity [1]. As the bearing become faulty from healthy condition, it reflects changes in the vibration signal which results variation in statistical parameters and its spectral characteristics. This variation in signal and its characteristic is an important indication for fault identification in bearing. Therefore, in order to identify the health of induction motor, different techniques are adopted which are capable to identify the variation in characteristics of the signal. The signals which are recorded from induction motor are often nonstationary in nature due to time-varying nature of machine operation. The fast Fourier transform (FFT) which is a frequency domain analysis technique is applicable only to stationary signals, so it is unable to identify the time information in relation to the changes in frequency in the signal. The acoustic signals and vibration signals of the machine are nonstationary in nature which is required to be analyzed for induction motor health monitoring. In the present work, the faults related to the induction motor bearing are categorized in four classes together with its location, i.e., healthy, inner raceway defect, rolling ball defect and outer raceway defect.

The extraction of features from bearing vibration signal is an active research subject for induction motor condition monitoring [2–4]. The aim is to obtain prominent features from the recorded signal which leads to fault detection correctly. This goal is achieved by collecting information in time, frequency and time–frequency domain which perceives real-time behavior of the machine in various operating conditions. As vibration signals are often acclaimed to explore the bearing faults, so a great deal

R. K. Patel (✉) · V. K. Giri
Rajkiya Engineering College, Sonbhdra 231206, India

S. Agrawal · V. K. Giri
Madan Mohan Malaviya University of Technology, Gorakhpur, Uttar Pradesh 273010, India

V. K. Giri et al. (eds.), *Computing Algorithms with Applications in Engineering*, Algorithms for Intelligent Systems, https://doi.org/10.1007/978-981-15-2369-4_23

of database is required for the analysis of condition monitoring of the machine [5, 6]. But, the fault detection is very well inclined to false positive rate due to large interrelationship of the data set, so correct identification of most characteristic features is required that would lead to assist in diagnosis of fault.

2 Feature Extraction and Selection Methods

In this present work, the induction motor bearing vibration signal data has been taken from the bearing data center of Case Western Reserve University (CWRU) laboratory, USA. The recording is done to frame the data set related to bearing installed at motor-driven system. A three-phase, 2 HP induction motor is attached with a dynamometer adaptively coupled with torque sensor. The controlling of dynamometer is done in order to achieve the required torque load. An accelerometer having bandwidth of 5000 Hz was fixed on the housing of motor at the driving end to gather the information of vibration signals of bearing. The data assemblage system is comprised of a high bandwidth amplifier especially intended for the vibration signal recording having a sampling frequency of 12 kHz.

From the literature, it is found that the task of classification does not perform accurately on the data obtained through sensors since it requires feature variables. The construction of features is the process where the signals are manipulated in such a way it produces a set of characteristic features of the signal. The features can be selected in such a way that is able to discriminate the data into classes or groups. Different computed features extracted from the data may distinctively vary differently for different systems. The aspect of extraction of features is to obtain relevant and covenant facts of the original one which can be efficiently applied to identify an object with assessment value as compared to objects of the same class and distinct for objects in distinct class. Hence, extracted features' quality is going to decide the success rate of identification of class. Therefore, extraction of features leads to an idea of searching distinct attributes of data that are unvarying to inappropriate transformation of the input. In the present work, the statistical features are calculated as Table 1, i.e., (F1–F11) [7, 8]. The problem of classification after analysis is always challenging to obtain superior accuracy from the available data. So, to reduce the intricacy, time and to improve the precision, the feature selection is critically important. The feature selection is the process of searching the meaningful subset of features automatically from all features. For evaluation by any classification algorithms, it is necessary to remove irrelevant features from the feature set just to reduce over fitting so that accuracy is improved as well as reduce the training time [9].

Table 1 Features extracted from data in time domain

Feature number	Features extracted	Measures
F1	Mean	Measure of central tendency
F2	Median	
F3	Mode	
F4	Standard deviation	Measure of variability
F5	Variance	
F6	Range	
F7	Kurtosis	Measures of dispersion
F8	Skewness	
F9	Crest factor	Dimensionless parameter
F10	Impulse factor	
F11	Clearance factor	

3 Principal Component Analysis (PCA)

PCA is an analytical procedure which converts the set of observation into a set of values of unrelated variables from linear correlated variables by using an orthogonal transformation [10, 11]. This is performed by using eigenvalue analysis of the correlation matrix. Depending on eigenvalues, the related eigenvectors are principal components. The largest eigenvalues will be the first component, and the smallest is the last one, as eigenvalues are associated with components in descending order. Mathematical depiction of a random variable is the basis of PCA, and the population of random vector can be represented by the following equations:

To obtain the principal component, the statistical information of a random variable is:

$$x = (x_1, \ldots, x_n)^{\mathrm{T}} \tag{2}$$

The mean (μ_x) and the covariance matrix $\left(C_{ij}\right)$ are given by

$$\mu_x = \text{mean}(x_n) \tag{3}$$

$$C_{ij} = \frac{1}{N}\sum_{n=1}^{N}(x_n - \mu_x)(x_n - \mu_x)^{\mathrm{T}} \tag{4}$$

where C_{ij} represent the covariance between the random variable component and x_i, i.e., variance of x_i the component. The covariance matrices obtained as:

$$C_x e_i = \lambda_i e_i, \quad i = 1, \ldots, n \tag{5}$$

where e_i are the eigenvector and λ_i is the corresponding eigenvalues. The first eigenvalue has the largest variance in the data, i.e., energy amount is most significant, since it is sorted in descending order. The matrix (A) having the eigenvectors of the covariance matrix represented as the row vectors and data vector (x) can be transformed by:

$$y = A(x - \mu_x) \tag{6}$$

This point in the perpendicular coordinate system is defined as eigenvectors. Components of y are the coordinates in the orthogonal base. The original data vector x can be reconstructed from y using Eq. (6):

$$x = A^T y + \mu_x \tag{7}$$

The primary vectors x, were planned on the organize axis defined by the principal component, are only few basic vectors of orthogonal basis, besides eigenvectors of the covariance matrix. The matrix which has the first K eigenvectors as rows is defined by A_K, and a same conversion can be made as given in Eq. (7):

$$\begin{aligned} y' &= A_K(x - \mu_x) \\ x' &= A_K^T y' + \mu \end{aligned} \tag{8}$$

On the coordinate axes, the original data vectors are projected with dimensions K, and then, the basic vectors are linearly combined to transform it back to vector. Due to this, the mean square error is minimized between data. Now, the principal component is represented as a linear arrangement of the original variables. PCA reveals combinations of the primary variables describing the dominant patterns and the main trends of the data. This is performed by decomposing eigenvector of the covariance matrix of the primary variables. The abstracted potential variables are orthogonal and are being sorted as per their eigenvalues. The percentage of variance (PV) for the principal components can be considered as:

$$\mathrm{PV} = \frac{\lambda_k}{\lambda_1 + \lambda_2 + \cdots + \lambda_n} \times 100\% \tag{9}$$

PCA is an effective approach to reduce the dimension by selecting that statistical pattern whose percentage variance are maximum, and its associated component magnitudes are analyzed only. These components statistical features variable are deliberated as significant features. In the present work, the variance obtained in percentage by different principal components for scheme 1 and scheme 2 are represented in Fig. 1.

The first two PCs (principal components 1 and 2) for scheme 1 accounted as 83% (PC1 = 74% and PC2 = 9%) and for scheme 2 accounted as 78% (PC1 = 53% AND PC2 = 25%) [11]. The relationship between the bearing data and its eleven

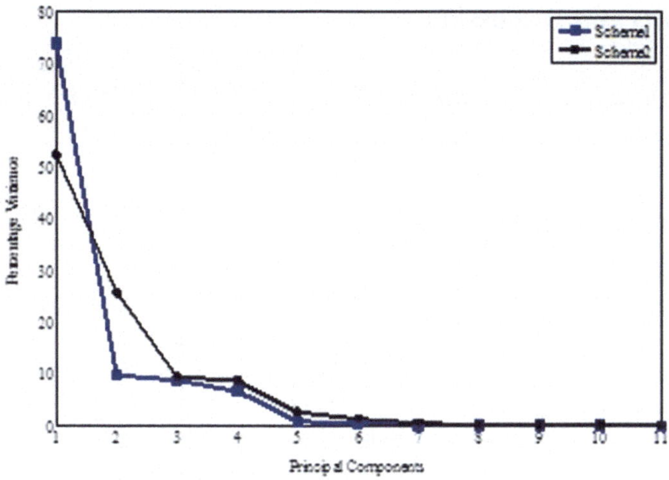

Fig. 1 Percentage of variance obtained by various principal components

statistical features has been analyzed by principal component analyzer using bi-plot. Bi-plot analysis is employed to study the correlation between the statistical features and the recorded vibration signal features for the proposed schemes. Bi-plot of PCA is shown in Fig. 2a for scheme 1 and in Fig. 2b for scheme 2. In Fig. 2a, it is clear that F4, F5, F7, F8 are highly correlated, and its variability across four classes (H, IRD, BD, ORD) is accounted mainly by the principal component. Features F4, F5, F7, F8 are grouped on the positive principal component 1 axis of the bi-plot, suggesting strong relationships among them, for scheme 1. Therefore, this will be significant features which are able to discriminate the type of fault effectively. Similarly, in Fig. 2b, features F4, F9, F10, F11 are found significant features for scheme 2.

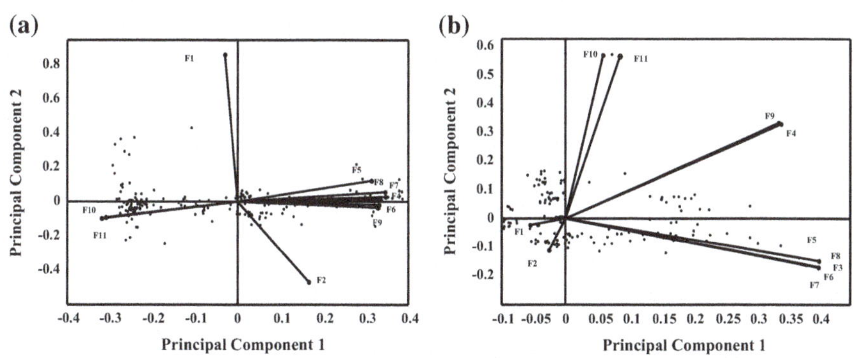

Fig. 2 PCA bi-plot summarizing the relationships among variables of the vibration signal for **a** scheme 1, **b** scheme 2

4 Results and Discussion

4.1 Training, Testing and Validation of ANN with Statistical Parameters on Experimental Data for Scheme 1

The present thesis aims to develop a method of bearing condition estimation using ANN [12, 13]. It is possible to classify the signals by using the characteristic vector as input. The input–output relationship of each node is determined by a set of connection weight, a threshold parameter and a node activation function. The activation function most commonly used in neural networks is sigmoid function. The characteristic vector developed affirms the number of nodes in the input layer. The decision of the number of nodes in the hidden layer is done by using trial and error method. Beginning with five, the number of hidden layer is increased by five in each trial. In this thesis, an example of a feedforward multi-layer perceptron (MLP), consisting of three layers, has been considered. The network is trained using a scale conjugate gradient backpropagation (SCGBP) [14, 15] training algorithm. SCG is a supervised learning algorithm for feedforward neural networks and is a member of the class of conjugate gradient methods. The minimization of the mean square error (MSE) between the real network output and the corresponding target in the training set is one of the most widely used performance criteria. A backpropagation is performed for the MSE because of which the network classification performance improves.

In the present work, the 70% data is used for training and 15% for validation and the rest 15% has been used for testing purpose. A MSE of 10E−5, a minimum gradient of 10E−6 and maximum iteration (epochs) of 1000 have been used. This training process is finished if any of these conditions are achieved. The initial weights and biases of the network have been generated automatically by the program. The four output nodes are exactly same in number of bearing conditions that is required to be classified. These four-digit output target nodes required to be mapped by the ANN are distinguished as: (1, 0, 0, 0) for a healthy bearing (H), (0, 1, 0, 0) for a bearing with inner race defect (IRD), (0, 0, 1, 0) for a rolling ball defect fault (RBD) and (0, 0, 0, 1) for an outer race defect (ORD). As explained in Sect. 4.2 and Fig. 1, the bearing vibration data has been divided into two schemes, i.e., scheme 1 and scheme 2. All the eleven extracted features, as explained in Table 1, are used to train the neural network for the data of scheme 1 and scheme 2. First, the analysis is carried out by taking 11 parameters for training with different numbers of neurons in hidden layer.

For the training and testing of neural network, four cases are considered. Each case consists of normal and faulty condition. Table 1 contains all eleven features which are calculated from the prepared data set. A matrix of 200 × 11 features is used for classification, to identify four cases related to identification of bearing condition. Table 2 shows the details of data which is used for the classification for scheme 1 and scheme 2 both. From each data set, statistical features have been calculated. These

Table 2 Data preparation for the classification

Defect type	Load (HP)	Fault size (in.)	Total data set
Normal (N)	0, 1, 2, 3	–	20
Inner race defect (IRD)	0, 1, 2, 3	0.007, 0.014, 0.021	60
Ball defect (BD)	0, 1, 2, 3	0.007, 0.014, 0.021	60
Outer race defect (ORD)	0, 1, 2, 3	0.007, 0.014, 0.021	60

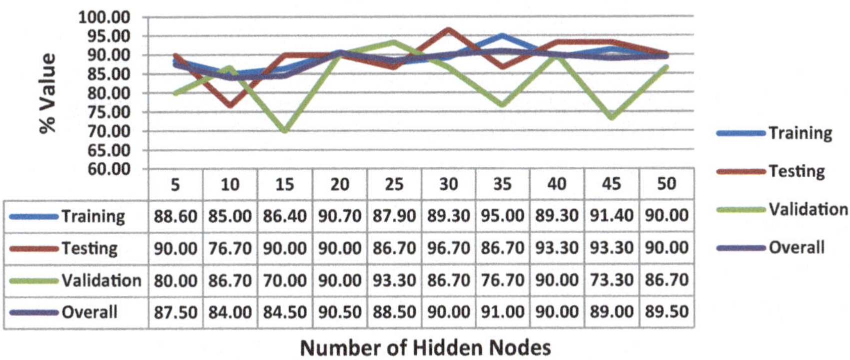

	5	10	15	20	25	30	35	40	45	50
Training	88.60	85.00	86.40	90.70	87.90	89.30	95.00	89.30	91.40	90.00
Testing	90.00	76.70	90.00	90.00	86.70	96.70	86.70	93.30	93.30	90.00
Validation	80.00	86.70	70.00	90.00	93.30	86.70	76.70	90.00	73.30	86.70
Overall	87.50	84.00	84.50	90.50	88.50	90.00	91.00	90.00	89.00	89.50

Fig. 3 Performance of neural network with different numbers of neuron in hidden layers with all features as an input

cases are combination of level of faults (0.007, 0.014 and 0.021 inch) and loading (0 HP, 1 HP, 2 HP and 3 HP) of the machine because both affect the performance of the machine.

Figure 3 shows the percentage value of training, testing, validation and overall success of the neural network performance for scheme 1. For discrimination, the quantity of neurons in the hidden layer is selected as 5, 10, 15, 20, 25, …, 0.50. It has been observed that the lowest training rate obtained is 85.00% for 10 neurons in hidden layers, and the highest rate is 95.00% for 35 neurons in hidden layers. For testing, the lowest and highest rates obtained are 76.70%, 96.7% for 10 and 30 neurons in hidden layers, respectively. For validation, the lowest and highest rates obtained are 70 and 93.3% for 15 and 25 neurons in the hidden layers, respectively. The best overall performance with 91.00% accuracy is observed with 35 neurons in numbers.

Figure 4 shows the accuracy achieved in percentage (%) by confusion matrix for training, testing, validation and overall performance with 35 numbers of neurons in hidden layers, having 11 input parameters of scheme 1. Figure 5 shows the best validation performance curve of the ANN for fault discrimination with 11 parameters, 35 neurons in hidden layer and 28 epochs.

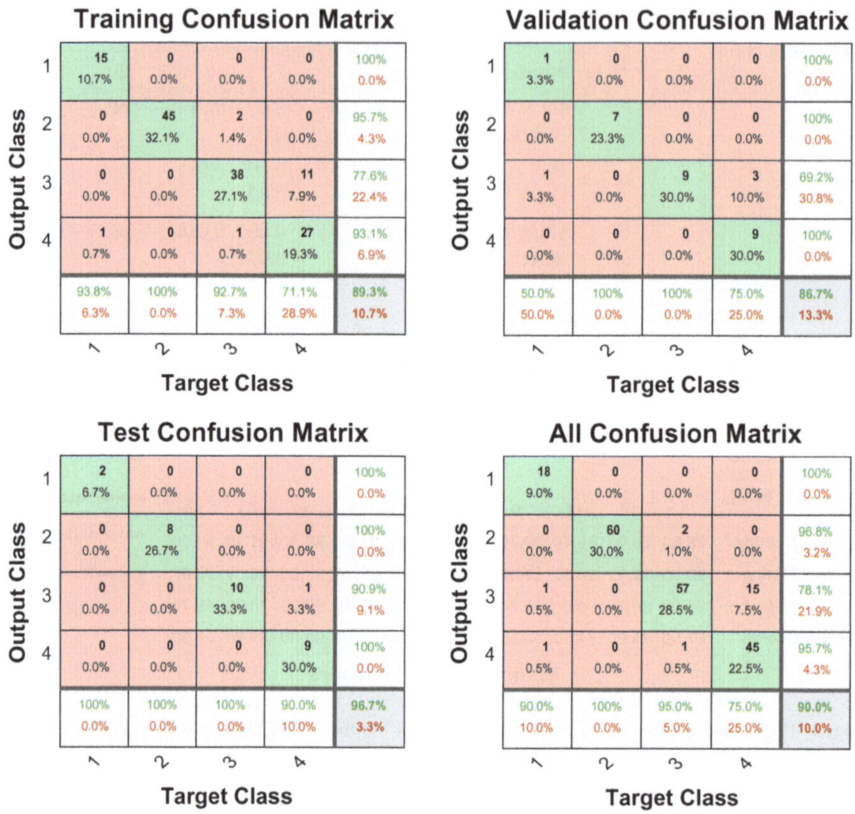

Fig. 4 Confusion matrix for the scheme 1 with all features as an input

4.2 Training and Testing of ANN with Statistical Parameters on Experimental Data for Scheme 2

As in case of scheme 1, the ANN is trained with different numbers of input parameters and varying numbers of hidden neurons. Figure 6 shows the confusion matrix including training and validation and testing for 11 statistical input parameters with 5 neurons in hidden layers in neural network for scheme 2. The NN has been trained with various neurons in hidden layers, i.e., 5, 10, 15, 20, …, 50.

It has been observed that for all number of neuron, 100% performance has been achieved for training, testing and validation, that is, better than the scheme 1 data. Figure 7 shows the best validation performance curve of the ANN for fault discrimination. The number of parameters is 11, and the number of neurons in hidden layer is 5, and it is achieved on 84 epochs. The error in performance obtained for scheme 2 is lower as compared with the performance achieved in case of scheme 1 data.

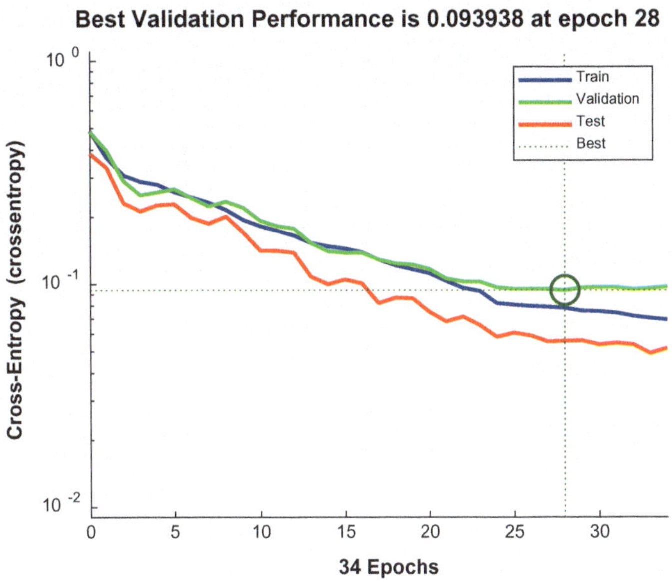

Fig. 5 Validation performance curve for scheme 1 with all features as an input

4.3 Training and Testing of ANN with Reduced Statistical Parameters on Experimental Data of Scheme 1

In the previous section, all eleven statistical features were considered as input parameters for training, testing and validation for scheme 1 and scheme 2; now, a features reduction technique known a PCA is adopted. The details of PCA are described in Sect. 3. Figure 2b shows that features F4, F5, F7, F8 are relevant for which is further used for classification purpose for scheme 1. So, the ANN is trained with this limited number of input parameters and performance recorded with various numbers of neurons in hidden layer, i.e., 5, 10, 15, …, 0.50.

Figure 8 represents the obtained percentage value of training, testing, validation and overall success of the neural network performance for various hidden layer networks. From the figure, it has been observed that the lowest training rate obtained is 75.7% for 15 neurons in hidden layer, and the highest rate is 81.4% for 40 neurons. In case of testing, the lowest and highest rates are 66.7 and 90% for 45 and 50 neurons in hidden layer, respectively. For validation, the lowest and highest rates are 70 and 83.3% for 10 and 5 neurons, respectively. The best overall performance with 81% accuracy is observed with 20 neurons in hidden layer. If the obtained results are compared with the results obtained in case of all input features, it is found that the performance decreases by 10% for scheme 1.

Figure 9 represents the confusion matrix with accuracy achieved in percentage (%) for training, testing and validation. Figure 10 represents the best validation

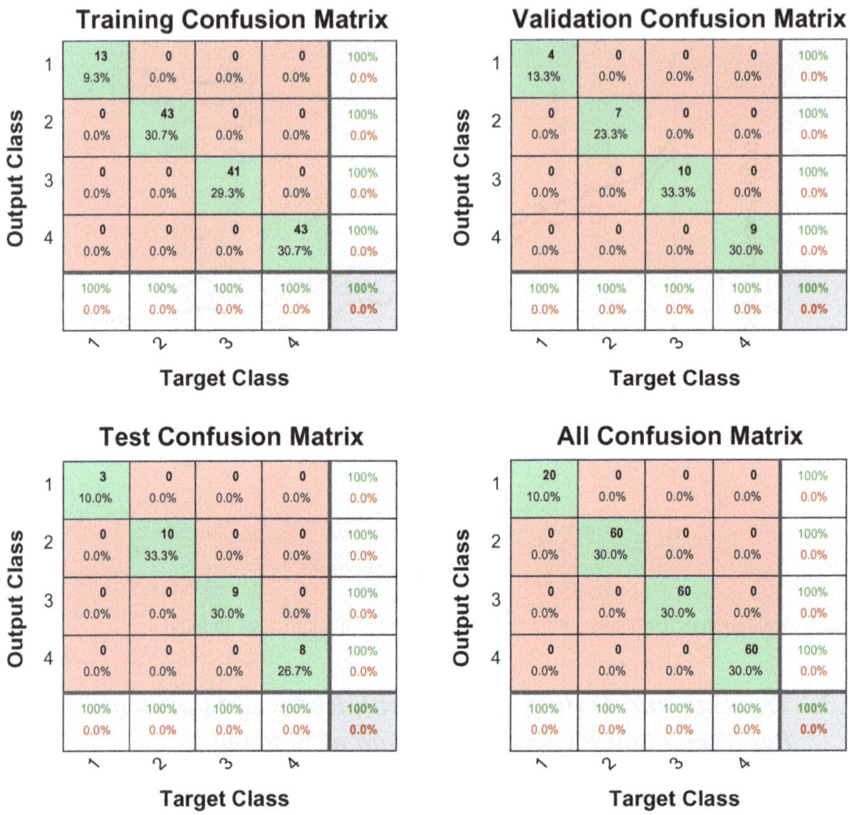

Fig. 6 Confusion matrix for scheme 2 with all features as an input

performance curve by ANN for fault discrimination with 4 parameters, 20 neurons in hidden layer at 66 epochs.

4.4 Training and Testing of ANN with Reduced Statistical Parameters on Experimental Data of Scheme 2

In Sect. 4.2, all eleven statistical features as input parameters were considered for training, testing and validation. Now the same process is adopted but with reduced statistical features which are significant and relevant as explained in Sect. 3. Figure 2b represents that features like F4, F9, F10, F11 are the only four features that are significant and relevant for classification purpose for scheme 2 as all these significant features lie in positive principal component 1 axis of the bi-plot, suggesting strong relationship among them. So, for scheme 2, the ANN is now trained with these four input features and various numbers of neurons in hidden layer. Figure 11 represents

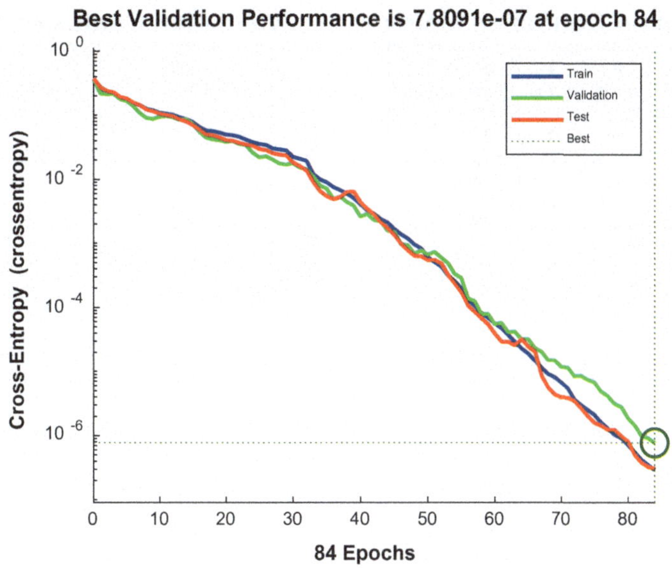

Fig. 7 Validation performance curve for scheme 2 with all features as an input

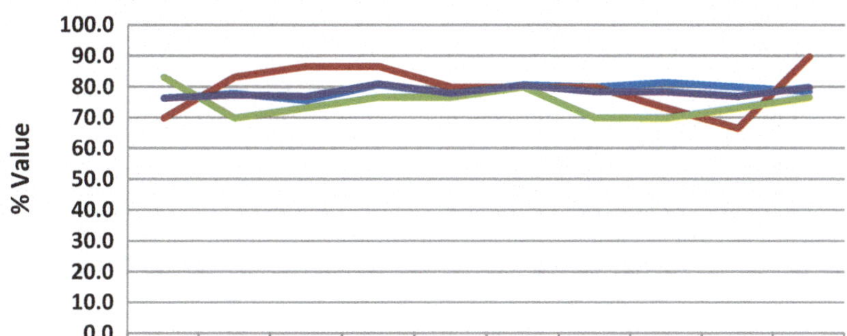

	5	10	15	20	25	30	35	40	45	50
Training	76.4	77.9	75.7	80.7	77.9	80.7	80.0	81.4	80.0	78.6
Testing	70.0	83.3	86.7	86.7	80.0	80.0	80.0	73.3	66.7	90.0
Validation	83.3	70.0	73.3	76.7	76.7	80.0	70.0	70.0	73.3	76.7
Overall	76.5	77.5	77.0	81.0	78.0	80.5	78.5	78.5	77.0	80.0

Fig. 8 Performance of neural network with different numbers of neuron in hidden layers with reduced features as an input

Fig. 9 Confusion matrix for the scheme 1 with reduced features as an input

the confusion matrix for training, testing and validation for 4 statistical input parameters with 5 neurons in hidden layer of the neural network. The neural network has been trained with different numbers of neurons in hidden layer, i.e., 5, 10, 15, 20, …, 50, and it has been observed that with all number of neurons, 100% performance is achieved for training, testing and validation. Figure 12 shows the best validation performance curve of ANN for fault discrimination. The number of input parameter is 4, and the number of neurons in hidden layer is 5, and it is achieved on 80 epochs. The error in performance obtained for scheme 2 is same as compared with performance achieved in case of scheme 1 for reduced features.

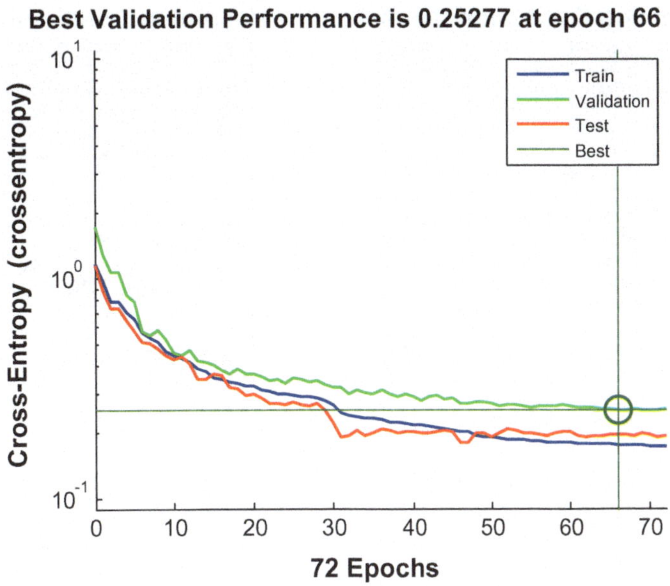

Fig. 10 Validation performance curve for scheme 1 with reduced features as an input

5 Conclusions

This present work applies one of the intelligent approaches, i.e., the artificial neural network (ANN). MATLAB function has been utilized for the training and testing of ANN. The network was trained and tested with different numbers of input parameters and varying number of neurons in hidden layer. The training and testing has been performed on experimental data. Sample plots of performance curve along with target and actual outputs have been presented. The accuracy of ANN is found to be quite good for both the health discrimination as well as fault classification for scheme 1 and scheme 2 which is 91% and 100%, respectively. It is also noted that as number of input parameters decreases, the accuracy of ANN also decreases from 81 and 99.5% for scheme 1 and scheme 2, respectively.

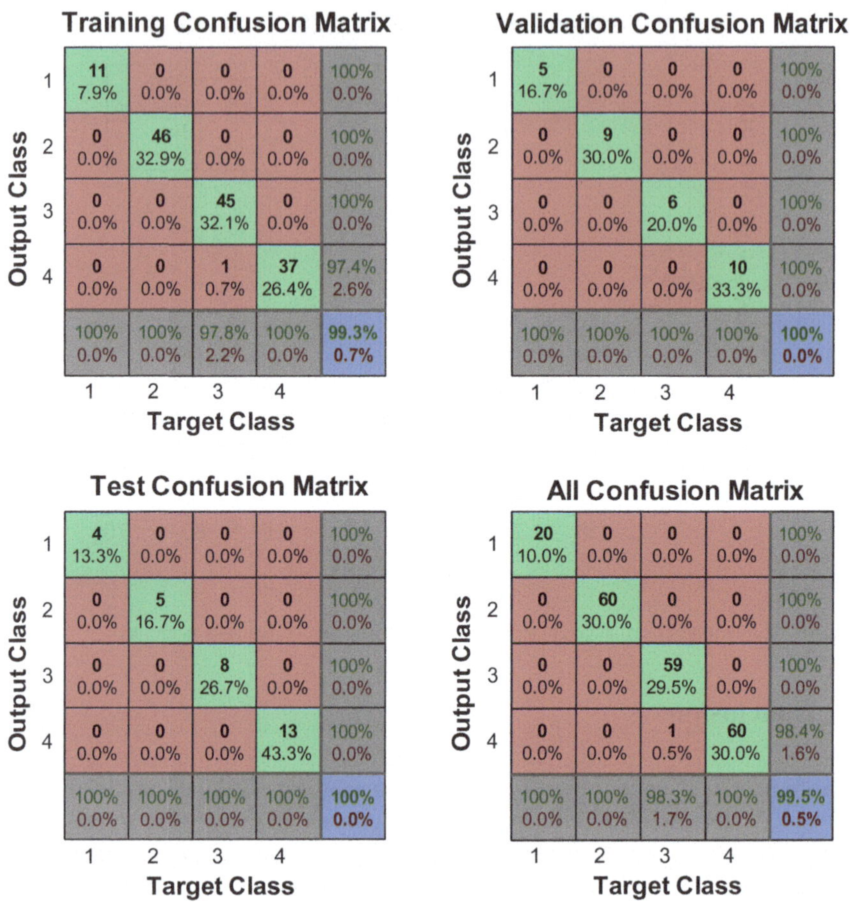

Fig. 11 Confusion matrix for the scheme 2 with reduced features as an input

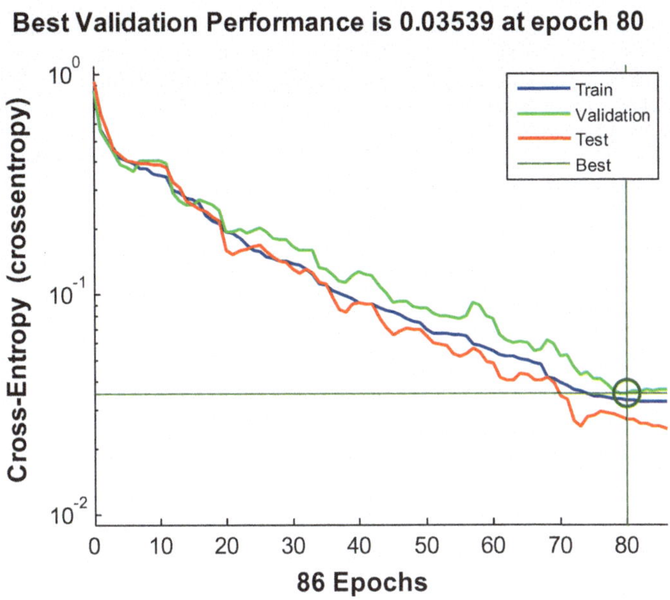

Fig. 12 Validation performance curve for scheme 2 with reduced features as an input

References

1. Jiang F, Li W, Wang Z, Zhu Z (2012) Fault severity estimation of rotating machinery based on residual signals. Adv Mech Eng 2012:1–8
2. Jiang H, Wang Z, He Z (2006) Wavelet design for extracting weak fault feature based on lifting scheme. Front Mech Eng China 2:199–203
3. Ruijuan J, Chunxia X (2008) Mechanical fault diagnosis and signal feature extraction based on fuzzy neural network. Transform 234–237
4. Yang H (2004) Automatic fault diagnosis of rolling element bearings using wavelet based pursuit features
5. Rauber TW, de Assis Boldt F, Varejao FM (2015) Heterogeneous feature models and feature selection applied to bearing fault diagnosis. IEEE Trans Ind Electron 62(1):637–646
6. Smith WA, Randall RB (2015) Rolling element bearing diagnostics using the Case Western Reserve University data: a benchmark study. Mech Syst Signal Process 64–65:100–131
7. Zhang C, Li B, Chen B, Cao H, Yanyang Z, He Z (2014) Periodic impulsive fault feature extraction of rotating machinery using dual-tree rational dilation complex wavelet transform. J Manuf Sci Eng 136(5):051011
8. Romero JFA, MS D'angelo, Saotome O, Müller SF (2005) Wavelet packet feature extraction for vibration. In: Proceedings of COBEM, 2005, no. 1992, pp 1–8
9. Patel RK, Chandel A (2012) Induction machine bearing fault diagnosis with discrete wavelet transform using vibration signal. J Electr Eng 5(4):1–6
10. Aguado D, Montoya T, Borras L, Seco A, Ferrer J (2008) Using SOM and PCA for analysing and interpreting data from a P-removal SBR. Eng Appl Artif Intell 21(6):919–930
11. Dong S, Sun D, Tang B, Gao Z, Yu W, Xia M (2014) A fault diagnosis method for rotating machinery based on PCA and Morlet Kernel SVM. Math Probl Eng 2014:1–8
12. Patel RK, Giri VK (2016) ANN based performance evaluation of BDI for condition monitoring of induction motor bearings. J Inst Eng Ser B 98(3):267–274

13. Onel I, Benbouzid M (2008) Induction motors bearing failures detection and diagnosis using a RBF ANN park pattern based method. Int Rev Electr Eng 3(1):159–165
14. Vijay GS, Kumar HS, Srinivasa PP, Sriram NS, Rao RBKN (2012) Evaluation of effectiveness of wavelet based denoising schemes using ANN and SVM for bearing condition classification. Comput Intell Neurosci, vol 2012
15. Kumar HS, Pai PS, Sriram NS, Vijay GS (2013) ANN based evaluation of performance of wavelet transform for condition monitoring of rolling element bearing. Procedia Eng 64:805–814

Chapter 24
Fixed Final Time and Fixed Final State Linear Quadratic Optimal Control Problem of Fractional Order Singular System

Tirumalasetty Chiranjeevi, Raj Kumar Biswas and Shashi Kant Pandey

1 Introduction

Most of the researchers showing interest on singular systems because these systems give more accurate physical behavior [1]. Singular systems having some special features which are not observed in classical systems are nonproperness of transfer matrix, consistent initial conditions, noncausality between input and state or input and output, impulse terms and input derivatives in the state response, etc., because of these features singular systems have attracted many researchers from last few decades [1]. Researchers also showing interest on fractional order systems because modeling of these systems gives accurate results [2, 3]. Very limited work is presented in the literature regarding FOSS [4–6] particularly in optimal control problem (OCP). Therefore, in this work, we consider LQOCP of FOSS in the sense of Caputo FD.

Abounded work has been done in the area of OCP but limited work has been done in the area of fractional optimal control problem (FOCP). In this regard, general formulation of FOCP and different numerical schemes are presented in [7–10]. Biswas and sen in [11, 12] presented formulation and numerical schemes for solving continuous-time FOCP with different endpoint conditions. Discrete-time FOCP formulation and solution scheme is presented in [13–15]. Authors in [16] presented an analytical numerical scheme for FOCP in the sense of conformable fractional derivative. Formulation of both continuous-time and discrete-time FOCP with constraints on control is presented in [17]. In literature, authors presented different numerical schemes like combination of epsilon penalty and variational methods [18], fixed point approach [19], Ritz variational method [20], semidefinite programming approach

T. Chiranjeevi (✉) · R. K. Biswas
Electrical Engineering Department, NIT Silchar, Silchar, Assam, India

T. Chiranjeevi
Rajkiya Engineering College, Sonbhadra, Uttar Pradesh, India

S. K. Pandey
Electrical Engineering Department, Rajkiya Engineering College, Sonbhadra, UP, India

V. K. Giri et al. (eds.), *Computing Algorithms with Applications in Engineering*,
Algorithms for Intelligent Systems, https://doi.org/10.1007/978-981-15-2369-4_24

[21], Ritz method [22], modified Adomian decomposition method [23], shifted Legendre orthonormal polynomials [24] and reflection operator based numerical method [25] for solving FOCPs. Optimal control of FOSS by pseudo-state space approach has been considered in [26, 27]. In [28], authors presented formulation and GLA based solution scheme for fixed final time and free final state OCP of FOSS. LQOCP of FOSS with fixed final time and free final state endpoint condition in terms of Caputo FD has been presented in [29].

The present work is extension of the work done in [29]. In this work, we have considered fixed final time and fixed final state endpoint condition. PI is considered in general form. System dynamics is expressed in terms of Caputo FD. Necessary conditions are obtained from the formulation of FOCP. GLA based numerical scheme is used for solving necessary conditions. In order to demonstrate the efficiency of numerical scheme, an example is illustrated.

2 Mathematical Background

In the literature [6], we can found several definitions of FDs. The most important and commonly using FDs are Riemann–Liouville (RL), Caputo and Grünwald–Letnikov.

The left and right RL derivatives of order α are given as

$$
{}_{0}^{\mathrm{RL}}D_{t}^{\alpha} z(t) = \frac{1}{\Gamma(1-\alpha)} \frac{\mathrm{d}}{\mathrm{d}t} \int_{0}^{t} (t-\tau)^{-\alpha} z(\tau) \mathrm{d}\tau
$$

$$
{}_{t}^{\mathrm{RL}}D_{1}^{\alpha} z(t) = \frac{1}{\Gamma(1-\alpha)} \left(\frac{-\mathrm{d}}{\mathrm{d}t} \right) \int_{t}^{1} (\tau-t)^{-\alpha} z(\tau) \mathrm{d}\tau
$$

The left and right Caputo derivatives of order α are given as

$$
{}_{0}^{C}D_{t}^{\alpha} z(t) = \frac{1}{\Gamma(1-\alpha)} \int_{0}^{t} (t-\tau)^{-\alpha} \left(\frac{\mathrm{d}}{\mathrm{d}\tau} \right) z(\tau)\, \mathrm{d}\tau
$$

$$
{}_{t}^{C}D_{1}^{\alpha} z(t) = \frac{1}{\Gamma(1-\alpha)} \int_{t}^{1} (\tau-t)^{-\alpha} \left(\frac{-\mathrm{d}}{\mathrm{d}\tau} \right) z(\tau)\, \mathrm{d}\tau
$$

The left and right Grünwald–Letnikov derivatives of order α are given as

$$
{}_{0}^{GL}D_{t}^{\alpha} z(t) = \underset{\substack{h\to 0\\ kh=t}}{Lt} \frac{1}{h^{\alpha}} \sum_{i=0}^{k} w_{i}^{(\alpha)} z(t-ih)
$$

$$ {}_{t}^{GL}D_{1}^{\alpha}z(t) = \underset{\substack{h\to 0\\ kh=t}}{Lt}\frac{1}{h^{\alpha}}\sum_{i=0}^{N-k} w_i^{(\alpha)} z(t+ih) $$

where $w_i^{(\alpha)} = (-1)^i \binom{\alpha}{i}$.

3 Problem Definition and Numerical Scheme

The main aim of this work is defined as follows. In the first step, transformation is carried out from FOSS to standard fractional order state space system. In the second step, we obtain necessary conditions from the formulation of FOCP by considering transformed system dynamics and transformed PI. In the third step, we use GLA based numerical scheme for approximation of state and co-state equations of necessary conditions.

First step:

The problem is defined as follows:

Find an optimal control $u(t)$ which minimizes the PI

$$ J(u) = \frac{1}{2}\int_{0}^{1}[x^T(t)Qx(t) + u^T(t)Ru(t)]\mathrm{d}t \tag{1} $$

subject to the dynamics of the singular system

$$ E\,{}_{0}^{C}D_{t}^{\alpha}x(t) = Ax(t) + Bu(t) \tag{2} $$

with the boundary conditions $x(0) = x_0$ and $x(1) = x_f$.

Where $x \in \Re^{m\times 1}$, $u \in \Re^{n\times 1}$, $Q \in \Re^{m\times m} \geq 0$ and $R \in \Re^{n\times n} > 0$, $A \in \Re^{m\times m}$ is the state matrix, $B \in \Re^{m\times n}$ is the input matrix, $E \in \Re^{m\times m}$ is the singular matrix and ${}_{0}^{C}D_{t}^{\alpha}x(t)$ is the Caputo derivative.

Choose two non-singular matrices S, M and gain matrix K in order to satisfy the following relations, respectively as [1, 29]

$$ \mathrm{SEM} = \mathrm{diag}(I_{m_1}, O),\;\; S(A+BK)M = \mathrm{diag}(\vec{A}, I_{m_2}) \tag{3} $$

$$ \deg\,(|sE - (A+BK)|) = \mathrm{rank}(E) \tag{4} $$

where $\vec{A}$ is a new state matrix of order $m_1 \times m_1$, O is a nilpotent matrix of order $m_2 \times m_2$, $m_1 = rank(E)$ and $m_1 + m_2 = m$.

Consider a feedback control law

$$u(t) = Kx(t) + w(t) \tag{5}$$

where $w(t)$ is the new control vector of order $n \times 1$ and K is the gain matrix of order $n \times m$.

Substituting Eq. (5) in Eq. (2), we get

$$E\,{}_0^C D_t^\alpha x(t) = (A + BK)x(t) + Bw(t) \tag{6}$$

Choose the coordinate transformation in order to convert the FOSS to standard fractional order state space system as [1, 28, 29]

$$x(t) = M\begin{bmatrix} x_1(t) \\ x_2(t) \end{bmatrix}, x_1(t) \in \Re^{m_1}, x_2(t) \in \Re^{m-m_1} \tag{7}$$

We can write Eq. (6) by using Eqs. (3) and (7) as

$${}_0^C D_t^\alpha x_1(t) = \vec{A}x_1(t) + B_1 w(t) \tag{8}$$

$$0 = x_2(t) + B_2 w(t) \tag{9}$$

Combining Eq. (7) with Eq. (9), we get

$$\begin{bmatrix} x(t) \\ u(t) \end{bmatrix} = \begin{bmatrix} I & 0 \\ K & I \end{bmatrix}\begin{bmatrix} x(t) \\ w(t) \end{bmatrix} = \begin{bmatrix} M & 0 \\ KM & I \end{bmatrix}\begin{bmatrix} x_1(t) \\ x_2(t) \\ w(t) \end{bmatrix} = \begin{bmatrix} M & 0 \\ KM & I \end{bmatrix}\begin{bmatrix} I & 0 \\ 0 & -B_2 \\ 0 & I \end{bmatrix}\begin{bmatrix} x_1(t) \\ w(t) \end{bmatrix} \tag{10}$$

Substituting Eq. (10) in Eq. (1) and further simplifying, we get

$$J(v) = \frac{1}{2}\int_0^1 \left[x_1^T(t)\breve{Q}x_1(t) + v^T(t)\hat{R}v(t)\right]dt \tag{11}$$

where $\breve{Q} = \hat{Q} - P\hat{R}^{-1}P^T$, $v(t) = w(t) + \hat{R}^{-1}P^T x_1(t)$ and

$$\begin{bmatrix} \hat{Q} & P \\ P^T & \hat{R} \end{bmatrix} = \begin{bmatrix} I & 0 \\ 0 & -B_2 \\ 0 & I \end{bmatrix}^T \begin{bmatrix} M & 0 \\ KM & I \end{bmatrix}^T \begin{bmatrix} Q & 0 \\ 0 & R \end{bmatrix}\begin{bmatrix} M & 0 \\ KM & I \end{bmatrix}\begin{bmatrix} I & 0 \\ 0 & -B_2 \\ 0 & I \end{bmatrix}.$$

Substituting $w(t) = v(t) - \hat{R}^{-1}P^T x_1(t)$ in Eq. (8), we get

$${}_0^C D_t^\alpha x_1(t) = (\vec{A} - B_1\hat{R}^{-1}P^T)x_1(t) + B_1 v(t) = \breve{A}x_1(t) + B_1 v(t) \tag{12}$$

Therefore, FOSS (2) has been transformed into standard fractional order state space system (12).

Second step:

Now, we formulate the FOCP by considering the transformed system dynamics given by Eq. (12) and transformed PI given by Eq. (11) in order to obtain the new control vector $v(t)$.

The augmented PI using Lagrange multiplier $\lambda(t)$ can be written as

$$J_a(v) = \int_0^1 \left\{ \frac{1}{2}\left(x_1^{\mathrm{T}}(t)\breve{Q}x_1(t) + v^{\mathrm{T}}(t)\hat{R}v(t) \right) + \lambda^{\mathrm{T}}(t)\left(\breve{A}x_1(t) + B_1 v(t) - {}_0^C D_t^{\alpha} x_1(t) \right) \right\} \mathrm{d}t \tag{13}$$

The first variation of $J_a(v)$ becomes

$$\delta J_a(v) = \int_0^1 \left\{ \begin{array}{l} \left[\breve{Q}x_1(t) + \breve{A}^{\mathrm{T}}\lambda(t) - {}_t^{RL}D_1^{\alpha}\lambda(t) \right]^{\mathrm{T}} \delta x_1(t) + \left[\hat{R}v(t) + B_1^{\mathrm{T}}\lambda(t) \right]^{\mathrm{T}} \delta v(t) \\ + \left[\breve{A}x_1(t) + B_1 v(t) - {}_0^C D_t^{\alpha} x_1(t) \right]^{\mathrm{T}} \delta\lambda(t) \end{array} \right\} \mathrm{d}t - \lambda(1)\delta x(1) \tag{14}$$

For optimum, $\delta J_a(v) = 0$. Therefore, the coefficients of $\delta x_1(t)$, $\delta w(t)$ and $\delta\lambda(t)$ in Eq. (14) become zero. This leads to

$${}_0^C D_t^{\alpha} x_1(t) = \breve{A}x_1(t) + B_1 v(t) \tag{15}$$

$${}_t^{RL} D_1^{\alpha}\lambda(t) = \breve{Q}x_1(t) + \breve{A}^{\mathrm{T}}\lambda(t) \tag{16}$$

$$v(t) = -\hat{R}^{-1}B_1^{\mathrm{T}}\lambda(t) \tag{17}$$

Finally, the first variation becomes

$$\lambda(1)\delta x(1) = 0 \tag{18}$$

In this work, we have considered fixed final time and fixed final state endpoint condition. Therefore, the transversality condition becomes as $\delta x(1) = 0$.

Third step:

We can approximate the necessary conditions given by Eqs. (15) and (16) by using GLA as

$$\frac{1}{h^{\alpha}}\sum_{i=0}^{k} w_i^{(\alpha)} x_1(kh - ih) = \breve{A}x_1(kh) - B_1\hat{R}^{-1}B_1^{\mathrm{T}}\lambda(kh) + \frac{(kh)^{-\alpha}}{\Gamma(1-\alpha)}x(0),\ k = 1, 2, \ldots, N \tag{19}$$

$$\frac{1}{h^{\alpha}}\sum_{i=0}^{N-k} w_i^{(\alpha)}\lambda(kh+ih) = \breve{Q}x_1(kh) + \breve{A}^{\mathrm{T}}\lambda(kh), \quad k = N-1, N-2, \ldots, 0 \tag{20}$$

Equations (19) and (20) represent the set of 2 N equations. In order to solve these equations, we divide entire time domain into N equal sub-domains. Size of each sub-domain is $h = 1/N$ and the time at node k is $t_k = kh$. We can solve these equations by using any linear equation solver. Ones $x_1(t)$ and $\lambda(t)$ are known, by using Eqs. (5) and (9), we obtain $u(t)$ and $x_2(t)$.

4 Numerical Simulation

In this section, an example is considered to demonstrate the applicability of the formulation and efficiency of solution scheme.

Consider a FOSS described by

$$\begin{bmatrix} 1 & 0 \\ 0 & 0 \end{bmatrix} {}_0^C D_t^{\alpha} x(t) = \begin{bmatrix} 0 & 1 \\ -1 & 0 \end{bmatrix} x(t) + \begin{bmatrix} 1 \\ 1 \end{bmatrix} u(t) \tag{21}$$

Consider a quadratic PI

$$J = \frac{1}{2}\int_0^1 \left(x_1^2(t) + x_2^2(t) + u^2(t)\right)\mathrm{d}t \tag{22}$$

Initial and final conditions are

$$x_1(0) = 1 \text{ and } x_1(1) = 0 \tag{23}$$

The matrices K, S and M are chosen as

$$K = \begin{bmatrix} 1 & -1 \end{bmatrix},\ S = \begin{bmatrix} 1 & 0 \\ 0 & 1 \end{bmatrix},\ M = \begin{bmatrix} 1 & 0 \\ 0 & -1 \end{bmatrix}$$

Using foregoing considerations, we obtain necessary conditions as

$$\begin{aligned} {}_0^C D_t^{\alpha} x_1(t) &= x_1(t) - \lambda(t) \\ {}_t^{RL} D_1^{\alpha} \lambda(t) &= 2x_1(t) + \lambda(t) \\ v(t) = w(t) &= -\lambda(t) \end{aligned}$$

We can solve the above set of necessary conditions by using GLA based numerical method discussed in the previous section. Results are produced for different values of α. Figures 1, 2, and 3 show the optimal states $x_1(t)$, $x_2(t)$ and the optimal control

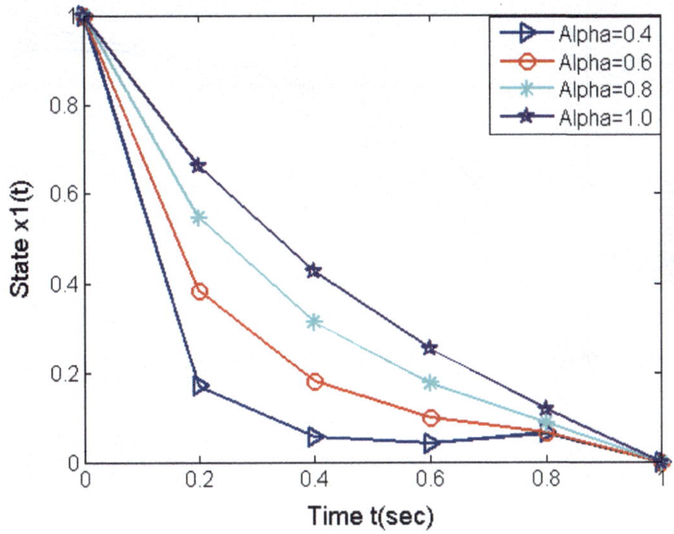

Fig. 1 State $x_1(t)$ for different α, $N = 5$

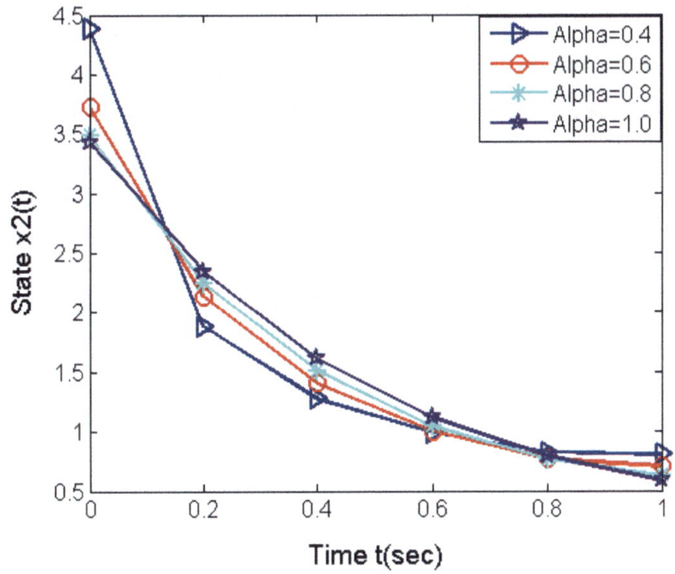

Fig. 2 State $x_2(t)$ for different α, $N = 5$

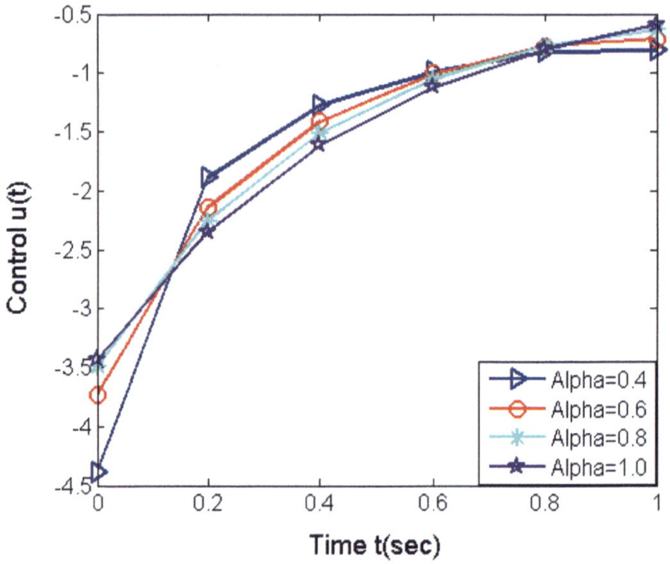

Fig. 3 Optimal control $u(t)$ for different α, $N = 5$

$u(t)$ for different α and $N = 5$. From these results, we can observe that control effort decreases as α decreased. We can also observe that, like in [7, 11, 28], amplitude of optimal states and optimal control increases as α is increased. Furthermore, the results of OCP of integer order system are recovered from the results of OCP of fractional order system at $\alpha = 1$.

5 Conclusions

Formulation and solution scheme of fixed final time and fixed final state LQOCP of FOSS in the sense of Caputo FD has been presented in this paper. Singular systems are complex in nature, so both numerical and analytical solutions are difficult. Therefore, using coordinate transformation, first we have converted FOSS into standard fractional order state space system and then necessary conditions are obtained. GLA based numerical scheme is used for solving necessary conditions. An example is considered to demonstrate the efficiency of solution scheme. For different values of α, optimal states and optimal control results are obtained. From the results, we conclude that when α decreases, amplitude of both states and control decreased and it demands small control effort.

References

1. Dai L (1989) Singular control systems. In: Thoma M, Wyner A (eds) Lecture notes in control and information sciences. Springer, Berlin, pp 1–340
2. Suri babu G, Chiranjeevi T (2016) Implementation of fractional order PID controller for an AVR system using GA and ACO optimization techniques. In: IFAC-ACODS conference, 49(1), NIT Trichy, India, pp 456–461
3. Dzielinski A, Sierociuk D, Sarwas G (2010) Some applications of fractional order calculus. Bull Pol Ac: Tech 58(4), 583–592
4. Kaczorek T (2011) Singular fractional linear systems and electrical circuits. Int J Appl Math Comput Sci 21(2):379–384
5. Wei J (2010) The constant variation formulae for singular fractional differential systems with delay. Comput Math Appl 59(3):1184–1190
6. Barbosa KA, de Souza CE, Coutinho D (2018) Admissibility analysis of discrete linear time—varying descriptor systems. Automatica 91:136–143
7. Agrawal OP (2004) A general formulation and solution scheme for fractional optimal control problems. Nonlinear Dyn 38:323–337
8. Baleanu D, Agrawal OP (2009) A central difference numerical scheme for fractional optimal control problems. J Vib Control 15(4):583–597
9. Agrawal OP, Baleanu D (2007) A Hamiltonian formulation and a direct numerical scheme for fractional optimal control problems. J Vib Control 13(9–10):1269–1281
10. Agrawal OP (2008) A quadratic numerical scheme for fractional optimal control problems. ASME J Dyn Sys, Meas, Control 130, 0110101–0110106
11. Biswas RK, Sen S (2011) Fractional optimal control problems with specified final time. ASME J Comput Nonlinear Dyn 6:0210091–0210096
12. Biswas RK, Sen S (2014) Free final time fractional optimal control problems. J Franklin Inst 351:941–951
13. Dzielinski A, Czyronis PM (2013) Fixed final time and free final state optimal control problem for fractional dynamic systems-linear quadratic discrete-time case. Bull Pol Ac: Tech 61(3), 681–690
14. Dzielinski A, Czyronis PM (2012) Fixed final time optimal control problem for fractional dynamic systems-linear quadratic discrete-time case. In: Mikolaj B, Krzysztof (eds) Advances in control theory and automation, Printing House of Bialystok University of Technology, Poland, pp 71–80
15. Chiranjeevi T, Biswas RK (2017) Discrete-time fractional optimal control. Mathematics 5(2):1–12
16. Chiranjeevi T, Biswas RK (2019) Closed-form solution of optimal control problem of a fractional order system. J King Saud Univ—Sci. https://doi.org/10.1016/j.jksus.2019.02.010
17. Chiranjeevi T, Biswas RK (2018) Formulation of optimal control problems of fractional dynamic systems with control constraints. J Adv Res Dyn Control Syst 10:201–212
18. Lotfi A (2017) A combination of variational and penalty methods for solving a class of fractional optimal control problems. J Optim Theory Appl 174(1):65–82
19. Zeid SS, Kamyad AV, Effati S et al (2017) Numerical solutions for solving a class of fractional optimal control problems via fixed-point approach. SeMA 74(4):585–603
20. Lotfi A, Yousefi SA (2017) A Generalization of Ritz-variational method for solving a class of fractional optimization problems. J Optim Theory Appl 174(1):238–255
21. Dehghan R, Keyanpour M (2017) A semidefinite programming approach for solving fractional optimal control problems. Optimization 66(7):1157–1176
22. Nemati A, Yousefi SA (2017) A numerical scheme for solving two-dimensional fractional optimal control problems by the Ritz method combined with fractional operational matrix. IMA J Math Control Inf 34(4):1079–1097
23. Alizadeh A, Effati S (2018) Modified Adomian decomposition method for solving fractional optimal control problems. T I Meas Control 40(6):2054–2061

24. Ezz-Eldien SS, Hafez RM, Bhrawy AH, Baleanu D, El-Kalaawy AA (2017) New Numerical Approach for Fractional Variational Problems Using Shifted Legendre Orthonormal polynomials. J Optim Theory Appl 174(1):295–320
25. Biswas RK, Sen S (2009) Numerical method for solving fractional optimal control problems. In: ASME IDETC/CIE conference, San Diego, California, USA, pp 1–4
26. Biswas RK, Sen S (2010) Fractional optimal control problems: a pseudo-state-space approach. J Vib Control 17(7):1034–1041
27. Biswas RK, Sen S (2011) Fractional optimal control within Caputo's derivative. In: ASME IDETC/CIE conference, Washington, DC, USA, pp 1–8
28. Chiranjeevi T, Biswas RK, Chudamani S (2019) Optimal control of fractional order singular system. Int J Electr Eng Edu. https://doi.org/10.1177/0020720919833031
29. Moubarak MRA, Ahmed HF, Khorshi O (2018) Numerical solution of the optimal control for fractional order singular systems. Differ Equ Dyn Syst 26:279–291

Chapter 25
Fuzzy Controller for DTC-SVM of Induction Motor Using Sample Reference Phase Voltages

Y. Laxmi Narasimha Rao and G. Ravindranath

1 Introduction

Variable speed drives have empowered extraordinary execution in electric drives and conveyed emotional vitality reserve funds by coordinating engine speed and torque to the determined burden prerequisites. Most VSDs in the market depend on a modulator organize that conditions voltage and recurrence contributions to the drive, yet aims intrinsic time delay in handling control signals. In contrast, direct torque control (DTC) extraordinarily expanding drive torque reaction. DTC technology likewise gives different advantages running up to framework-level highlights [1].

DTC of induction motor directly decouples flux and torque using parks. Transformation of direct torque control technique using different switching state which in elaborate has eight states out of which two are redundant states. A proper suitable on/off operation of switches produces the flux and torque independent similar to that of DC motor for variable speeds [2]. Vector control of induction motor [3] which requires current parameters for controlling the speed requires shaft encoders making it more complex can be avoided with DTC.

SVPWM induction motor combined with DTC produces less ripples in torque and current [4] at the same time trying to improve the utilization of DC bus bars. Space vector pulse width modulation requires switching states which are to be operated in a sequence and also angular position of reference phasor [5], and hence, faster computation devices are required. DTC-SVPWM requires PI controller to achieve faster response in terms of transient states (i.e.,) rise time, delay time and settling time. PI controller has drawbacks of sensitivity parameter variation. DTC-SVM of induction motor is shown in Fig. 1.

Y. Laxmi Narasimha Rao (✉)
M.V.S.R Engineering College, Nadergul, Hyderabad, Telangana 501510, India

G. Ravindranath
Matrusri Engineering College, Saidabad, Hyderabad, Telangana 501510, India

V. K. Giri et al. (eds.), *Computing Algorithms with Applications in Engineering*,
Algorithms for Intelligent Systems, https://doi.org/10.1007/978-981-15-2369-4_25

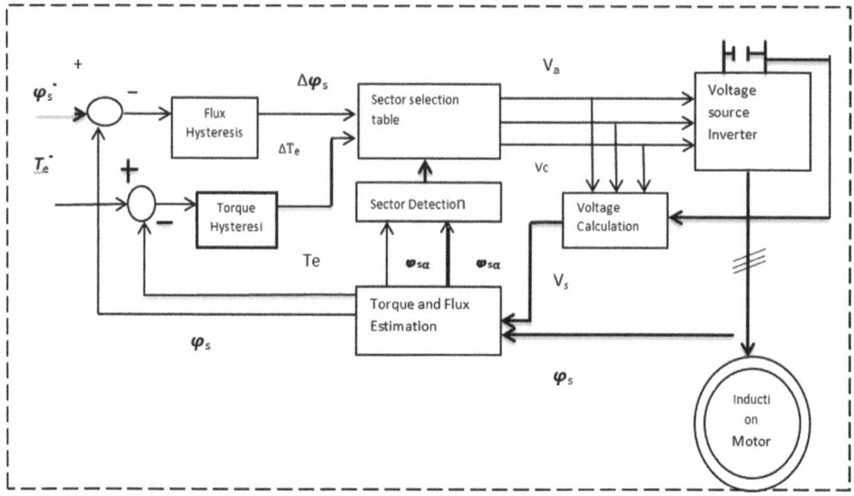

Fig. 1 DTC-SVM of induction motor

To overcome the effect of parameter, variation fuzzy logic controllers are used where the effect of parameters on the performance of induction motor can be eliminated [5, 6]. DTC-SVM, fuzzy logic which a rule-based decision invokes the exact parameter in terms of linguistic variables, thereby producing the appropriate results [7, 8]. In the proposed method, using sample reference phase voltage of DTC-SWM with fuzzy logic inputs are chosen to produce accurate output by looking into the decision based on linguistic variables.

2 DTC-SVM Model

DTC-SVPWM plots for two-level inverters, each reference stage voltage is differentiated and the triangular carrier and the individual stage voltages are made, self-governing of each other. To activate the most outrageous top plentifulness of the basic stage voltage, in direct guideline, abnormal mode voltage, V_{offset1}, is added to the reference stage voltages, where the degree of V_{offsset1} is given by:

$$V_{\text{offset1}} = -(V\text{max} + V_{\text{min}})/2 \tag{1}$$

In Eq. (1), $V_{\max}$ is the greatest extent of the three examined reference stage voltages, while $V_{\min}$ is the base size of the three tested reference stage voltages in an assessing interval, Eq. (1) relies upon the route that in an investigating interval, the reference stage which has most diminished degree crosses the triangular carrier first, and causes the essential advancement in the inverter switch in state, notwithstanding the way that the reference organize, which has the superlative significance, crosses the carrier last and causes the last trading change in the inverter changing states. Along these lines, the changing times of the space vectors can be obtained from the most extreme stage and min arrange example mention stage voltage amplitudes in a two-level inverter conspire.

2.1 Algorithm for DTC-SVM Using Sample Reference Phase Voltages

Step 1: Get the operating voltages of V_{an}, V_{bn} and V_{cn} from the errors produce due to flux and torque
Step 2:

$$T_{\text{as}} = V_{\text{an}} * T_s / V_{\text{dc}} \tag{2}$$

$$T_{\text{bs}} = V_{\text{bn}} * T_s / V_{\text{dc}} \tag{3}$$

$$T_{\text{cs}} = V_{\text{cs}} * T_s / V_{\text{dc}} \tag{4}$$

Step 3:

$$T_{\text{offset1}} = -\max(T_{\text{as}}, T_{\text{bs}}, T_{\text{cs}}) + \min(T_{\text{as}}, T_{\text{bs}}, T_{\text{cs}})/2 \tag{5}$$

Step 4:

$$T^*_{\text{as}} = T_{\text{as}} + T_{\text{offset1}} \tag{6}$$

$$T^*_{\text{bs}} = T_{\text{bs}} + T_{\text{offset1}} \tag{7}$$

$$T^*_{\text{cs}} = T_{\text{as}} + T_{\text{offset1}} \tag{8}$$

Once the calculation of Eqs. (6–8) is done, triangular wave is compared with the reference signal to obtain the switching states of each inverter leg (Fig. 2).

3 DTC-SVM of Induction Motor with Fuzzy Logic Controller

The Takagi–Sugeno Fuzzy Controller takes as data sources the stator flux $E\varphi s$ and the operating motor torque $E\tau$ produces the direct and quadrature components of voltage vector in turn produces the reference phase voltage. The above controller considers

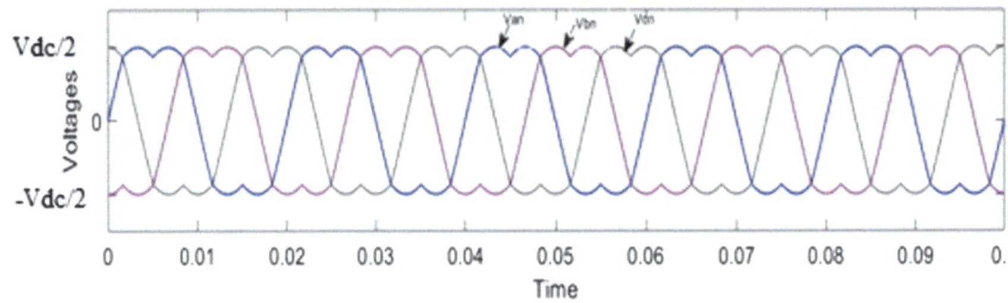

Fig. 2 Comparison of phase voltages with triangular wave

the error of operating variable so that it reproduces the quadrature components of operating voltages from U_{dc} so that appropriate stage is obtained, thereby eliminating the switching time calculations compared to the conventional DTC-SVM of induction motor. Exact reference voltages are obtained by changing the coefficients as explained in the next section (Fig. 3).

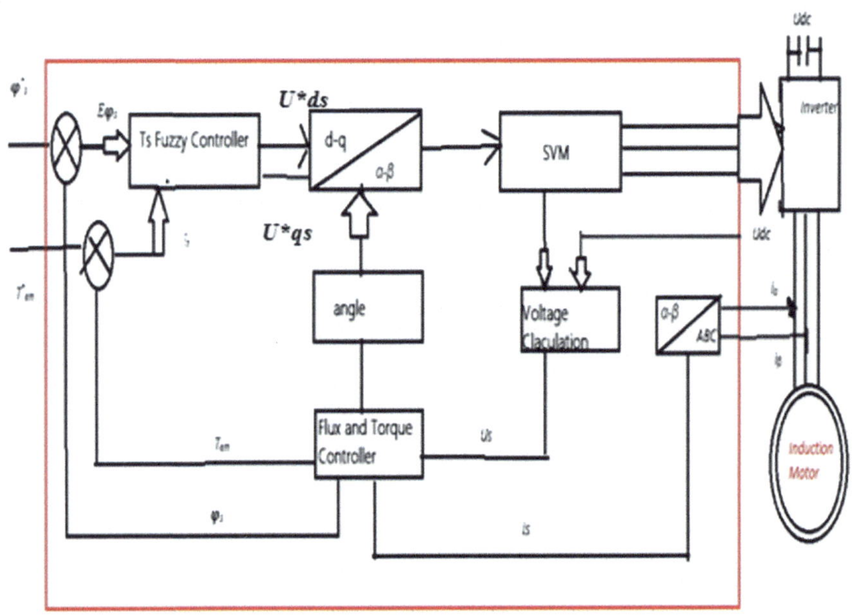

Fig. 3 FLC for DTC-SVM of induction motor using sample reference phase voltages

4 Membership Functions

The block diagram representation of fuzzy controller is shown in Fig. 4. The member function of the motor namely produced by motor flux and electromagnetic torque is shown in Figs. 5 and 6, respectively. These MFs shape and parameters were found through simulation procedure with different multiplications and with the knowing rules of linguistic variables for each test.

The minimum and maximum limits of the control parameter for the flux produced by the stator are fixed and lie between the limits of [−0.5 0.5]. The prospectus MFs have trapezoidal shapes anyway the middle one acknowledges triangular shape as is showed up in Fig. 5. In a similar fashion, the membership function for torque is also fixed and lies within the range as shown in Fig. 6, the limit lies in the range of [−20 20]. For the two data sources, the semantic names *n*, ze and *p* means negative, zero and positive, respectively.

The immediate segment of the stator voltage U^*_{dc} is dictated by the guidelines of the structure

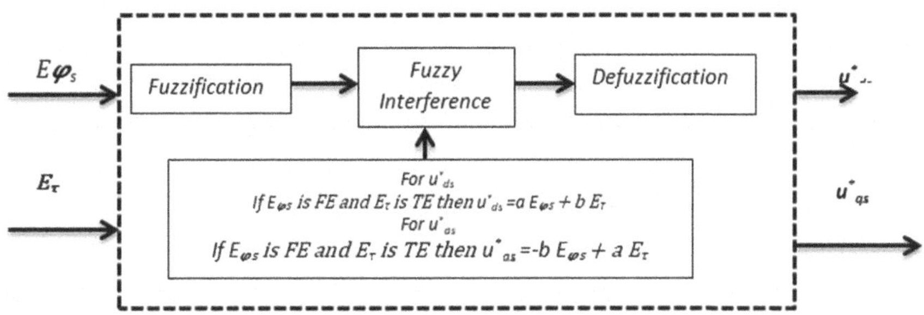

Fig. 4 Block representation of fuzzy logic controller

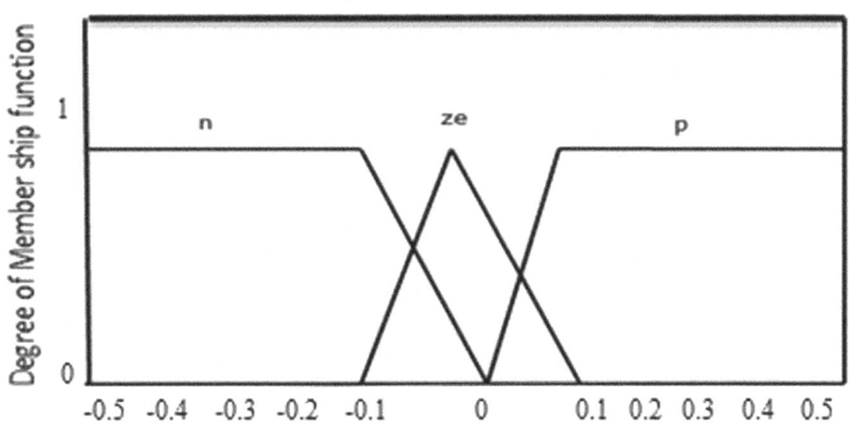

Fig. 5 Stator flux error 'e' as membership function

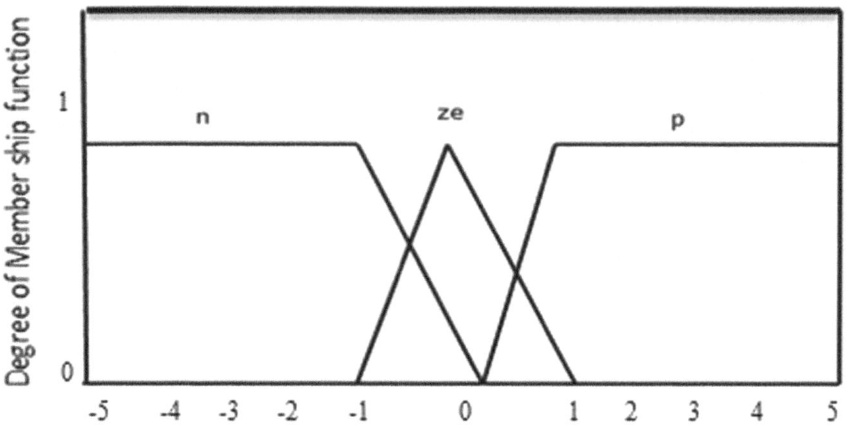

Fig. 6 Torque error '$E\tau$' its membership function

Rx: on the off chance that $E\varphi s$ is Fe and Te is TE at that point

$$U_{\text{ds}}^{*} = aE\varphi s + bE\tau \tag{9}$$

In any case, the quadrature part of the stator voltage $U^{*}qs$ is controlled by the guidelines of the structure

$$U_{\text{qs}}^{*} = -bE\varphi s + aE\tau \tag{10}$$

where $\{n, ze, p\}$ are the linguistic variables for the fuzzy arrangements of the data sources and, a and b are coefficients of the primary request polynomial capacity. For example, when the resulting capacity of the standard Ri is genuine number, the ensuing capacity is a zero-request polynomial and we have a zero request (Table 1).

Table 1 Fuzzy rule base computations of U_{ds}^{*} and U_{qs}^{*}

$E\varphi s/E\tau$	n	ze	p
n	$U_{\text{ds}}^{*} = a\,\text{Fe} + b\,\text{Te}$ $U_{\text{qs}}^{*} = b\,\text{Fe} + a\,\text{Te}$	$U_{\text{ds}}^{*} = a\,\text{Fe} + b\,\text{Te}$ $U_{\text{qs}}^{*} = -b\,\text{Fe} + a\,\text{Te}$	$U_{\text{ds}}^{*} = a\,\text{Fe} + b\,\text{Te}$ $U_{\text{qs}}^{*} = -b\,\text{Fe} + a\,\text{Te}$
ze	$U_{\text{ds}}^{*} = a\,\text{Fe} + b\,\text{Te}$ $U_{\text{qs}}^{*} = -b\,\text{Fe} + a\,\text{Te}$	$U_{\text{ds}}^{*} = a\,\text{Fe} + b\,\text{Te}$ $U_{\text{qs}}^{*} = -b\,\text{Fe} + a\,\text{Te}$	$U_{\text{ds}}^{*} = a\,\text{Fe} + b\,\text{Te}$ $U_{\text{qs}}^{*} = -b\,\text{Fe} + a\,\text{Te}$
p	$U_{\text{ds}}^{*} = a\,\text{Fe} + b\,\text{Te}$ $U_{\text{qs}}^{*} = -b\,\text{Fe} + a\,\text{Te}$	$U_{\text{ds}}^{*} = a\,\text{Fe} + b\,\text{Te}$ $U_{\text{qs}}^{*} = -b\,\text{Fe} + a\,\text{Te}$	$U_{\text{ds}}^{*} = a\,\text{Fe} + b\,\text{Te}$ $U_{\text{qs}}^{*} = -b\,\text{Fe} + a\,\text{Te}$

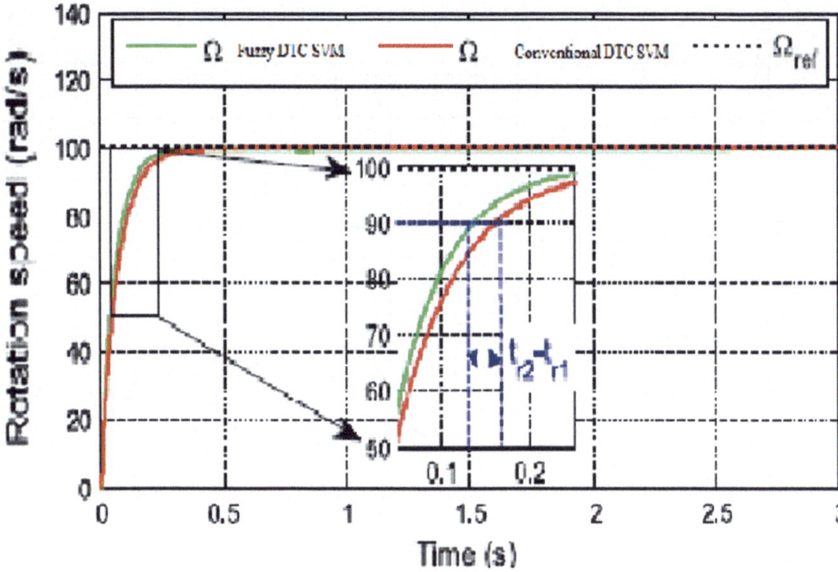

Fig. 7 Speed comparison of conventional DTC-SVM and fuzzy DTC-SVM

5 Results

The recreations were performed utilizing MATLAB through Simulink which incorporates the fuzzy implementation, fuzzy rationale tool compartment. The exchanging recurrence of PWM inverter was set to be 10 kHz, the stator reference transition considered was 0.9 wb and the coefficients considered were $a = 90$ and $b = 2$. So as to explore the adequacy of the proposed control framework and so as to check the circle solidness of the total framework, we tried out few tests. We utilized distinctive powerful working conditions, for example, step change in the load with variable loads. The tests were initially carried out on PI controller and then on the proposed fuzzy controller and then observed and also the comparison made for the currents flowing in stator (Figs. 7 and 8).

6 Conclusions

This part presents the DTC-SVM scheme with T-S fuzzy controller for two-level inverters connected to three-phase induction motor. The standard DTC-SVM plan takes two PI controllers to create the reference stator voltage vector. To improve the impediment of this customary, DTC-SVM plan is proposed with the T-S fuzzy controller to substitute both PI Controllers. The proposed controller processes the quadrature and direct axis parts of the reference stator voltage vector in the stator

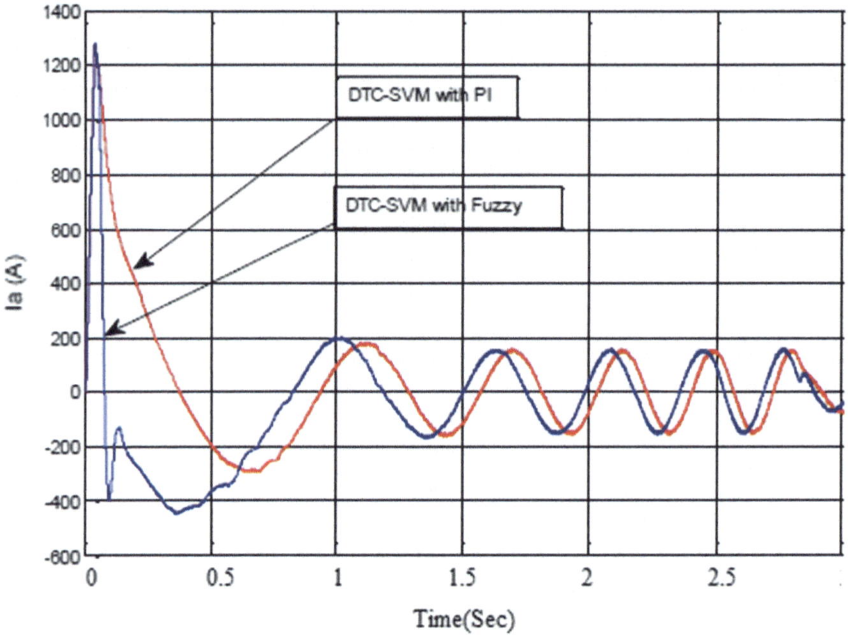

Fig. 8 Current drawn by two different controllers

reference layout. The standard base for the proposed controller is described in limit of the flux error produced by stator and similar for the torque variable used in the speed control parameter using trapezoidal and triangular membership function within prescribed limits. The quick fragment of the stator voltage takes a straight blend of its commitments as a consequent bit of the principles; in any case, the quadrature part of the stator voltages takes the near direct mix used in the fundamental yield anyway with the coefficients traded, not to be low torque swells are acquired using space vector modulation technique.

The proliferation results show that the proposed DTC-SVM with fuzzy controller achieved extraordinary performance in terms of rise time, settling time reduced, errors in flux and torque are also reduced when compared to PI controller of DTC-SVM. The above work can be implemented practically with DSP processor or FPGA. The above work can also be extended using artificial intelligence control like ANFIS, genetic algorithm and particle swarm optimization.

References

1. Lin G, Xu Z (2010) Direct torque control of induction motor base on fuzzy logic. In: 2010 2nd International conference on computer engineering and technology (ICCET), vol 4, pp V4-65–V4-655
2. Depenbrock M (1998) Direct self control of induction machine. IEEE Trans PE 4:420–429
3. Kennel R, El-rafaei A, Elkady F, Mahmoud S, Elkholy E, (2003) Torque ripple minimization for induction motor drives with direct torque control. In: Proceedings of 5th international conference on power electronics and drive systems, vol 1, pp 210–215
4. Takada N, Tanaka K (1992) Vector control of induction motor without shaft encoder. IEEE Trans IA 28(1):157–164
5. Bacha F, Dhifaoui R, Buyse H (2001) Real–time implementation of direct torque control of an induction machine by fuzzy logic controller. In: International conference on electrical machines and systems (ICEMS), vol 2, pp 1244–1249
6. Qu X, Song B, Li H (2010) DTC with adaptive stator flux observer and stator resistance estimator for induction motors. In: Paper presented at the 8th world congress on intelligent control and automation, pp 2460–2463
7. Brahim M, Farid T, Ahmed A, Nabil T, Toufik R (2011) A new fuzzy direct control strategy for induction machine based on indirect matrix converter. Int J Res Rev Comput Eng 1:18–22
8. Khanna R, Singla M, Kaur G (2009) Fuzzy logic based direct torque control of induction motor. In: Conference of power and energy society general meeting, Calgary, AB, pp 1–6

Chapter 26
Efficiency and Cost Optimization of Three-ϕ Induction Motor Using Soft Computing Techniques

Niraj Kumar Shukla, Shashi Kant Pandey and Rajeev Srivastava

1 Introduction

Three-phase induction motors are broadly used in industries, and with the invent of power electronic converters, induction motors can be used in variable speed applications. The three-phase induction motor performance can be controlled by the use of different controllers. For electrical drives, high performance is essential, and this is achieved with the help of robust controllers. Performance of the three-phase induction motor can be enhanced with the help of artificial intelligence systems. This paper proposes two such robust controller for the induction motor, and their performance analysis is done [1, 2]. Conventional controllers show poor response when unpredictability of the drive such as load disturbances and mechanical parameter variations are considered. Therefore, artificial intelligence strategies such as artificial neural network (ANN), expert system (ES), fuzzy logic control (FLC) have been introduced to control the performance of electrical drives. These controllers are termed as intelligent controllers. In this paper, the load torque of motor is used as the input for both the controllers, and the output of the controller is optimal current which is producing optimal flux for the motor. In this paper, the concept of artificial neural network and concept of neuro-fuzzy inference system are used for the field-oriented controlled induction motor drive. The use of ANFIS controller combines the best features of both fuzzy logic and as well as artificial neural network [3, 4]. The performance of adaptive neural-fuzzy interface system-based controller is compared with that of artificial neural network controller. In [6–8], the concept of LMC and SC has applied for obtaining the optimal flux value, but the performance of the model suffers when load variation is considered. This problem is overcome by

N. K. Shukla (✉) · R. Srivastava
Department of Electronics & Communication Engineering, University of Allahabad, Prayagraj, India

S. K. Pandey
Department of Electrical Engineering, Rajkiya Engineering College, Sonbhadra, India

V. K. Giri et al. (eds.), *Computing Algorithms with Applications in Engineering*, Algorithms for Intelligent Systems, https://doi.org/10.1007/978-981-15-2369-4_26

using proposed method. The results of simulated model show a great improvement in power consumption, and optimization of efficiency is achieved.

2 System Configuration and Mathematical Modelling of the Drive

Figure 1 represents the vector control induction motor drive with ANFIS-based controller which produces optimal current. To control the torque and rotor flux of the motor of a vector control motor, the synchronously rotating vector components of stator currents Iqs and Ids are independently controlled.

In a vector-controlled drive operation, the stator current is adjusted by using its frame of reference, i.e. in direct axis and quadrature axis components. A time-varying three-phase induction motor describes the dynamic behaviour of torque and voltages which is expressed by Eqs. 1, 2 and 9. A change of variables can be used to minimize the complication of these mathematical equations by removing all time-varying inductance [5]. The direct and quadrature equivalent circuit of the induction motor in a synchronous reference frame is represented in Fig. 2. The motor rating and parameters of three phase induction motor is given by Table 1.

The synchronously rotating reference frame mathematical equation of induction machine is given as [7].

$$v_{qs} = r_s i_{qs} + \frac{d}{dt}\lambda_{qs} + \omega_e \lambda_{ds} \tag{1}$$

$$v_{ds} = r_s i_{ds} + \frac{d}{dt}\lambda_{ds} - \omega_e \lambda_{qs} \tag{2}$$

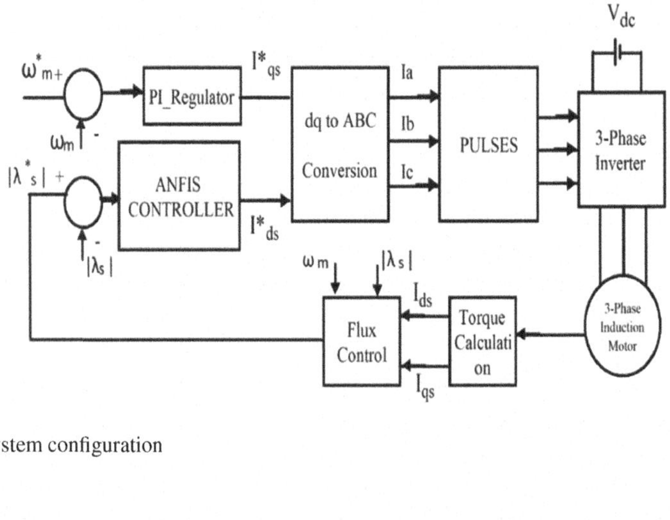

Fig. 1 System configuration

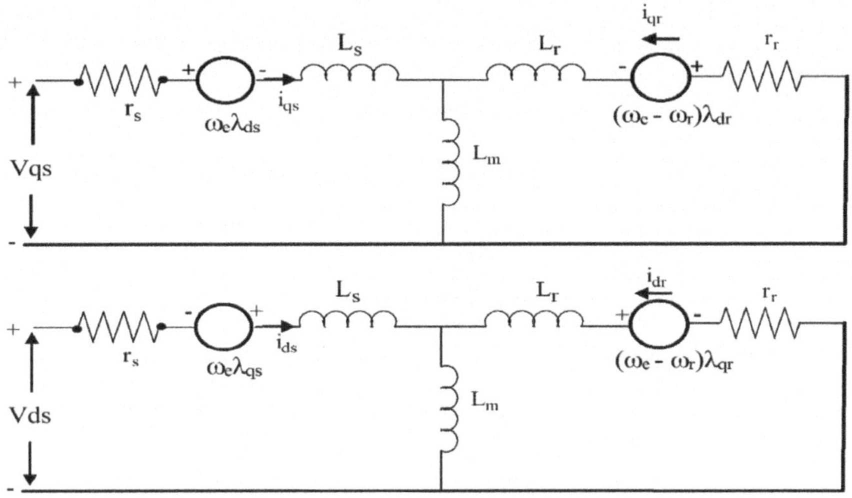

Fig. 2 Induction motor *d-q* equivalent circuit in synchronous frame

Table 1 Ratings and parameter of three-phase induction motor

Parameters	Values
Nominal power	50 HP
Voltage applied	460 V
Frequency	60 Hz
Stator resistance Rs and stator inductance Ls	0.087 Ω and 0.8 mH
Rotor resistance Rr and Rotor inductance Lr	0.228 Ω and 0.8 mH
Mutual inductance Lm (H)	34.7 mH
Inertia, friction factor, pole	1.662 (kg m^2), 0.01(Nm s), 4
Synchronous speed	1800 rpm
Angular frequency	188.45 rad/s

$$v_{qr} = r_r i_{qr} + \frac{d}{dt}\lambda_{qr} + (\omega_e - \omega_r)\lambda_{dr} = 0 \tag{3}$$

$$v_{dr} = r_r i_{dr} + \frac{d}{dt}\lambda_{dr} - (\omega_e - \omega_r)\lambda_{qr} = 0 \tag{4}$$

where

$$\lambda_{qs} = L_s i_{qs} + L_m i_{qr} \tag{5}$$

$$\lambda_{ds} = L_s i_{ds} + L_m i_{dr} \tag{6}$$

$$\lambda_{qr} = L_r i_{qr} + L_m i_{qs} \tag{7}$$

$$\lambda_{dr} = L_r i_{dr} + L_m i_{ds} \tag{8}$$

The electromagnetic torque equation is represented in terms of stator current and rotor flux linkage as:

$$T_e = \frac{3}{2}\frac{P}{2}\frac{L_m}{L_r}\left[\lambda_{dr} i_{qs} - \lambda_{qr} i_{ds}\right] \tag{9}$$

Input power is calculated using only DC voltage and current.

$$P_{in} = V_{dc} * I_{dc} \tag{10}$$

In this study, the motor torque and speed are kept constant, and hence, the output power of motor is constant.

Output power is given as

$$P_{out} = T_e * \omega_m \tag{11}$$

The efficiency is written as

$$\eta = \frac{P_{out}}{P_{in}} \tag{12}$$

3 Controller Design

3.1 Artificial Neural Network (ANN) Controller

Artificial neural network (ANN) is an information processing system which has capability to study and upgrade its operation using biological neural network training. It consists of a set of highly interrelated simple nonlinear processing elements such as neurons, units, cells or nodes. Every neuron is attached to the other neurons with the help of direct communication links where each has an analogous weight. The weight represents information being used by the network to solve the complication [4]. By using weights, the relation between input and output can be adjusted. The mathematical expression of neuron is given by:

$$y = \sum_{i=0}^{n} W_i . X_i + b \tag{13}$$

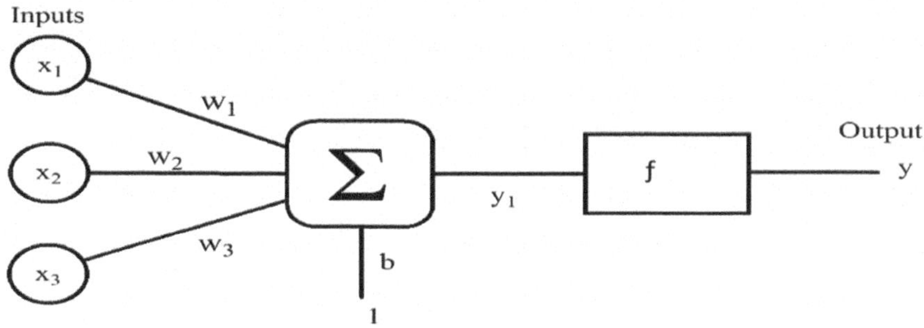

Fig. 3 A detailed view of one neuron model

where X_1, X_2, …, X_n are the input signals of the neuron, W_1, W_2, …, W_m are the weights, and b is the biasing parameter. Y represents the output signal of neuron. Figure 3 shows the detailed view of one neuron.

3.2 Adaptive Neuro-Fuzzy Inference System (ANFIS) Controller

ANFIS is an intelligent technique which is very effective in problems like modelling and control of complex systems. It is a group of adaptive networks that are utilitarian identical to fuzzy interference system. The fuzzy neural network is formulated to combine fuzzy inference mechanism and neural network which implements a fuzzy inference [7]. In ANFIS, the fuzzy inference widely used is the first-order Sugeno fuzzy model because of its intelligibility, high explainable, computational efficiency and adaptive approach techniques. ANFIS prototypical structure is represented in Fig. 4.

It incorporates the features of learning capability of neural network with proper knowledge characterization of fuzzy logic. A neuro-fuzzy system is trained by a learning algorithm obtained from neural network theory (Figs. 5 and 6).

Figures 7, 8, 9 and 10 depict the membership function and trained values of input and output variables of ANFIS model. The trained values are obtained from neural network theory.

4 Simulation Results of the Model

In this study, the performance analysis of the suggested ANFIS controller is assessed at different conditions of load, and the results are compared with the ANN controller.

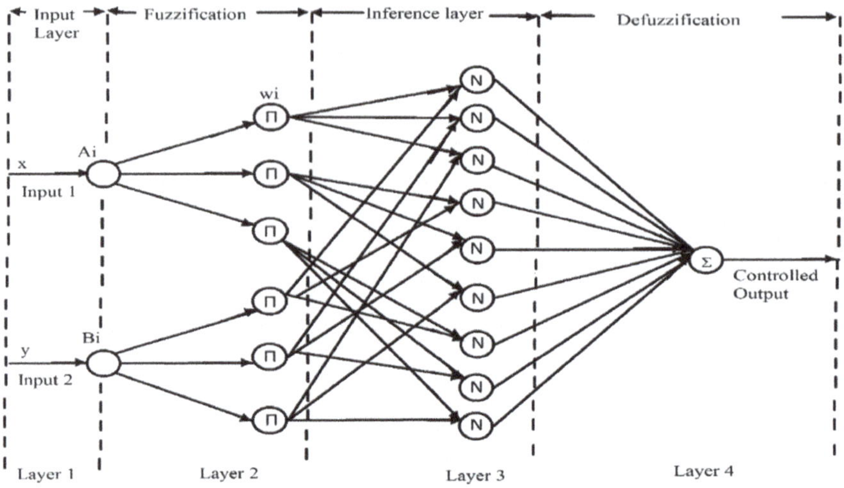

Fig. 4 Structure of ANFIS controller

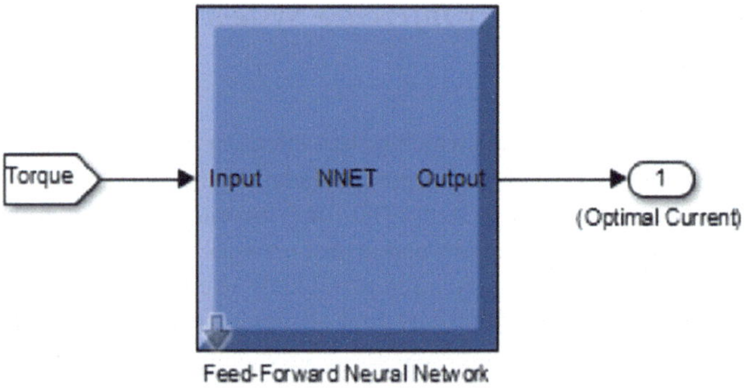

Fig. 5 ANN model

Fig. 6 ANFIS model

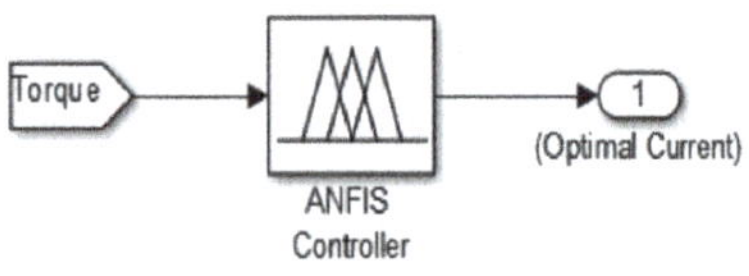

4.1 Case Study I: At 150 Nm Load Torque and Different Motor Speed at 90 rps, 120 rps, 150 rps and 180 rps

The efficiency and input power of the three-phase induction motor drive for the same load torque and at different speeds are studied, and its performance is compared and shown in Figs. 11, 12, 13 and 14. The optimal value of current is derived by

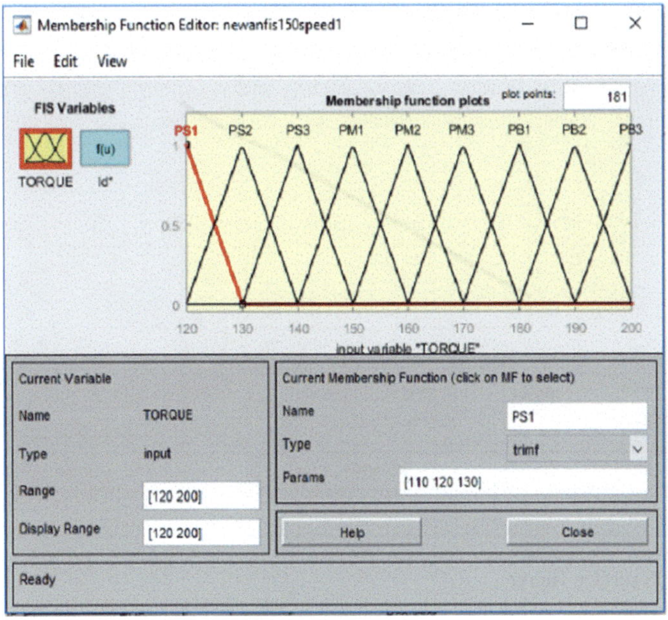

Fig. 7 Membership function for input and output variables

Fig. 8 ANFIS trained values

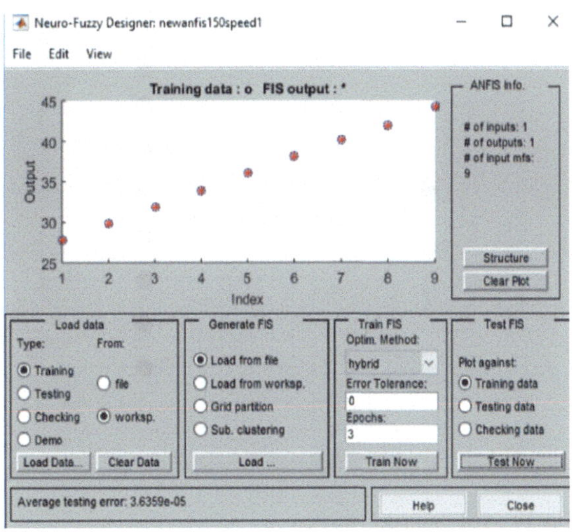

mathematical modelling of motor; it is given to the controllers. The output of the controller is fed to the current generation block which generates PWM pulses for the inverter operated drive resulting in efficient operation of drive in terms of maximum efficiency.

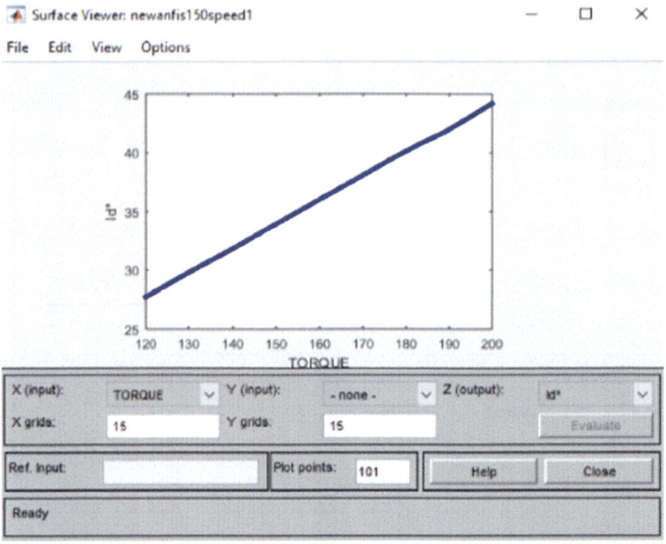

Fig. 9 ANFIS surface viewer

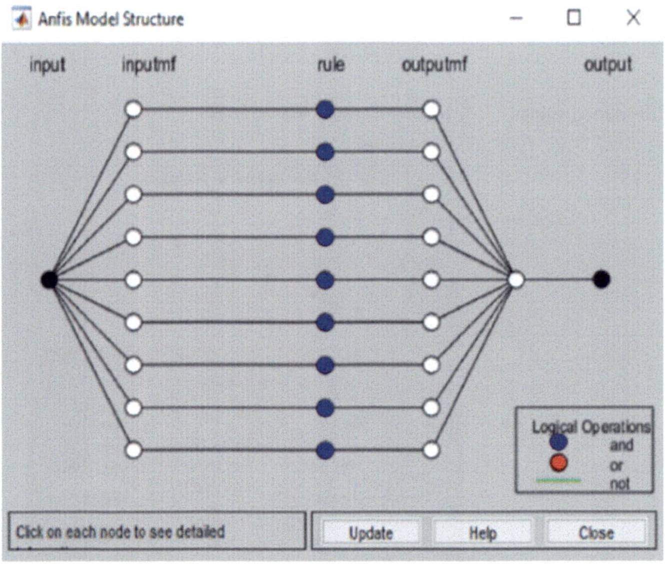

Fig. 10 ANFIS model structure

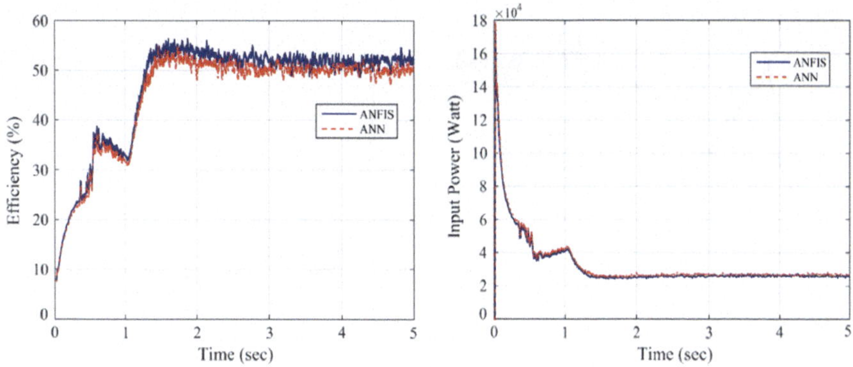

Fig. 11 Efficiency and input power comparison of ANN and ANFIS controllers at 150 Nm and 90 rps

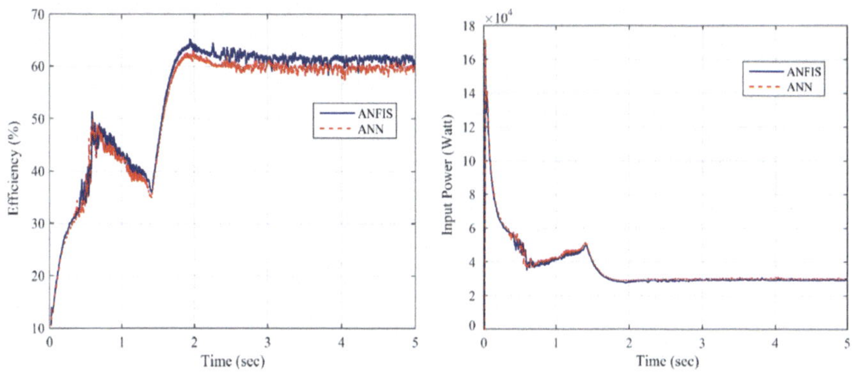

Fig. 12 Efficiency and input power comparison of ANN and ANFIS controllers at 150 Nm and 120 rps

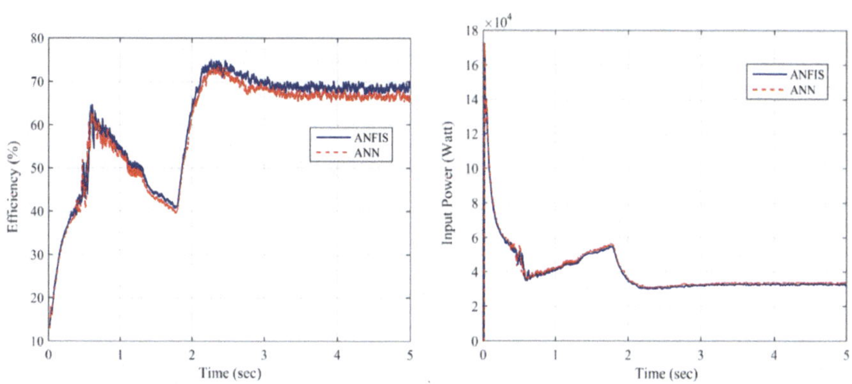

Fig. 13 Efficiency and input power comparison of ANN and ANFIS controllers at 150 Nm and 150 rps

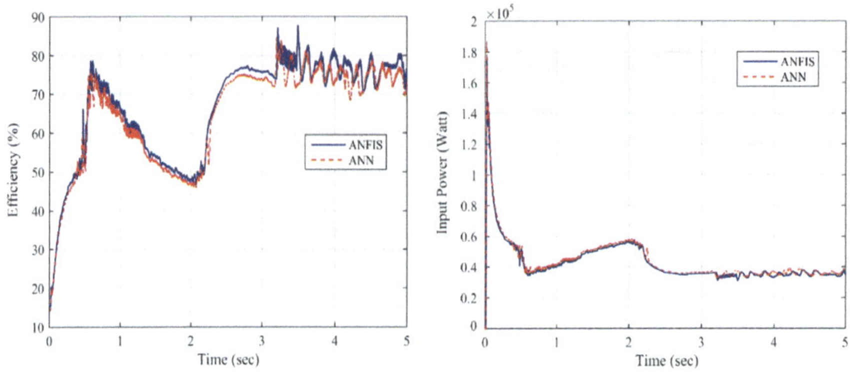

Fig. 14 Efficiency and input power comparison of ANN and ANFIS controllers at 150 Nm and 180 rps

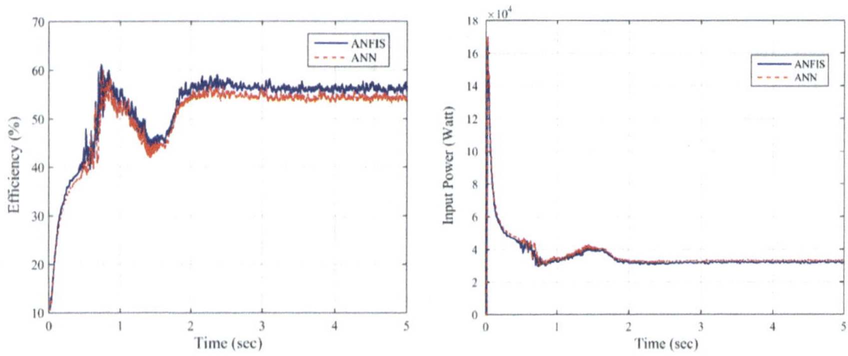

Fig. 15 Efficiency and input power comparison of ANN and ANFIS controllers at 200 Nm and 90 rps

4.2 *Case Study II: At 200 Nm Load Torque and Different Motor Speed at 90 rps, 120 rps, 150 rps and 180 rps*

The result of this case study is shown in Figs. 15, 16, 17, 18 and result comparison of input power, saved power and efficiency of the two controller are shown in Figs. 19, 20 and 21.

5 Result Analysis

Tables 2 and 3 show a comparative analysis of the results of two controllers, and it shows that better results are obtained from ANFIS controller and energy is saved up

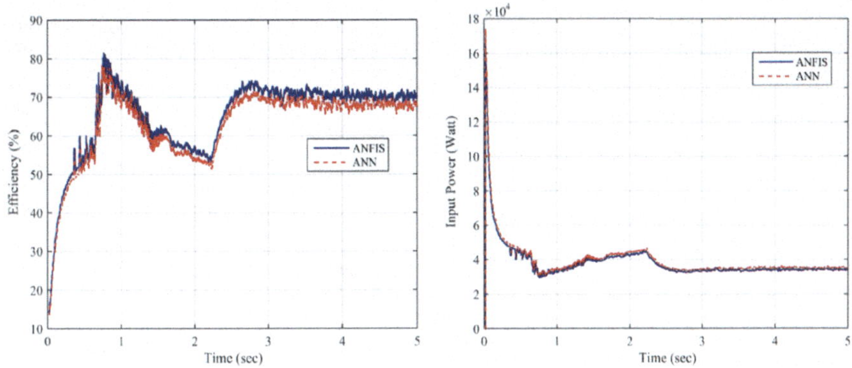

Fig. 16 Efficiency and input power comparison of ANN and ANFIS controllers at 200 Nm and 120 rps

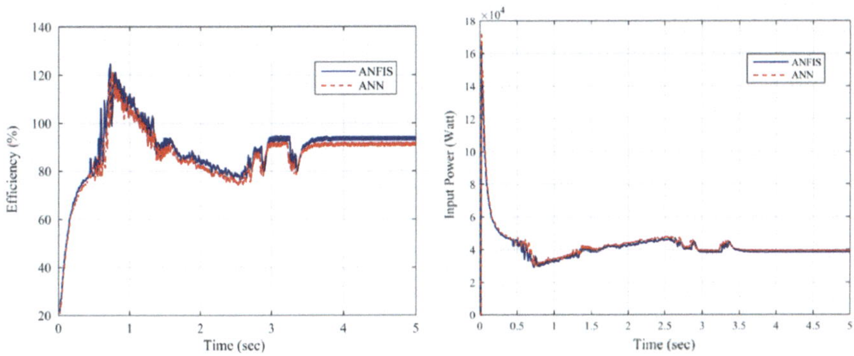

Fig. 17 Efficiency and input power comparison of ANN and ANFIS controllers at 200 Nm and 150 rps

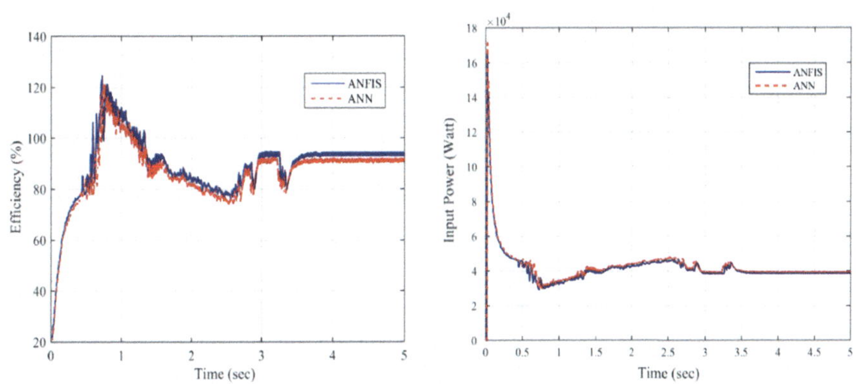

Fig. 18 Efficiency and input power comparison of ANN and ANFIS controllers at 200 Nm and 180 rps

Fig. 19 Comparison of input power and speed

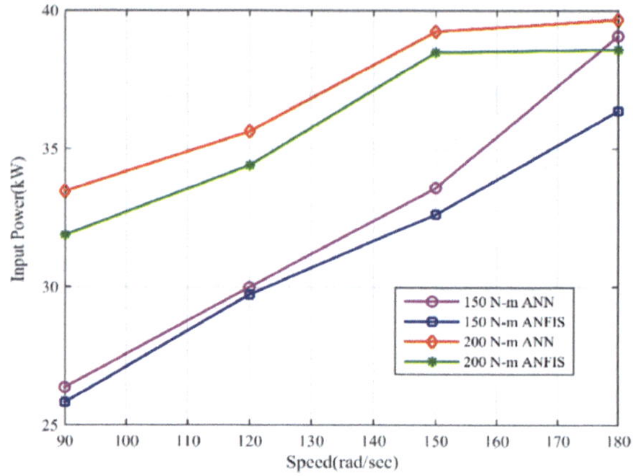

Fig. 20 Comparison of saved power and speed

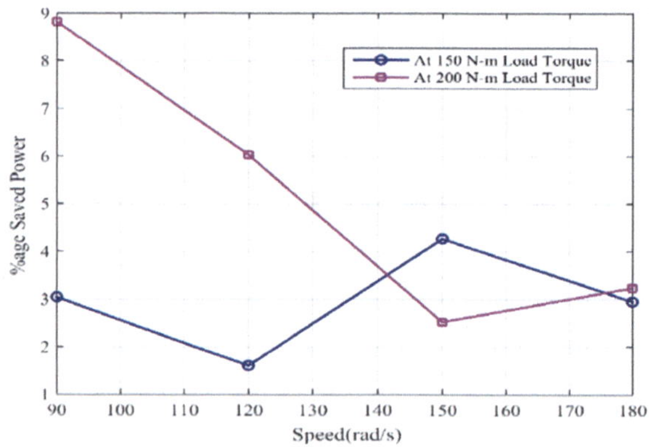

Fig. 21 Comparison graph of efficiency at different load torques

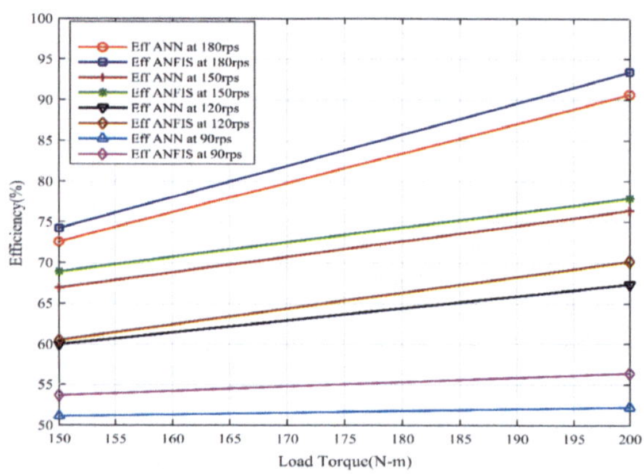

Table 2 Comparison of ANN controller and ANFIS controller results at 150 Nm

Speed	90 rad/s		120 rad/s		150 rad/s		180 rad/s	
Controller	ANN	ANFIS	ANN	ANFIS	ANN	ANFIS	ANN	ANFIS
System efficiency at 150 Nm (%)	51.16	52.25	59.96	60.54	66.96	68.93	72.63	74.24
Input power (kW)	26.39	25.84	30.02	29.73	33.6	32.64	37.17	36.37
Electric demand saving (kW)	0.55		0.29		0.96		0.8	
Savings/output %age	3.05		1.61		4.27		2.96	
Annual saving of energy for 5840 h @.1052 $	337.90 $		178.16 $		589.79 $		491.49 $	

Table 3 Comparison of ANN controller and ANFIS controller results at 200 Nm

Speed	90 rad/s		120 rad/s		150 rad/s		180 rad/s	
Controller	ANN	ANFIS	ANN	ANFIS	ANN	ANFIS	ANN	ANFIS
System efficiency at 200 Nm (%)	53.74	56.43	67.34	70.20	76.42	77.93	90.71	93.45
Input power (kW)	33.49	31.9	35.64	34.19	39.26	38.5	39.69	38.52
Electric demand saving (kW)	1.59		1.45		0.76		1.17	
Savings/output %age	8.83		6.041		2.533		3.25	
Annual saving of energy for 5840 h @.1052 $	976.84 $		890.83 $		466.91 $		718.81 $	

to a great extent. The operating cost and efficiency of the complete drive are much improved by ANFIS controller.

6 Conclusion

In this paper, a comparative analysis between two optimization techniques is done, and performance evaluation is done. These optimization techniques generate optimal value of current which minimizes the losses of induction motor and improved the efficiency. From the above tables, it is concluded that there is reasonable reduction in input power due to which the cost optimization of electric power is achieved. From the present study, it is concluded that ANFIS controller scheme is proved to be a promising technique as compared to ANN controller technique under different load conditions.

References

1. Satish Kumar H, Parthasarathy SS (2017) A novel neural network intelligent controller for vector controlled induction motor drive. In: 2017 International conference on alternative energy in developing countries and emerging economics, Bangkok, Thailand, pp 692–697
2. Kilic E, Ozcalik HR, Yilmaz S, Sit S (2015) A comparative analysis of FLC and ANFIS controller for vector controlled induction motor drive. In: IEEE international conference on ACEMP-OPTIM-ELECTROMOTION 2015, Side, Turkey, 2015, pp 102–106
3. Hussain S, Bazaz MA (2014) ANFIS implementation on a three phase vector controlled induction motor with efficiency optimisation. In: IEEE international conference on circuits, systems, communication and information technology (CSCITA), Mumbai, India, pp 391–396
4. Aakanksha T, Naveen A (2015) Artificial neural network controller for induction motor drive. Int J Sci Res (IJSR) 4(6):805–812
5. Bose BK (2003) Modern power electronics and AC drives, 1st edn. Pearson Education, Inc., London
6. Malkhandale UB, Bawane NG (2017) Comparison of current controllers induction machine 1 HP based on ANN, Fuzzy, ANFIS and PI. IOSR J Electr Electron Eng (IOSR-JEEE) 12(4):36–42. ISSN: 2278-1676
7. Naga Sujatha K, Vaisakh K (2010) Implementation of adaptive neuro fuzzy inference system in speed control of induction motor drives. J Intell Learn Syst Appl 2:110–118. https://doi.org/10.4236/jilsa.2010.22014. Published Online May 2010
8. Choudhary PK, Dubey SP (2016) Efficiency optimization of induction motor drive in steady-state using artificial neural network. In: 2016 International conference on computation of power, energy information and communication (ICCPEIC), Chennai, pp 295–302

Chapter 27
Optimal Placement and Sizing of Distributed Generations Using Soft Computing Approach in Radial Distribution Network

Umesh Kumar Gupta, Shashi Kant Pandey and Ram Ishwar Vais

1 Introduction

The period after 1950 marked the beginning of the interconnection of the electricity grids. The extension of synchronous zones has led to an overall improvement in reliability through exchanges and mutual assistance between large geographical areas. The production of electricity was largely centralized and the electrical system managing companies were maintained great monopolies. The liberalization of electricity markets combined with the integration of renewable energies has literally upset this old pattern to arrive at a new organizational model [1]. Wind turbines first and then photovoltaic production has gradually introduced the notion of dispersed production [2]. Local production of electricity should be further developed. From the perspective of diminishing CO_2 emissions, and in order to increase the efficiency of fuel consumption, cogeneration systems are an interesting solution. Unlike the transport of electrical energy, the heat generated by these systems and used as hot water or steam cannot be transported over long distances. This requires local use of these types of generator. Under these conditions, the ideal positioning of the heat-electricity cogeneration systems is in the distribution system and at the load sites [3].

The principle of the electronic connection of a source is now very well mastered. Converters for wind turbines, photovoltaic panels and storage devices are produced in large series [4]. However, the coexistence between conventional sources and sources connected with a power electronics converter in the context of a microarray remains a real challenge and constitutes a still wide scope for research on electrical systems in terms of control but also in terms of protection.

Most developing countries are experiencing significant growth in energy demand. This growth, linked to that of the population and to the improvement of the standard

---

U. K. Gupta (✉) · S. K. Pandey · R. I. Vais
Department of Electrical Engineering, Rajkiya Engineering College Sonbhadra, Sonbhadra, UP, India

V. K. Giri et al. (eds.), *Computing Algorithms with Applications in Engineering*, Algorithms for Intelligent Systems, https://doi.org/10.1007/978-981-15-2369-4_27

of living, is estimated at an annual average of about 3.8% for the next three decades [5]. In order to satisfy this expansion of energy needs, many investments associated with an adequate energy policy are needed, both for the construction of new production units and for the improvement and extension of the current transmission and distribution networks. The basic question is how this indispensable development should be carried out. Under current economic conditions, public subsidies are steadily decreasing, making it more difficult to mobilize the financing needed to build coherent energy systems. The World Commission on Environment and Development characterizes feasible improvement [6] as: 'advancement that addresses the issues of the present without trading off the capacity of future eras to address their issues.' Thus, the search for a path of sustainable development imposes three conditions to fulfill simultaneously:

- Improving the quality of life.
- The maintenance of permanent access to natural resources, whether renewable or non-renewable.
- Reduction of emissions of waste or pollutants in order to avoid any persistent environmental damage.

Therefore, decentralization of electricity production based on renewable energy resources should be considered [7], as the increase in production is not without consequences on the environment. However, the basis for such an energy strategy must be economically efficient, socially equitable, environmentally sustainable and contribute to reducing disparities.

The Kyoto Protocol creates the Clean Development Mechanism (CDM) [8]. The CDM will enable developing countries (PEDs) either to initiate CO_2 emission reduction projects themselves or in partnership with industrialized countries and in return to sell to them the emission reductions certified. If the CDM is well established in the energy sector, it will contribute to reversing or slowing down the trend toward increased use of fossil fuels. It should be pointed out that the success of this mechanism will depend on the ability of the governments of the developing countries to establish criteria for judging the sustainability of projects, the benefits of technology transfer and the reduction of pollution.

2 Distribution Network

Distribution networks (DNs) can be characterized as which are worked at the most minimal voltage levels in the general power systems and serve residential, commercial and industrial loads. Classically, most DNs are outspread by nature, having a single source of power [9]. The present day power DNs are continuously facing problem due to increasing load demand, these entire causes reduced voltage and increased burden on the networks also results on the process, scheduling, methodical and security concerns of distribution networks. This causes excessive power losses in distribution system. Classification of the DGs can be done as [10]:

Type-I: DG, which can inject real power only to DNs, for example photovoltaic, fuel cells, etc.
Type-II: DG, which can inject reactive power only, hence increase the voltage profile of the corresponding power network, for example; capacitors, synchronous compensator, Kvar compensator, etc.
Type-III: DG, which can inject both real and reactive power to DNs, for example synchronous machines.
Type-IV: DG, which can inject real power but withdraws reactive power from DNs, for example induction generators utilized for the electricity production in the wind mills.

2.1 Distribution Network Power Losses

According to studies, 70% of the total power networks losses are occurring in the distribution networks, whereas transmission networks signify only 30% of the total losses [9]. Power delivery proficiency can be enhanced by reducing the power loss, specifically at the distribution stage.

The succeeding methods are adopted for reducing the DNs losses [9].

- Reinforcement of the feeders.
- Reactive power compensation.
- Conductor grading.
- Reconfiguration of feeder.
- Placement of distributed generator.

2.2 Optimization Methods Applied to Distribution Network

The strategies used to tackle the arranging issue can be separated into two classes: mathematical programming methods (deterministic methods) and non-deterministic (or stochastic) methods, in particular those using evolutionary algorithms. Deterministic methods are effective when one has an idea of the global optimum (in fact, it converges toward the optimum closest to the starting point). More complex cases (many local optima, non-differentiable functions, etc.) are often treated by non-deterministic methods. However, these can lead to longer computing times [11].

The mathematical programming of a techno-economic problem consists in finding, among all the points x satisfying certain conditions, one which makes a certain criterion $f(x)$ maximal (or minimal, depending on the case), which will be interpreted as a gain in the first case (and as a cost in the second). When the variable x is of finite dimension, and its components $(x_1, \ldots, x_n)$ can only take integer values, we speak of programming in integers; when it is continuous, that is to say, when x

describes R_n or another vector space, one speaks of linear, convex or non-convex programming according to the properties of the functions *f*, *g* and *h*. Finally, the variable *x* may present itself as a function of other more primitive variables, notably time; then dynamic programming is used. Mathematical programming can therefore be defined as the numerical analysis of optimization and control problems. It was of great historical importance, for at a time when the means of calculation were far from being what they now, only high-performance numerical methods can link theory and practice, i.e., demonstrate to users the relevance of models adopted by researchers, and feed them into concrete problems [11].

Strategies for tackling mathematical issues rely on the objective function and on the set of constraints. The following major sub domains exist: linear programming (LP), mixed-integer programming (MIP), quadratic programming (QP), nonlinear programming (NLP), simulated annealing (SA), tabu search, genetic algorithms (GA), dynamic programming, etc.

3 Problem Formulation

Objective Function

Considering N bus distribution system, the formula for real power loss minimization is expressed as [12]:

$$\text{Min } P_L = \sum_{i=1}^{N}\sum_{j=1}^{N}\left[\alpha_{ij}\left(P_iP_j + Q_iQ_j\right) + \beta_{ij}\left(Q_iP_j + P_iQ_j\right)\right] \tag{1}$$

where

$$\alpha_{ij} = \frac{r_{ij}}{V_iV_j}\cos\left(\delta_i - \delta_j\right)$$

$$\beta_{ij} = \frac{r_{ij}}{V_iV_j}\sin\left(\delta_i - \delta_j\right)$$

with

$$Z_{ij} = r_{ij} + jx_{ij}$$

where

Z_{ij} is the impendence of the line between bus *i* and bus *j*.
r_{ij} is the resistance of the line between bus *i* and bus *j*.
x_{ij} is the reactance of the line between bus *i* and bus *j*.
V_i is the voltage magnitude at bus *i*.
V_j is the voltage magnitude at bus *j*.

δ_i is the voltage angle at bus i.
δ_j is the voltage angle at bus j.
P_i and Q_i are the active and reactive power injections at bus i.
P_j and Q_j are the active and reactive power injections at bus j.

Constraint

The objective function in Eq. (1) is subjected to following constraints [12]:

1. ***System Power Flow Conditions Need to be Fulfilled***:

$$P_{Gi} - P_{Di} = \sum_{j=1}^{N} V_i V_j \left[G_{ij}\cos(\delta_i - \delta_j) + B_{ij}\sin(\delta_i - \delta_j)\right] \quad \forall i = 1, 2, 3, \ldots, N \tag{2}$$

$$Q_{Gi} - Q_{Di} = \sum_{j=1}^{N} V_i V_j \left[G_{ij}\sin(\delta_i - \delta_j) - B_{ij}\cos(\delta_i - \delta_j)\right] \quad \forall i = 1, 2, 3, \ldots, N \tag{3}$$

where G_{ij} is the conductance of the line between bus i and bus j.
B_{ij} is the susceptance of the line between bus i and bus j.

$$P_i = P_{Gi} - P_{Di}$$
$$Q_i = Q_{Gi} - Q_{Di}$$

P_{Gi} and Q_{Gi} are power generations of generators at bus i. P_{Di} and Q_{Di} are the loads at bus i.

2. ***Voltage Constraint at Each Bus (± 5% of Rated Voltage) Need to be Fulfilled*** [13]:

$$V_{\min} \le V_i \le V_{\max} \tag{4}$$

where $i = 1, 2, 3, \ldots, N$.

3. ***Line Current Constraint Need to be Fulfilled***:

$$I_i \le I_i^{\text{Rated}} \quad \forall i \in \{\text{branches of the network}\} \tag{5}$$

where I_i^{Rated} is current admissible for branch i inside safe point of confinement of temperature.

4 Proposed Method

Numerous optimization algorithms are available for sizing and location of DG in radial distribution networks [14–20]. Out of those several algorithms, PSO and GWO algorithms are very prominent and effective which will be used here to decide the optimal size and location of Type-I DG and Type-IV DG units and results for these algorithms will be compared with each other.

4.1 Particle Swamp Optimization

James Kennedy and Russell C. Eberhart proposed a PSO approach in 1995. This approach is a heuristic method [21]. The evaluation of candidate solution of current search space is done on the basis of iteration process. The minima and maxima of objective function are determined by the candidate's solution as it fits the task's requirements. Since PSO algorithm does not accept the objective function data as its inputs, therefore the solution is randomly away from minimum and maximum (locally/globally) and also unknown to the user. The speed and position of candidate's solution is maintained and at each level, fitness value is also updated. The best value of fitness is recorded by PSO for an individual record. The other individuals reaching this value are taken as the individual best position and solution for given problem. The individuals reaching this value are known as global best candidate solution with global best position. The upgradation of global and individual best fitness value is carried out and if there is a requirement then global and local best fitness values are even replaced. For PSO's optimization capability, the updation of speed and position is necessary. Each particle's velocity is updated with the help of subsequent formula:

$$v_i(t+1) = wv_i(t) + c_1 r_1 \left[\hat{x}_i(t) - x_i(t)\right] + c_2 r_2 [g(t) - x_i(t)] \tag{6}$$

PSO Algorithm for DG Placement and Sizing

Step 1: Start
Step 2: Initialization of variables: Read the various data, which includes bus data, line data, population size, dimension of the problems, maximum number of iterations, DG size limitations, local information (c_1), global information (c_2), etc.
Step 3: Initializing swarm and velocities: Create random population of DG size and random population of location of DG considering the limit of DG size and its locations, respectively.
Step 4: Evaluation of initial cost of all candidates: Calculate the objective function (OF), i.e., cost by using load flow analysis (LFA), of all particles considering the DG size and its location which is generated in previous step.
Step 5: Find out the best candidate among initial population: According to the obtained cost of all particles, find the optimum cost which is called global minima or local minima (in the initial condition, both local and global minima are same). Find the particle position corresponding to optimum cost, i.e., global best (gb), and also find the personal best (pb). In the initial condition, personal best position (pb) is same as initial population.

Step 6: Starting iteration:
Set iteration counter, $iter = 1$
Step 7: Update candidates velocity and position.
Step 8: Authentication of particles: Check the validity of candidate according to the candidate's particular conditions, i.e., validate the candidate's new positions according to the limitations of DG size and its valid locations. If new position of any particle is not valid, then randomly regenerate that candidate according to their conditions.
Step 9: Evaluate the cost function (CF), i.e., real power loss minimization using LFA, corresponding to new particles.
Step 10: Update the personal_best (pb) and global_best (gb) value.
Step 11: Increment the iteration count (iter = iter + 1)
Step 12: If stopping criteria is not satisfied (i.e., if iter ≤ maxiter && repeat < maxrepeat) then go to **Step 7**, otherwise continue to next step.
Step 13: Print the results and plot graphs.
Step 14: Stop.

4.2 Grey Wolf Optimization

Grey wolf optimization (GWO) algorithm was first proposed in 2014 by Mirjalili et al. [22]. GWO is an iterative search technique, which is based on swarm intelligence. GWO algorithm was simulated by the self-governing behavior and the hunting mechanism of grey wolves for a prey in the forest. Grey wolves usually prefer to live in a pack. A pack of grey wolves consists of four different categories of wolves according to their ranking, α, β, δ and ω. In a pack, they abide themselves by the harsh social leadership hierarchical structure as shown in the Fig. 1.

The leaders of a grey wolf pack are called as alphas (α), which may be a male or female. Prominently these alphas take decision like hunting of prey, dozing place, etc. Their decisions are binding to the pack. In other words, alpha wolves are also called as dominant wolves among the pack because pack followed the decision taken by them. The advisors to the alpha wolves are designated as betas (β) (second ranking in hierarchy), which work as subordinate to the alpha and help them while making any decision. The β wolf can be a male or female, he/she should respect alphas, as well commands other low-ranking wolves in a pack. The β imposes the alpha's command to the pack also provides feedback to α. The third level of grey wolves named as deltas (δ), which have to accede to α and β but they command lowest rank

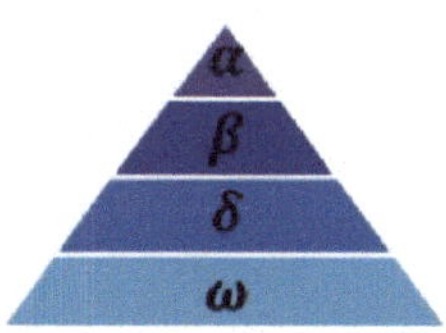

Fig. 1 Grey wolf social leadership hierarchy (superiority increases from bottom to top) [22]

grey wolves named as omegas (ω). The custodians, predators, elders and vanguards belong to δ category wolves. Omega wolves forever have to accede to all the other superior wolves.

Mathematical Model of GWO Algorithm

Grey wolf while hunting takes certain steps which are as follows:

1. Tracking the prey
2. Encircling the prey
3. Attacking the prey.

In this section, social hierarchy and hunting steps are presented in mathematical models.

Social hierarchy: While modeling social hierarchy of the grey wolves in mathematical form, the alpha (α) is considered as the fittest solution. Thus, the beta (β) and delta (δ) are considered as second best and third best solution, respectively. The remaining candidate (wolves) solutions are concluded as omega (ω). The alpha (α), beta (β) and delta (δ) escort the hunting (optimization) process in GWO algorithm. The omega (ω) wolves follow them during hunting process.

Encircling Prey: During hunting (optimization) grey wolves encircle the pray until it stops moving. The modeling of encircling behavior of the grey wolves in mathematical form is represented by following equations:

$$\vec{D} = \left| \vec{C} . \vec{X}_{pt} - \vec{X}(t) \right| \tag{7}$$

$$\vec{X}(t+1) = \vec{X}_{pt} - \vec{A} . \vec{D} \tag{8}$$

where $\vec{X}_P$ is the position vector of the prey, $\vec{X}$ denotes the position vector of a grey wolf, t denotes the current iteration, and $\vec{A}$, $\vec{C}$ represent coefficient vectors.

The computation of coefficient vectors $\vec{A}$ and $\vec{C}$ is carried out as:

$$\vec{A} = 2.\vec{a}.\vec{r}_1 - \vec{a} \tag{9}$$

$$\vec{C} = 2.\vec{r}_2 \tag{10}$$

where $\vec{a}$ = value decreases linearly from 2 to 0 during iteration.

$\vec{r}_1, \vec{r}_2$ are random vectors in [0, 1] and allow wolves to reach any position in the search space around the prey to obtain best solution.

Optimization The following equations are formulated for the candidates to update their position.

$$\begin{Bmatrix} \overrightarrow{D}_\alpha = \left| \overrightarrow{C}_1 . \overrightarrow{X}_\alpha - \overrightarrow{X} \right| \\ \overrightarrow{D}_\beta = \left| \overrightarrow{C}_2 . \overrightarrow{X}_\beta - \overrightarrow{X} \right| \\ \overrightarrow{D}_\delta = \left| \overrightarrow{C}_3 . \overrightarrow{X}_\delta - \overrightarrow{X} \right| \end{Bmatrix} \tag{11}$$

$$\begin{Bmatrix} \overrightarrow{X}_1 = \overrightarrow{X}_\alpha - \overrightarrow{A}_1 . \overrightarrow{D}_\alpha \\ \overrightarrow{X}_2 = \overrightarrow{X}_\beta - \overrightarrow{A}_2 . \overrightarrow{D}_\beta \\ \overrightarrow{X}_3 = \overrightarrow{X}_\delta - \overrightarrow{A}_3 . \overrightarrow{D}_\delta \end{Bmatrix} \tag{12}$$

$$\overrightarrow{X}(t+1) = \frac{\overrightarrow{X}_1 + \overrightarrow{X}_2 + \overrightarrow{X}_3}{3} \tag{13}$$

Attacking Prey (Exploitation) and Search for Prey (Exploration)
The mathematical modeling of attacking the pray by grey wolves is carried out by linearly decreasing the value of $\vec{a}$ from 2 to 0 during iteration count, thus value of $\overrightarrow{A}$ will also be decreasing during each iteration given in Fig. 2. The '$\overrightarrow{A}$' will take any random values in between [−2a, 2a]. The random values '$\overrightarrow{A}$' are utilized to force the candidate (search agent) to move toward or away from the prey. When $|A| < 1$, the wolves are forced to attack the prey and when $|A| > 1$, the grey wolves are enforced to diverge from the prey. In this way, GWO algorithm searches for optimum globally and locally.

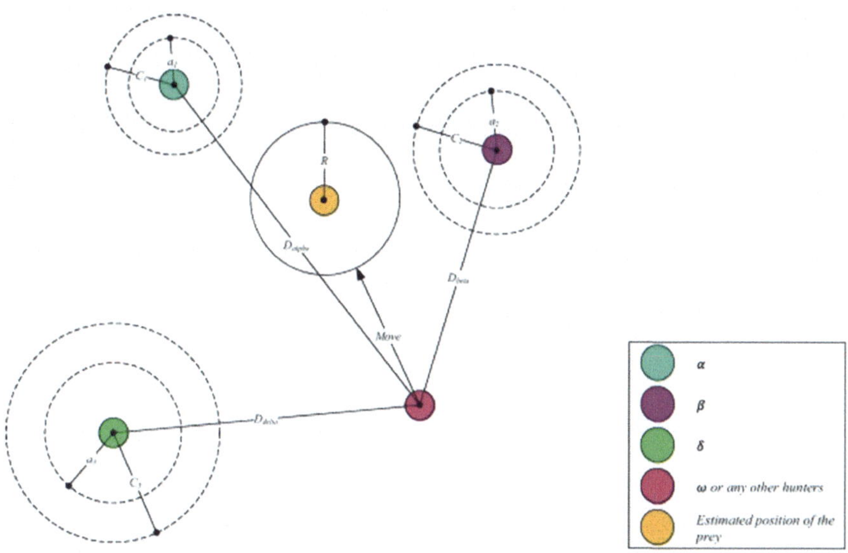

Fig. 2 Grey wolf position update in GWO [22]

GWO Algorithm for DG Placement and Sizing

Step 1: Start
Step 2: Initialization of variables: Load the various data, which includes bus data, line data, population size, dimension of the problems, maximum number of iterations, DG size limitations, parameter $\vec{a}$, coefficient vectors $\overrightarrow{A}$ and $\overrightarrow{C}$, etc.
Step 3: Initializing grey wolf population: Create random population of DG size and random population of location of DG considering the limit of DG size and its locations, respectively.
Step 4: Evaluation of initial cost of all candidates (search agents): Calculate the objective function (OF), i.e., cost by using load flow analysis (LFA), of all particles considering the DG size and its location which is generated in previous step.
Step 5: Finding the first three best solutions $\overrightarrow{X}_{\alpha}$, $\overrightarrow{X}_{\beta}$ and $\overrightarrow{X}_{\delta}$ according to the obtained cost of all candidates.
Step 6: Starting iteration:
Set iteration counter, *iter* = 1
Step 7: Update each candidate position: Update current search agent position according to equation:

$$\overrightarrow{X}(t+1) = \frac{\overrightarrow{X}_1 + \overrightarrow{X}_2 + \overrightarrow{X}_3}{3}$$

Step 8: Authentication of candidate: Check the validity of candidate according to the candidate's particular conditions, i.e., validate the candidate's new positions according to the limitations of DG size and its valid locations. If new position of any particle is not valid, then randomly regenerate that candidate according to their conditions.
Step 9: Updating the parameter $\vec{a}$, coefficient vectors $\overrightarrow{A}$ and $\overrightarrow{C}$ and calculate the objective function (OF), i.e., real power loss minimization using LFA.
Step 10: Update $\overrightarrow{X}_{\alpha}$, $\overrightarrow{X}_{\beta}$ and $\overrightarrow{X}_{\delta}$.
Step 11: Increment the iteration count.
Step 12: If stopping criteria is not satisfied (i.e., if iter $\leq$ maxiter && repeat < maxrepeat) then go to **Step 7**, otherwise continue to next step.
Step 13: Print the results and plot graphs.
Step 14: Stop.

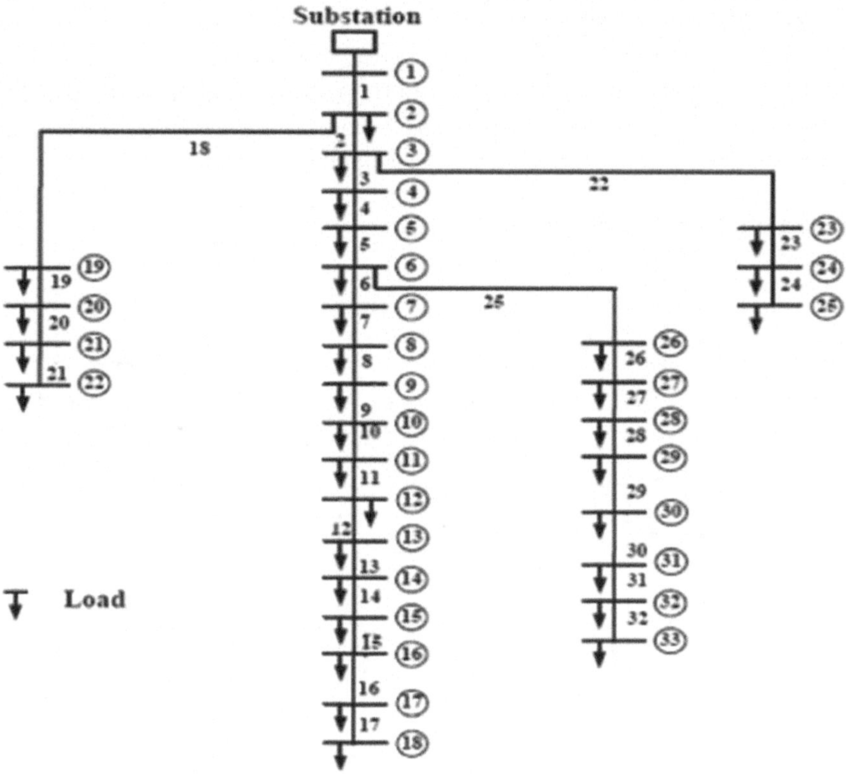

Fig. 3 IEEE-33 bus radial distribution system

Table 1 Comparison of active power loss in Type-I DG and Type-IV DG using PSO approach

Type of DG	Optimum DG location	Size of DG P (MW)	Power loss P (MW)
Type-I	Bus 6	2.611	0.112
Type-IV	Bus 30	1.513	0.071

Table 2 Comparison of active power loss in Type-I DG and Type-IV DG using GWO approach

Type of DG	Optimum DG location	Size of DG P (MW)	Power loss P (MW)
Type-I	Bus 6	2.465	0.110
Type-IV	Bus 30	1.159	0.070

5 Results and Analysis

The proposed methodology is tested on MATLAB 2014a using IEEE-33 Bus [23] test system is shown in Fig. 3. Tables 1 and 2 show the active power loss in Type-I DG and Type-IV DG using PSO approach and proposed approach

Results using PSO Optimization
See Figs. 4, 5 and 6.

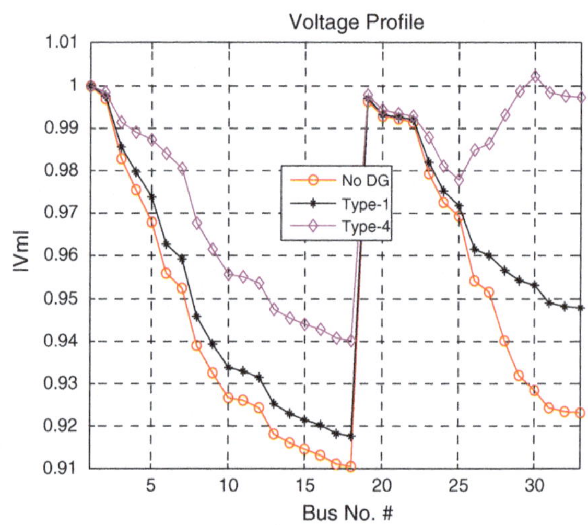

Fig. 4 Comparison of voltage profile for 33 bus radial distribution system

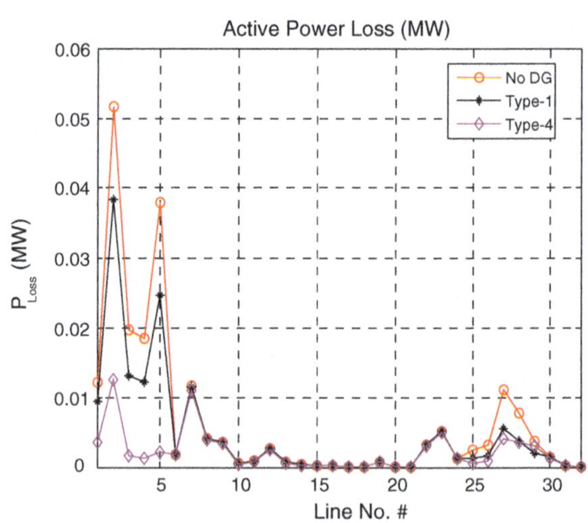

Fig. 5 Comparison of active power loss across each line of 33 bus radial distribution system

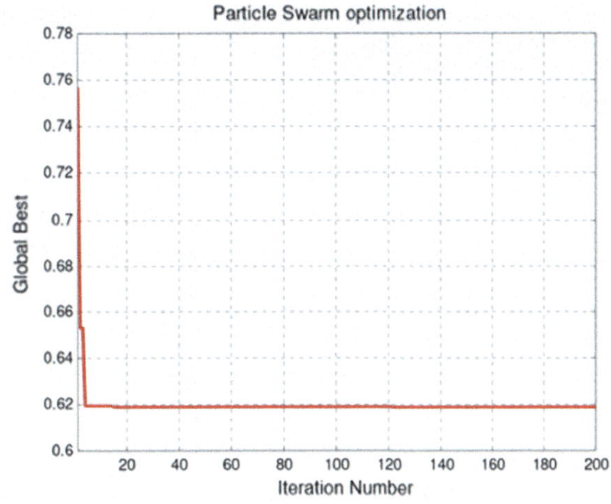

Fig. 6 Convergence characteristic of PSO algorithm

Results using GWO Optimization

Figure 4 shows comparison of voltage profile for 33 bus radial distribution system. Comparison of active power loss across each line of 33 bus radial distribution system is shown in Fig. 5.

Further, convergence characteristics are given in Fig. 6. The results of GWO optimization are given in Figs. 7, 8 and 9.

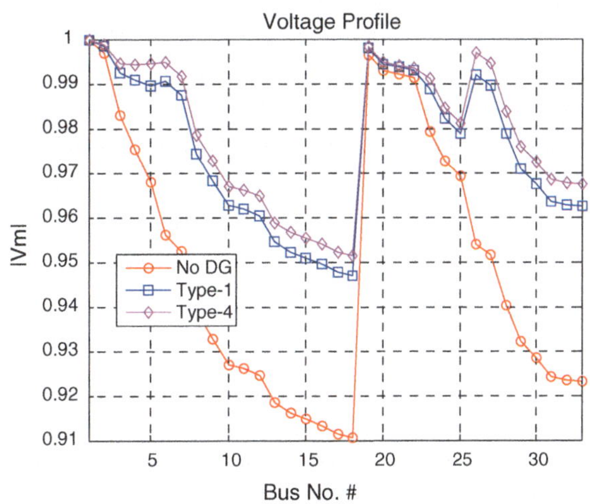

Fig. 7 Comparison of voltage profile for 33 bus radial distribution system

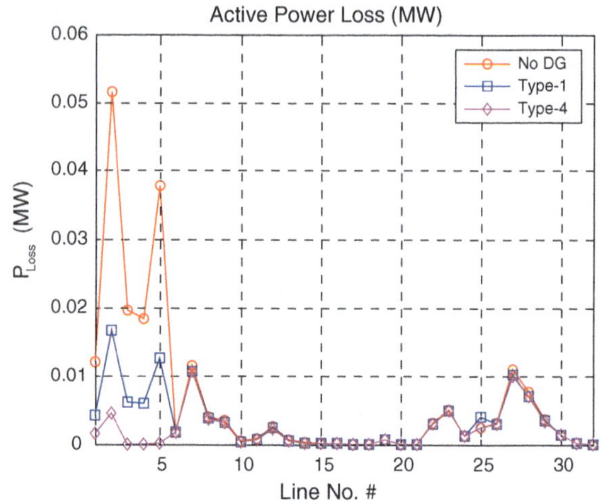

Fig. 8 Comparison of active power loss across each line of 33 bus radial distribution system

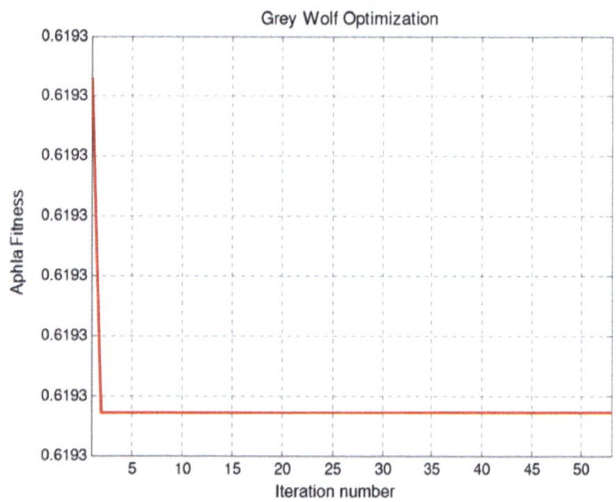

Fig. 9 Convergence characteristic of GWO algorithm

6 Conclusion

In this paper, allocation of DG unit in distribution network is carried out using PSO and GWO algorithms. Results obtained by these algorithms show that not only active power loss of the radial distribution network minimizes but also voltage at weak buses improved within limit, thus it is possible to find the optimal size of DG unit and its optimal location. Although both the algorithm gives competitive results but at the same time size of DG unit is further optimized by using GWO algorithm and it converges fast as compared to PSO algorithm (see Tables 1 and 2). Hence, GWO algorithm is most effective for the proposed work.

References

1. Michalik-Mielczarska G, Mielczarski W (2002) Quality of supply in liberalized electricity markets. In: 10th International conference on harmonics and quality of power, vol 1. IEEE, Rio de Janeiro, Brazil, pp 1–6
2. Philipson L (2000) Distributed and dispersed generation: addressing the spectrum of consumer needs. In: Power engineering society summer meeting, 2000, vol 3. IEEE, pp 1663–1665
3. Schellong W, Schmidla T (2013) Optimization of distributed cogeneration systems. In: 2013 International conference on industrial technology (ICIT). IEEE, Cape Town, South Africa, pp 879–884
4. Badwawi RA, Abusara M, Mallick T (2015) A review of hybrid solar PV and wind energy system. Smart Sci 3(3):127–138
5. https://www.theatlantic.com/technology/archive/2013/12/heres-why-developing-countries-will-consume-65-of-the-worlds-energy-by-2040/282006/
6. White paper, Our common future, Chapter 2: Towards sustainable development. UN Documents Gathering a Body of Global Agreements. http://www.un-documents.net/ocf-02.htm
7. Friedrichsen N, Klobasa M, Pudlik M (2015) Distribution network tariffs—the effect of decentralized generation and auto-consumption. In: 12th International conference on European energy market (EEM). IEEE, Lisbon, Portugal, pp 1–5
8. Sutter C, Parreño JC (2007) Does the current clean development mechanism (CDM) deliver its sustainable development claim? An analysis of officially registered CDM projects. Clim Change 84(1):75–90
9. Brown RE (2008) Electric power distribution reliability, 2nd edn. CRC Press, Boca Raton
10. Chowdhury S, Crossley P (2009) Microgrids and active distribution networks. The Institution of Engineering and Technology, London
11. Ganguly S, Sahoo NC, Das D (2013) Recent advances on power distribution system planning: a state-of-the-art survey. Energy Syst 4(2):165–193
12. Kansal S, Kumar V, Tyagi B (2013) Optimal placement of different type of DG sources in distribution networks. Int J Electr Power Energy Syst 53:752–760
13. Willis HL (2004) Power distribution planning reference book, 2nd edn. CRC Press, North Carolina
14. Ghosh S, Ghoshal SP, Ghosh S (2010) Optimal sizing and placement of distributed generation in a network system. Int J Electr Power Energy Syst 32(8):849–856
15. Arya LD, Koshti A, Choube SC (2012) Distributed generation planning using differential evolution accounting voltage stability consideration. Int J Electr Power Energy Syst 42(1):196–207
16. Singh SP, Rao AR (2012) Optimal allocation of capacitors in distribution systems using particle swarm optimization. Int J Electr Power Energy Syst 43(1):1267–1275
17. Keane A, Ochoa LF, Borges CL, Ault GW, Alarcon-Rodriguez AD, Currie RA, Pilo F, Dent C, Harrison GP (2013) State-of-the-art techniques and challenges ahead for distributed generation planning and optimization. IEEE Trans Power Syst 28(2):1493–1502
18. Murthy VVSN, Kumar A (2013) Comparison of optimal DG allocation methods in radial distribution systems based on sensitivity approaches. Int J Electr Power Energy Syst 53:450–467
19. Georgilakis PS, Hatziargyriou ND (2013) Optimal distributed generation placement in power distribution networks: models, methods, and future research. IEEE Trans Power Syst 28(3):3420–3428
20. Murty VVSN, Kumar A (2015) Optimal placement of DG in radial distribution systems based on new voltage stability index under load growth. Int J Electr Power Energy Syst 69:246–256

21. Kennedy J (2011) Particle swarm optimization. Encyclopaedia of machine learning. Springer, Berlin
22. Mirjalili S, Mirjalili SM, Lewis A (2014) Grey wolf optimizer. Adv Eng Softw 69:46–61
23. Kashem MA, Ganapathy V, Jasmon GB, Buhari MI (2000) A novel method for loss minimization in distribution networks. In: Proceedings of international conference on electric utility deregulation and restructuring and power technologies, DRPT2000. IEEE, London, UK, pp 251–256

Chapter 28
Maiden Application of Hybrid Shuffled Frog-Leaping Algorithm with Pattern Search Algorithm in AGC Studies of a Multi-area System

Naladi Ram Babu, Lalit Chandra Saikia, Dhenuvakonda Koteswara Raju and Tirumalasetty Chiranjeevi

1 Introduction

The mismatch between generated power and load tends to oscillations in tie-line power and frequency which is mitigated by the valve operation of the governor in automatic generation control (AGC) [1, 2]. The initial AGC works began with single area [1] and stretched out to multi-area thermal systems [3]. Thermal systems with generation rate constraints (GRC), governor dead band (GDB) and droop are considered for realistic approach [4]. During disturbances, AC tie-lines help to retain the system stability.

System performance gets deteriorated during faulty conditions. This can be achieved by the addition of high voltage direct current transmission (HVDC) tie-lines in parallel to the present AC lines. HVDC tie-lines have the power transfer capability through the longer distances. Authors in [5, 6] presented the control of frequency in HVDC tie-lines via voltage source converters. The aid of parallel AC-HVDC tie-lines was provided by Sharma et al. [7]. The frequency control study via parallel AC-HVDC tie-lines in two-area systems was demonstrated by authors in [6, 7]. However, the above frequency control studies with parallel AC-HVDC tie-lines are restricted to two-area only. This provides scope for further investigations. Several secondary controllers like FOPI [3], PID [8], PIDN [9], FOPIDN [10], fuzzy [11] and Tilt (T)-I-D (TID) [12] are used by the authors in the past. From the above literature, TID with filter (TIDN) is not yet utilized for the AGC studies which need further investigation.

N. R. Babu (✉) · L. C. Saikia · D. K. Raju · T. Chiranjeevi
Electrical Engineering Department, NIT Silchar, Silchar, Assam, India

T. Chiranjeevi
Rajkiya Engineering College, Sonbhadra, Uttar Pradesh, India

V. K. Giri et al. (eds.), *Computing Algorithms with Applications in Engineering*, Algorithms for Intelligent Systems, https://doi.org/10.1007/978-981-15-2369-4_28

Performance of secondary controllers will be the best when their parameters are optimum. Optimization techniques like firefly (FA) [9], biogeography-based optimization [11], hybrid algorithms [12], crow-search algorithm [13], pattern search [14], etc., are available in the literature. A new optimization technique named as shuffled frog-leaping algorithm (SFLA) [15] is available that works based on the elements for passing the information to next generations. The present trend refers to the hybridization of the algorithms with existing algorithms such as pattern search (PS) for better accuracy. Surprisingly, HSFLA-PS in not yet utilized for controller optimization in AGC studies. This needs further investigations.

Authors in [16, 17] presented the idea of testing the robustness of the controller by subjecting to variations in system loading and inertia conditions. How the same is not evaluated for TIDN controller using HSFLA-PS technique.

From the above-stated literature, the objectives are as follows:

- To compare the system dynamics with numerous secondary controllers like FOPI, PIDN and the proposed TIDN with HSFLA-PS technique.
- To compare the system dynamics among AC tie-line alone and parallel AC-HVDC tie-lines with HSFLA-PS technique and best controller in (a).
- To check the robustness of the TIDN controller with variations in inertia and loading conditions using HSFLA-PS technique.
- Performance comparison among HSFLA-PS, SFLA and FA optimization technique.

2 Study System

The thermal system of three-area with ±0.36% GDB, 3%/min GRC, 4% droop and HVDC in Fig. 1 is considered for investigation. The tie-line power equations with AC, HVDC tie-line and both parallel AC-HVDC tie-lines are given by (1), (2) and (3), respectively,

$$\Delta P_{\text{tie}\,ij\,\text{AC}} = \frac{2\Pi T_{jk}}{s}(\Delta F_j - \Delta F_k) \tag{1}$$

$$\Delta P_{\text{tie}\,ij\,\text{DC}} = \frac{K_{\text{DC}}}{1 + sT_{\text{DC}}}(\Delta F_j - \Delta F_k) \tag{2}$$

$$\Delta P_{\text{tie}\,ij} = \Delta P_{\text{tie}\,ij\,\text{AC}} + \Delta P_{\text{tie}\,ij\,\text{DC}} \tag{3}$$

The considered system is provided with TIDN controller and HSFLA-PS technique with SLP of 1% in area-1 subjecting to ISE in (4)

$$\eta_{\text{ISE}} = \int_0^T \left\{(\Delta F_j)^2 + (\Delta P_{k-m})^2\right\}.\mathrm{d}t \tag{4}$$

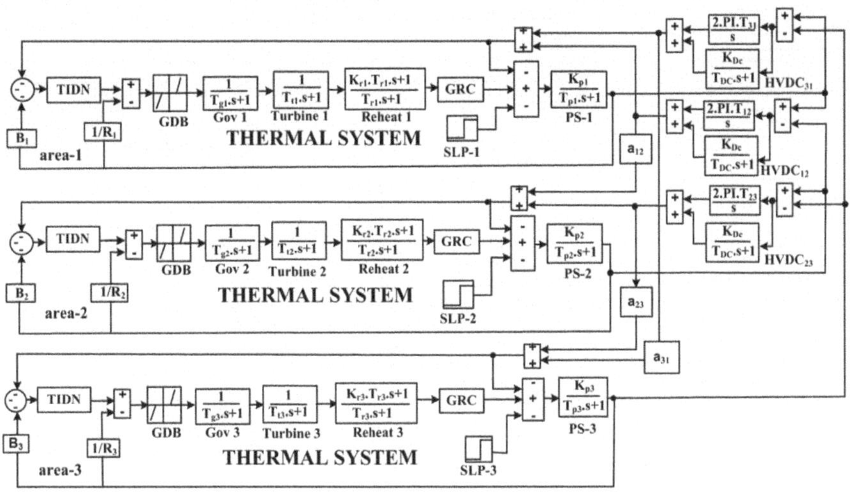

Fig. 1 Transfer function of unequal three-area thermal system with HVDC tie-lines, GRC, GDB and droop

where j, k and m are area number with $j = 1$, 2 and 3.

3 Tilt-Integral-Derivative with Filter (TIDN) Controller

The structure of TIDN is similar to PIDN controller and is shown in Fig. 2b. The only difference among PIDN and TIDN is the proportional term of PIDN is multiplied by "$(1/s)^n$." The transfer function of FOPI, PIDN and TIDN controller is given by (5), (6) and (7), respectively.

$$\text{TF}_{\text{FOPI}} = K_{Pj} + \left(K_{Ij}/s^{\lambda_j}\right) \tag{5}$$

$$\text{TF}_{\text{PIDN}} = K_{Pj} + \left(K_{Ij}/s\right) + K_{Dj}.s\left(\frac{N_j}{s+N_j}\right) \tag{6}$$

$$\text{TF}_{\text{TID}} = K_{Tj}.(1/s)^{n_j} + K_{Ij}/s + K_{Dj}.s\left(\frac{N_j}{s+N_j}\right) \tag{7}$$

where n is non-integer. The controller values are optimized by HSFLA-PS technique and ISE in (4) subjecting to constraints in (8)

$$\left.\begin{array}{l} 0 \le K_{Tj} \le 1, \quad 0 \le n_j \le 8, \\ 0 \le K_{Ij} \le 1, \quad 0 \le K_{Dj} \le 1 \ \text{and} \ 0 \le N_j \le 100 \end{array}\right\}. \tag{8}$$

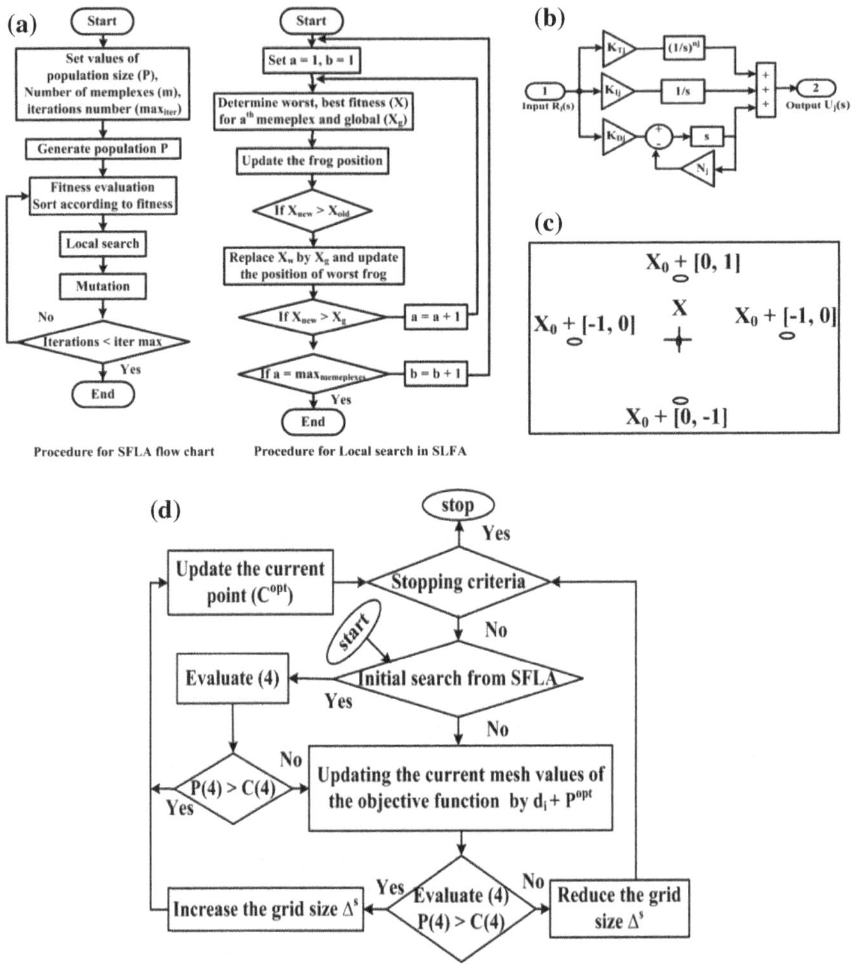

P- updated value of objective function C- current value of objective function

Fig. 2 Flow charts and controllers. **a** Flow chart of SFLA, **b** the proposed TIDN controller, **c** mesh points in pattern search, **d** flow chart of HSFLA-PS technique

4 Hybrid Shuffled Frog-Leaping Algorithm with Pattern Search (HSFLA-PS) Technique

4.1 Shuffled Frog-Leaping Algorithm (SFLA)

Eusuff [15] developed SFLA. It is a community-based algorithm with frogs as search elements. Group of frogs generally referred as memplexes that performs local search.

Each frog in a memplex retains information from their ancestors in a memetic evolution. This evolution mainly relies upon the survival of the best. Information exchanges take place during this process, and it continues till convergence is achieved. The frog position can be updated with fitness as given in (9)

$$\left.\begin{array}{l} D_i = \text{rand}.(X_b - X_w) \\ X_w^{\text{new}} = X_w^{\text{old}} + D_i \\ D_i^{\min} \leq D_i \leq D_i^{\max} \end{array}\right\} \tag{9}$$

where X_b and X_w are the fitness with the best and worst values. The frog with best fitness is saved and is updated by change in its position (D_i) with allowable step changes $\left(D_i^{\min}, D_i^{\max}\right)$. In SFLA, local search is achieved in a similar way to swarm optimization, and global exploitation is achieved by the periodically shuffling of frogs in memeplexes. The flow chart of SFLA is presented in Fig. 2a. The tuned values of SFLA are listed in Appendix.

4.2 Pattern Search (PS) Algorithm

Dolan et al. [14] developed PS technique. It comprises adaptable and adjustable operator that boosts the global values for tuning the local values. It starts with initial values from mesh that are given by SFLA. These mesh values are updated by an expansion of a scalar as a product to the present values. The scalar lies within the mesh shown in Fig. 2c, and it updates accordingly. If the updated value is having the best fitness, then it will be considered as a present point in the next iteration. This procedure is continued till the point with least fitness function is found. In next iteration, mesh size gets increased by multiple of 2, and procedure gets repeated. If it is found that the fitness value is getting increased, the mesh size should be reduced by a multiple of 0.5 till the stopping criteria are reached.

4.3 Hybrid Shuffled Frog-Leaping Algorithm with Pattern Search (HSFLA-PS) Technique

The optimum values obtained using SFLA are set as input to HSFLA-PS technique. These values are considered as s set of minimum values and maximum values are chosen from (8) and simulations are carried out. Thus, forming HSFLA-PS technique and its flow chart is depicted in Fig. 2d.

5 Results and Analysis

5.1 Hybrid Shuffled Frog-Leaping Algorithm with Pattern Search (HSFLA-PS) Technique

5.1.1 System Dynamic Comparison with Various Controllers like FOPI, PIDN and TIDN

The system in Fig. 1 (with AC tie-line alone) is considered for investigation. It is exposed to numerous secondary controllers like PIDN, FOPI and the proposed TIDN. The gains of controller are optimized by HSFLA-PS technique and SFLA. Its optimum values are given in Tables 1 and 2. The corresponding responses are given in Fig. 3a–d. Careful observations of these Figs reveal that the system with TIDN controller provides better settling time and less convergence that the other controllers.

5.1.2 System Dynamics with Parallel AC-HVDC Tie-Line

The system in Sect. 5.1.2 is provided with parallel AC-HVDC tie-lines. The system is investigated with the proposed TIDN controller, SFLA and HSFLA-PS technique. The obtained optimum values are given in Table 3. The corresponding responses are plotted in Fig. 4a–d. Careful observation reveals the parallel AC-HVDC tie-lines system with HSFLA-PS technique and TIDN controller gives better settling time, fewer oscillations in number.

5.2 Sensitivity Analysis of TIDN Controller with HSFLA-PS Technique

The optimum controller values obtained at nominal conditions are subjected to sensitivity analysis (SA) to analyze the robustness of proposed TIDN controller. SA is tested with system parameters like 50% loading condition and inertia constant ($H = 5$ s) with variations by $\pm 20\%$. The system is subjected with TIDN controller, and its values are optimized by HSFLA-PS technique. The corresponding dynamic responses with variation in H and loading conditions are shown in Fig. 5a–d. Critical observation of these figures explores that the responses with variation in system parameters are much same.

Table 1 Optimized controller gains and parameters of three-area thermal system using SFLA

Controller	FOPI			PIDN			TIDN		
Gains	Area-1	Area-2	Area-3	Area-1	Area-2	Area-3	Area-1	Area-2	Area-3
K_{Pj}	0.9050	0.3122	0.0500	0.5561	0.1978	0.3898	0.4725	0.4572	0.5180
n_j	–	–	–	–	–	–	7.2498	2.1747	3.9991
K_{Ij}	0.7071	0.2301	0.6850	0.2962	0.6986	0.5781	1.2841	1.7012	0.5963
λ_j	0.9452	0.1271	0.8340	–	–	–	–	–	–
K_{Dj}	–	–	–	0.7550	0.1352	0.6854	1.3601	0.4826	0.0159
N_j	–	–	–	35.780	54.265	40.161	36.458	57.245	74.5685

Table 2 Optimized controller gains and parameters of three-area thermal system using HSFLA-PS

Controller	FOPI			PIDN			TIDN		
Gains	Area-1	Area-2	Area-3	Area-1	Area-2	Area-3	Area-1	Area-2	Area-3
K_{Pj}	0.9174	0.3021	0.0501	0.5678	0.19878	0.3887	0.4823	0.4678	0.5210
n_j	–	–	–	–	–	–	7.2545	2.2015	4.0124
K_{Ij}	0.7178	0.2452	0.6985	0.2998	0.6585	0.5685	1.2867	1.7154	0.6547
λ_j	0.9685	0.1547	0.8152	–	–	–	–	–	–
K_{Dj}	–	–	–	0.7541	0.1484	0.6784	1.3501	0.4985	0.0199
N_j	–	–	–	36.458	55.645	44.251	38.542	58.625	78.054

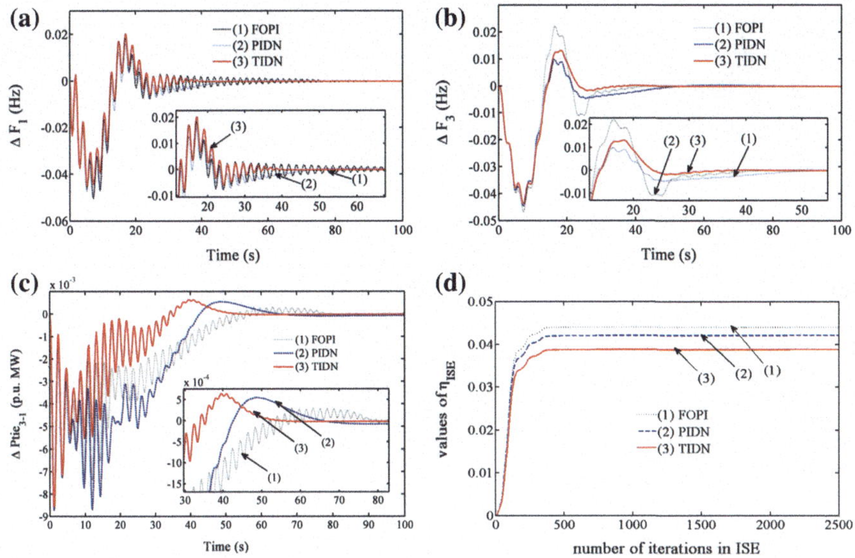

Fig. 3 Dynamic response and convergence characteristic comparison among FOPI, PIDN and TIDN controllers. **a** Frequency response in area-1, **b** frequency response in area-3, **c** tie-power deviation among area-3 and area-1 and **d** convergence characteristics of various controllers

Table 3 Optimized controller values of unequal three-area thermal system with parallel AC-HVDC tie-lines using SFLA and HSFLA-PS technique

TIDN	Using SFLA technique			Using HSFLA-PS technique		
	Area-1	Area-2	Area-3	Area-1	Area-2	Area-3
K_{Pj}	0.4963	0.1106	0.2274	0.5021	0.1502	0.2784
n_j	3.1781	3.2164	2.5192	3.1954	3.2897	2.6574
K_{Ij}	0.7309	0.5579	0.3015	0.7404	0.5687	0.3698
K_{Dj}	0.4590	0.6996	0.6936	0.4952	0.7087	0.6953
N_j	60.584	85.016	79.120	60.985	85.774	79.645

5.3 Dynamic Response and Convergence Characteristics Comparison with Various Algorithms

The TIDN controller values are optimized by FA, SFLA and HSFLA-PS techniques. The obtained responses are given in Fig. 6a–c. It is observed that the responses with HSFLA-PS technique are better in terms of settling time, oscillation magnitude and their corresponding values are listed in Table 4. Moreover, the convergence characteristic of HSFLA-PS technique in Fig. 6d converges faster than other.

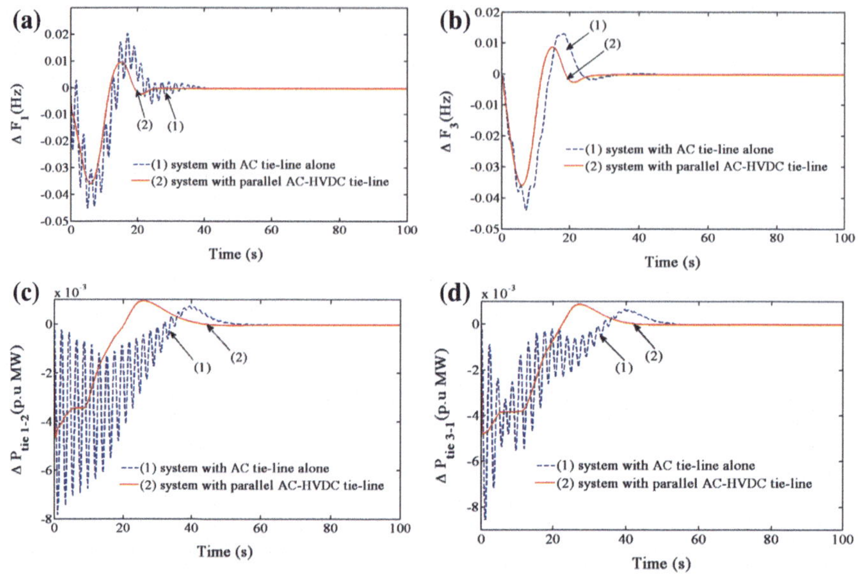

Fig. 4 Dynamic response comparison with AC and parallel AC-HVDC tie-line versus time. **a** Area-1 frequency response, **b** area-3 frequency response, **c** deviation in tie-power among area-1 and area-2 and **d** deviation in tie-power among area-3 and area-1

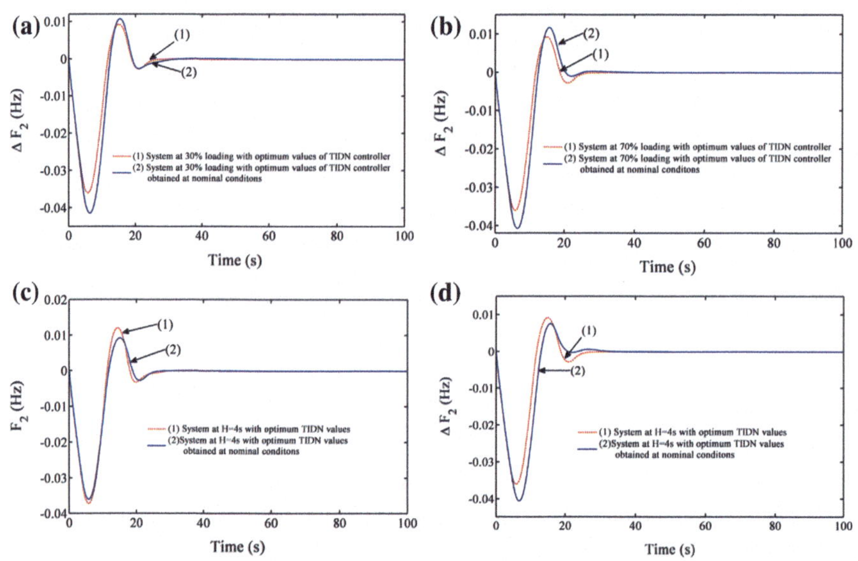

Fig. 5 Comparison of dynamic responses at varied and nominal conditions. **a** Area-2 frequency deviation with variation in 30% system loading condition from nominal values, **b** area-2 frequency deviation with variation in 70% system loading condition from nominal values, **c** area-2 frequency deviation with variation in H parameters at 4 s from nominal values and **d** area-2 frequency deviation with variation in H parameters at 6 s from nominal values

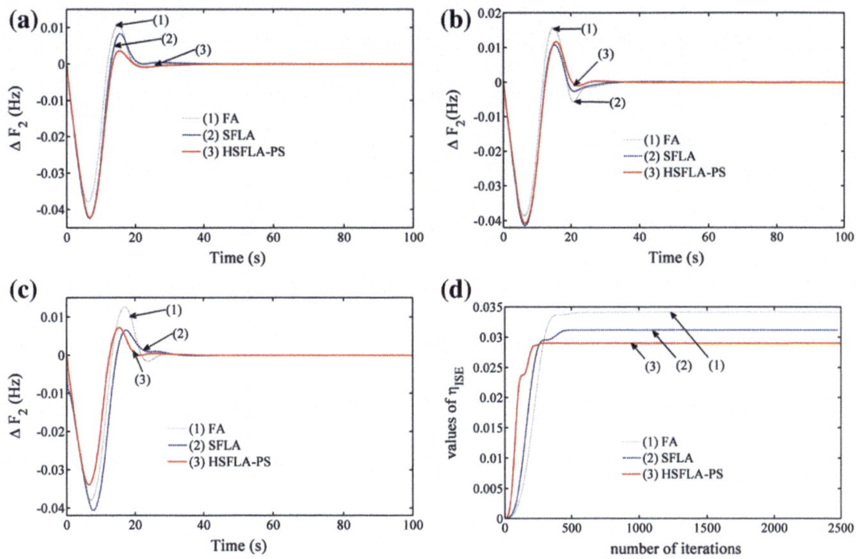

Fig. 6 Area-2 frequency deviation and convergence characteristics with various algorithms like FA, SFLA and HSFLA-PS technique. **a** With parallel AC-HVDC tie-lines, **b** with variation at 70% loading condition, **c** with inertia constant of 6 s and **d** convergence characteristics

6 Conclusions

The hybrid shuffled frog-leaping algorithm with pattern search (HSFLA-PS) technique is successfully utilized for simultaneous optimization of controller values under AGC Study. Investigations reveal that optimum values obtained with HSFLA-PS technique, and TIDN controller provides better settling time and convergences faster than FA and SFLA. Moreover, system dynamics with TIDN controller provides better dynamics over FOPI and PIDN controller in terms of oscillations magnitude and settling time. Further sensitivity analysis with variations in loading and inertia constant of TIDN controller with HSFLA-PS technique suggests that controller values at nominal conditions are robust. Furthermore, it is concluded that the parallel AC-HVDC tie-line system improves system dynamics than the latter.

Table 4 Various comparisons of the algorithm with generation size of 100 and iteration size of 50

Case study	Run time (s)			Settling time (s)			Magnitude of Peak overshoot (Hz) × 10^{-3}			Magnitude of Undershoot (Hz) × 10^{-3}			Cost function (ISE) × 10^{-3}		
	FA	SFLA	HSFLA-PS	FA	SFLA	HSFLA-PS	FA	SFLA	HSFLA-PS	FA	SFLA	HSFLA-PS	FA	SFLA	HSFLA-PS
ΔF_2 with parallel AC-HVDC tie-lines	705	635	535	35	28	25	11.1	8	4	−44	−37	−40	315	303	285
ΔF_2 at $H = 6$ s	715	665	545	30	27	23	10	5.1	5.2	−41	−36	−32	300	280	255
ΔF_2 at 70% loading	720	662	565	35	32	28	15	10	12	−42	−38	−40	303	285	255

FA firefly algorithm; *SFLA* shuffled frog-leaping algorithm; *HSFLA-PS* hybrid crow shuffled frog-leaping algorithm with pattern search technique

Appendix

1. System Parameters: $T_{jk,\mathrm{AC}} = 0.086$ pu MW/rad, $F = 60$ Hz, $H_j = 5$ s, $K_{pj} = 120$ Hz/MW pu, $B_j = 0.425$ pu MW/Hz $D_j = 8.33 \times 10^{-3}$ pu MW/Hz, P_{r1}: P_{r2}: $P_{r3} = 1000$ MW: 2000 MW: 4000 MW, $R_j = 2.4$ pu MW/Hz, $T_{pj} = 20$ s, $T_{\mathrm{DC}} = 0.03$ s, $T_{tj} = 0.3$ s, $K_{rj} = 0.5$ s, $T_{gj} = 0.08$ s, $K_{\mathrm{DC}} = 0.5$, $T_{rj} = 10$ s.
2. Algorithm parameters
   (a) Shuffled frog-leaping algorithm: Number of frogs = 50, maximum iterations = 100, number of memeplex = 5, memeplex size = 5, number of ofsprings (α) = 2, maximum number or iterations in a memeplex (β) = 3 and step size (σ) = 1
   (b) Firefly algorithm: maximum generation = 100, Firefly size = 50, gamma = 1, beta = 0.8, alpha = 0.25.

References

1. Elgerd OI (2007) Electric energy systems theory: an introduction. Tata McGraw-Hill, New Delhi
2. Ibraheem Kumar P, Kothari DP (2005) Recent philosophies of automatic generation control strategies in power systems. IEEE Trans Power Syst 20(1):346–357
3. Tasnin W, Saikia LC (2018) Performance comparison of several energy storage devices in deregulated AGC of a multi-area system incorporating geothermal power plant. IET Renew Power Gener 12(7):761–772
4. Mobarak Y (2015) Effects of the droop speed governor and automatic generation control AGC on generator load sharing of power. Int J Appl Power Eng 4(2):84–95
5. Adeuyi OD, Cheah-Mane M, Liang J et al (2015) Frequency support from modular multilevel converter based multi-terminal HVDC schemes. In: Power & energy society general meeting. IEEE, pp 1–5
6. Rakhshani E, Rodriguez P (2017) Inertia emulation in AC/DC interconnected power systems using derivative technique considering frequency measurement effects. IEEE Trans Power Syst 32(5):3338–3351
7. Sharma G, Nasiruddin I, Niazi KR (2016) Robust automatic generation control regulators for a two-area power system interconnected via AC/DC tie-lines considering new structures of matrix Q. IET Gener Transm Distrib 10(14):3570–3579
8. Jagatheesan K, Anand B, Samanta S, Dey N, Ashour AS, Balas VE, Design of a proportional-integral-derivative controller for an automatic generation control of multi-area power thermal systems using firefly algorithm. IEEE/CAA J Autom Sin. https://doi.org/10.1109/jas.2017.7510436
9. Pan I, Das S (2016) Fractional order AGC for distributed energy resources using robust optimization. IEEE Trans on Smart Grid 7(5):2175–2186
10. Sharma G, Ibraheem, Niazi KR, Bansal RC (2016) Adaptive fuzzy critic based control design for AGC of power system connected via AC/DC tie-lines. IET Gener Transm Distrib 11(2):560–569
11. Topno PN, Chanana S (2015) Tilt integral derivative control for two-area load frequency control problem. In: 2nd International conference on recent advances in engineering & computational sciences (RAECS), Chandigarh, pp 1–6

12. Sahu RK et al (2014) Optimal gravitational search algorithm for automatic generation control of interconnected power system. Ains Shams Eng J 5:721–733
13. Rambabu N, Saikia LC (2019) Automatic generation control of a solar thermal and dish-stirling solar thermal system integrated multi-area system incorporating accurate HVDC link model using crow search algorithm optimised FOPI Minus FODF controller. IET Renew Power Gener. https://doi.org/10.1049/iet-rpg.2018.6089
14. Dolan ED, Lewis RM, Torczon V (2003) On the local convergence of pattern search. SIAM J Optim 14:567–583
15. Eusuff M, Lansey K, Pasha F (2006) Shuffled frog-leaping algorithm: a memetic meta-heuristic for discrete optimization. Eng Optim 38(2):129–154
16. Rahman A, Saikia LC, Sinha N (2015) Load Frequency control of a hydro-thermal system under deregulated environment using biogeography-based optimised three-degree-of freedom integral-derivative controller. IET Gener Transm Distrib 9(15):2284–2293
17. Raju M, Saikia LC, Sinha N, Saha D (2017) Application of antlion optimizer technique in restructured automatic generation control of two-area hydro-thermal system considering governor dead band. In: Innovations in power and advanced computing technologies, Vellore, pp 1–6

Chapter 29
Multi-area AGC System Incorporating GTPP and Coyote Optimized PI Minus DN Controller

Naladi Ram Babu, Lalit Chandra Saikia, Dhenuvakonda Koteswara Raju and Tirumalasetty Chiranjeevi

1 Introduction

Automatic generation control (AGC) plays a crucial role in the control and operation of power system. The system's frequency and tie power deviate from its nominal values with change in load and other abnormal conditions. AGC is a supplementary control scheme which keeps system frequency within tolerable limits which in turn maintains the balance between the powers generated and the load. Power system is a combination of individual components or utilities that are coordinated together to form enormous complex systems. The concept of modeling an interconnected multi-area power system for AGC was demonstrated by Elgerd and Fosha [1]. Using the concept presented in [1], many studies on AGC have been found in the past. Authors in [2, 3] presented the idea of integrating solar thermal plant and geothermal power plant (GTPP) with thermal of two-area system. However, there are no studies where GTPP is considered in all the three areas which require further study.

Present AGC study focuses on the design of secondary controllers. Many studies include simple PI [4], PID [5, 6] and PIDN [4]. A new controller namely PI minus DN controller with a change of sign in PIDN structure is not yet utilized for AGC studies. This provides scope for further investigations. Performance of secondary controllers will be best when its values are optimized. Numerous optimization techniques namely gray wolf optimization [7], genetic algorithm [8], crow search algorithm [9], cuckoo search algorithm [10], symbiotic organisms search [11], sine–cosine [12], binary Jaya algorithm [13] and many more are utilized by the researchers in the past. A new optimization technique named coyote optimization algorithm (COA) [14] is available that works on the mating behavior of coyotes. Surprisingly, COA is not yet utilized for controller optimization in AGC studies. Many authors have performed

---

N. R. Babu (✉) · L. C. Saikia · D. K. Raju · T. Chiranjeevi
Electrical Engineering Department, NIT Silchar, Silchar, Assam, India

T. Chiranjeevi
Rajkiya Engineering College, Sonbhadra, Uttar Pradesh, India

V. K. Giri et al. (eds.), *Computing Algorithms with Applications in Engineering*, Algorithms for Intelligent Systems, https://doi.org/10.1007/978-981-15-2369-4_29

sensitivity to check the sturdiness of controller parameters [15, 16] with variations in system loading and inertia conditions. However, the same is not evaluated for PI-DN controller using COA technique.

From the above-stated literature, the objectives are as follows:

(a) To develop an unequal three-area thermal-GTPP systems.
(b) Application of COA for optimizing various controller values of such as PID, PIDN and PI-DN with 1% SLP and RLP while considering one at a time to determine the best.
(c) To study the effect of GTPP with best controller from (b).
(d) To explore the sensitivity analysis of the proposed PI-DN controller.

2 System Investigated

A three-area system is developed by comprising of thermal and GTPP systems as shown in Fig. 1a. The thermal systems are considered with GRC of 3%/min. The nominal system parameters are shown in appendix. The dynamics are evaluated by using 1% SLP in Area-1. COA technique is used for optimization of controllers such as PID, PIDN and PI-DN subjecting to ISE in (1).

$$J = \int_{0}^{t} \left\{ (\Delta F_i)^2 + \left(\Delta P_{\text{tie}i-j}\right)^2 \right\} \mathrm{d}t \tag{1}$$

where area number is i, j. For $i = 1, 2, 3$ and $j \neq i$. Simulink model and coding are done in MATLAB software.

3 Proportional–Integral Minus Derivative with Filter (PI-DN) Controller

The structure of PI-DN controller is similar to the PIDN controller, the only difference is the change in sign of derivative (D) component. The presence of D-component in forward path generated derivative kick. This is objectionable in electrical circuits [13, 14]. To overcome this, industrial engineer has redesigned and added D-component in feedback patch as shown in Fig. 1c. The equations of PID, PIDN and PI-DN controller are given in (2), (3) and (4).

$$\mathrm{TF}_{\mathrm{PID}} = K_{Pj} + \left(K_{Ij}/s\right) + \left(K_{Dj}.s\right) \tag{2}$$

$$\mathrm{TF}_{\mathrm{PIDN}} = K_{Pj} + (K_{Ij}/s) + K_{Dj}.s\left(\frac{N_j}{s + N_j}\right) \tag{3}$$

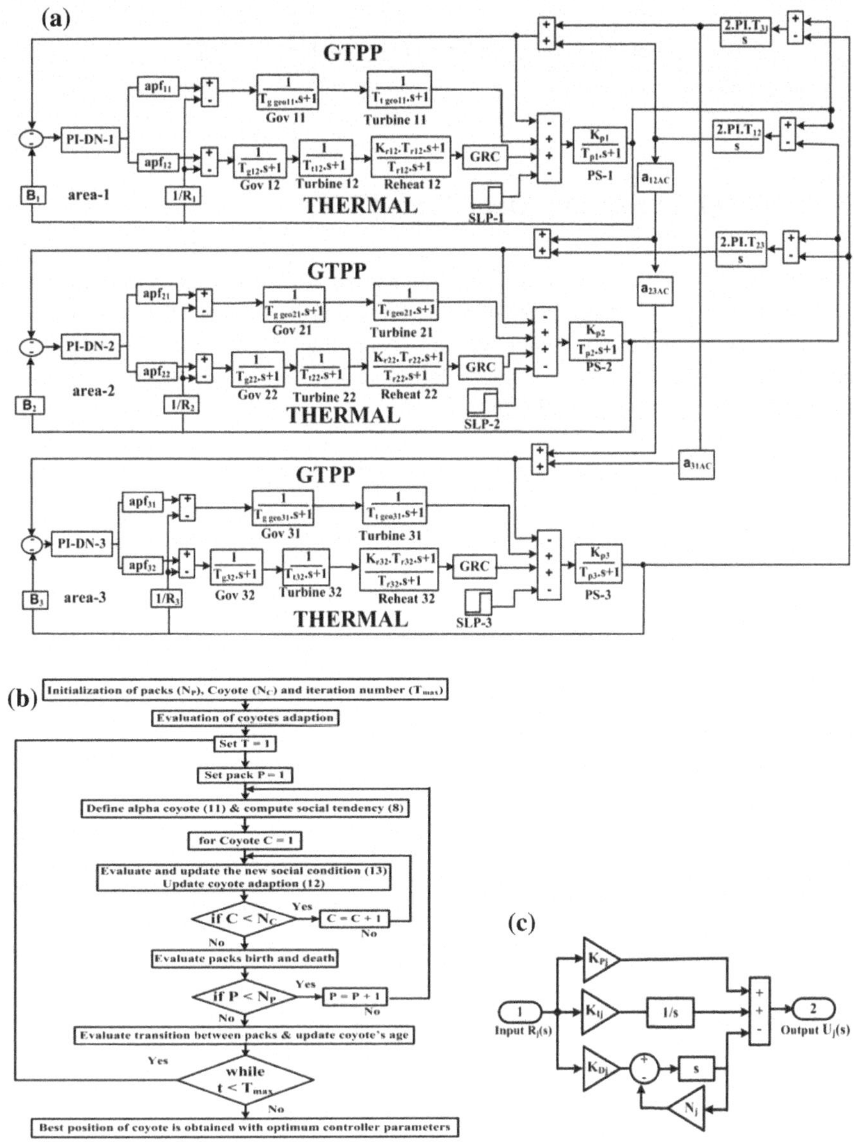

Fig. 1 System components, COA flowchart and controller. **a** Transfer function model of an unequal three-area thermal system with integration of GTPP, **b** flow chart of COA and **c** the proposed PI-DN controller

$$\mathrm{TF_{PI\text{-}DN}} = K_{Pj} + \left(K_{Ij}/s\right) - K_{Dj}.s\left(\frac{N_j}{s+N_j}\right) \tag{4}$$

4 Coyote Optimization Algorithm (COA)

It was proposed by Pierezan et al. [10]. It is a population-based algorithm inspired by Canis Latrans. The parameters of COA are listed in Table 1. The flowchart of COA is given in Fig. 1b. The total population is obtained by the multiplication of N_p and N_c. Each coyote (c) is a feasible solution and the social condition (soc) is the objective function and is given by (5).

$$\mathrm{SOC}_c^{p,t} = \overrightarrow{X} = (X_1, X_2, \ldots, X_D) \tag{5}$$

where time (t) and D is the search space dimension. The initial random soc of each coyote given by (6)

$$\mathrm{soc}_{c,j}^{p,t} = lb_j + r_j.(ub_j - lb_j) \tag{6}$$

where lb_j, ub_j are lower and upper limits of decision variable (j) and r_j is real random number in [0, 1]. After that, the coyote's adaptation with respect to the present social conditions is evaluated using (7)

$$\mathrm{fit}_c^{p,t} = f\left(\mathrm{soc}_c^{p,t}\right) \tag{7}$$

The alpha pack (P) is given by (8)

$$\mathrm{alpha}^{p,t} = \left\{\mathrm{soc}^{p,t} \mid \arg_{c=\{1,2,\ldots,N_c\}} \min f\left(\mathrm{soc}_c^{p,t}\right)\right\} \tag{8}$$

The cultural tendency of P is given by (9)

$$\mathrm{cult}_j^{p,t} = \begin{cases} O^{p,t}_{\frac{(N_c+1)}{2},j}, & N_c \text{ is odd} \\ \dfrac{O^{p,t}_{\frac{N_c}{2},j} + O^{p,t}_{\left(\frac{N_c}{2}+1\right),j}}{2}, & \text{otherwise} \end{cases} \tag{9}$$

where ranked social conditions ($O^{p,t}$) are in between [1, D]. The birth of new coyotes can be represented by (10).

Table 1 Nominal parameters of COA technique

Number of packs N_P = 20	Number of coyotes N_c = 10
stopping criteria $T_{\max}$ = 1000	rnd$_j$ [0,1]

$$\mathrm{pup}_j^{p,t} = \begin{cases} \mathrm{soc}_{r1,j}^{p,t}, & \mathrm{rnd}_j \prec P_\mathrm{s} \text{ or } j = j_1 \\ \mathrm{soc}_{r2,j}^{p,t}, & \mathrm{rnd}_j \geq P_\mathrm{s} + P_\mathrm{a} \text{ or } j = j_2 \\ R_j, & \text{otherwise} \end{cases} \quad (10)$$

where r is random with dimensions (j_1, j_2), scatter probability (P_s), association (P_a) and rnd_j is in range [0, 1] and are given by (11)

$$P_s = 1/D \quad \text{and} \quad P_a = (1 - P_s)/2 \quad (11)$$

The coyotes with alpha (δ_1) and pack (δ_2) influence are given by (12)

$$\delta_1 = \mathrm{alpha}^{p,t} - \mathrm{soc}_{cr1}^{p,t} \quad \text{and} \quad \delta_2 = \mathrm{cult}^{p,t} - \mathrm{soc}_{cr2}^{p,t} \quad (12)$$

The new social condition of coyote is updated and given by (13)

$$\mathrm{new\_soc}_c^{p,t} = \mathrm{soc}_c^{p,t} + r_1.\delta_1 + r_2.\delta_2 \quad (13)$$

where alpha weights (r_1) and pack weights (r_2) are used to update the social condition as global solution by using (14)

$$\mathrm{new\_fit}_c^{p,t} = f\left(\mathrm{new\_soc}_c^{p,t}\right) \quad (14)$$

5 Results and Discussions

5.1 System Dynamics with Various Controllers like PID, PIDN and the Proposed PI-DN Controller

5.1.1 Application of 1% SLP in Area-1

The system in Fig. 1a is equipped with PID, PIDN and PI-DN controller. In each case, the controller values are optimized by COA. The optimum values are tabulated in Table 2. Its corresponding dynamic responses are compared in Fig. 2a–d. Critical analysis of Fig. 2 explores that PI-DN controller performs better than others with regard to settling time, peak undershoot and much lesser oscillations.

5.1.2 Application of RLP in Area-1

The load on power system is random in nature. The RLP in Area-1 shown in Fig. 3a is considered for investigation. Under this random load change, the various controllers such as PID, PIDN and PI-DN are used one at a time. The controller values are

Table 2 COA optimized controller values of three-area thermal-GTPP system with considering 1% SLP in Area-1

Controller	PID			PIDN			PI-DN		
Gains	Area-1	Area-2	Area-3	Area-1	Area-2	Area-3	Area-1	Area-2	Area-3
K_{Pj}	0.7586	0.0465	0.2543	0.6476	0.1815	0.1024	0.4254	0.0260	0.4725
K_{Ij}	0.3845	0.0639	0.4638	0.9215	0.3053	0.1324	0.8467	0.0946	0.2841
K_{Dj}	–	–	–	0.8549	0.4524	0.6171	0.4234	0.5205	0.3601
N_j	–	–	–	79.681	52.479	16.126	10.074	93.564	85.458

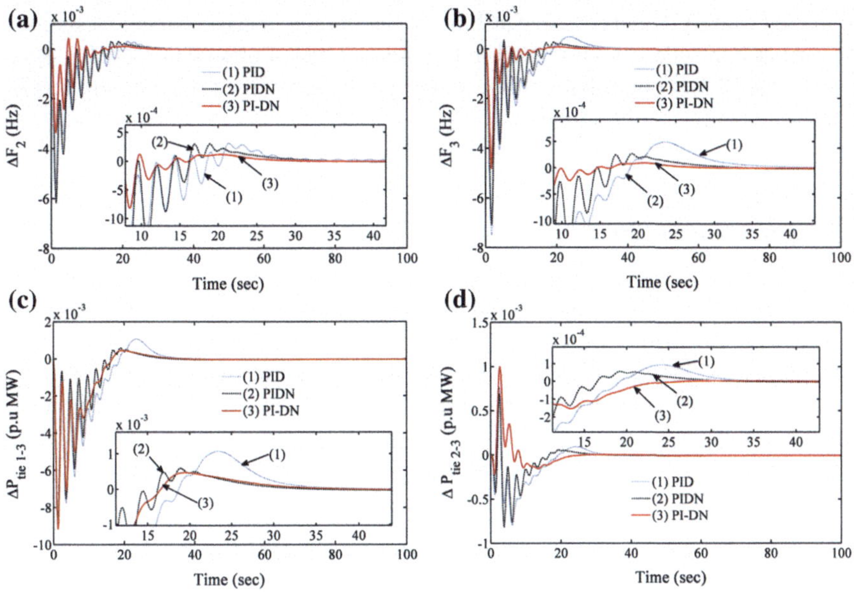

Fig. 2 Dynamic response comparison with various controllers like PI, PIDN and PI-DN considering 1% SLP in Area-1 versus time, **a** ΔF_2, **b** ΔF_3, **c** $\Delta P_{\text{tie 1-3}}$ and **d** $\Delta P_{\text{tie 2-3}}$

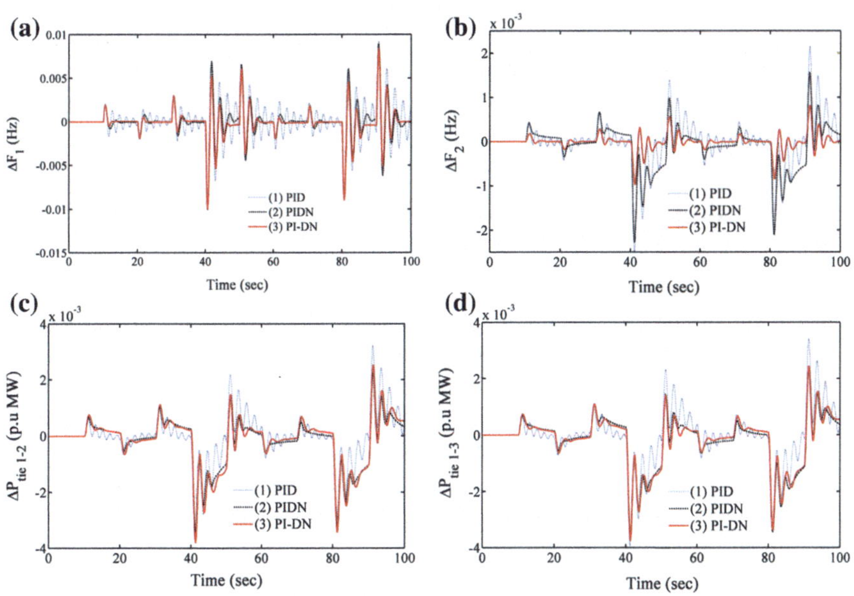

Fig. 3 System dynamic comparison with various controllers like PI, PIDN and PI-DN controller with RLP in Area-1, **a** ΔF_1, **b** ΔF_2, **c** $\Delta P_{\text{tie 1-2}}$ and **d** $\Delta P_{\text{tie 1-3}}$

optimized with COA and are listed in Table 3. The comparisons of dynamic responses for various controllers are provided in Fig. 4b–d. Critical observation of Fig. 4b–d reveals that PI-DN controller exhibits better performance over other controllers in terms of magnitudes of oscillations.

5.2 Effect of GTPP

The system in Fig. 1a is considered and the investigations are carried out with thermal systems and the proposed PI-DN controller. The obtained responses are analyzed with thermal-GTPP system shown in Fig. 4a–d. Careful investigations reveal that the system with GTPP provides better settling time with fewer oscillations. Optimum values are not shown.

5.3 Sensitivity Analysis of the Proposed PI-DN Controller

In this study, the loading condition is changed from nominal 50% to $\pm 25\%$ and inertia constant (H) is changed from nominal 5 s to $\pm 25\%$. In each changed condition, the PI-DN controller values are optimized by COA. The responses at changed conditions are compared with optimum values corresponding to nominal conditions in Fig. 5a–d. It is noted that both responses are same. Thus, the optimum PI-DN values obtained at nominal condition are robust when there is large variation in system parameters.

6 Conclusion

A maiden attempt has been made to apply PI-DN controller in AGC with a new metaheuristic optimization technique called coyote optimization algorithm for optimization of controller gains. Comparison of dynamic performance of PID, PIDN and the proposed PI-DN controller reveals the best performance over others in terms of settling time, peak undershoot and magnitude of oscillations under 1% SLP and RLP in Area-1. Moreover, a comparative study is carried out by considering GTTP in all the areas with the thermal system alone. Furthermore, sensitivity analysis explores that the optimum gains of PI-DN controller obtained at nominal conditions are robust for variations in system conditions.

Table 3 COA optimized controller values of three-area thermal system with GTPP considering RLP in Area-1

Controller	PID			PIDN			PI-DN		
Gains	Area-1	Area-2	Area-3	Area-1	Area-2	Area-3	Area-1	Area-2	Area-3
K_{Pj}	0.3517	0.2863	0.8047	0.7464	0.2988	0.5550	0.0885	0.1013	0.1919
K_{Ij}	0.6184	0.0772	0.7678	0.5076	0.8878	0.8941	0.8200	0.2494	0.0446
K_{Dj}	–	–	–	0.5966	0.0059	0.8729	0.0010	0.4383	0.3057
N_j	–	–	–	14.772	23.035	34.413	60.960	67.418	88.010

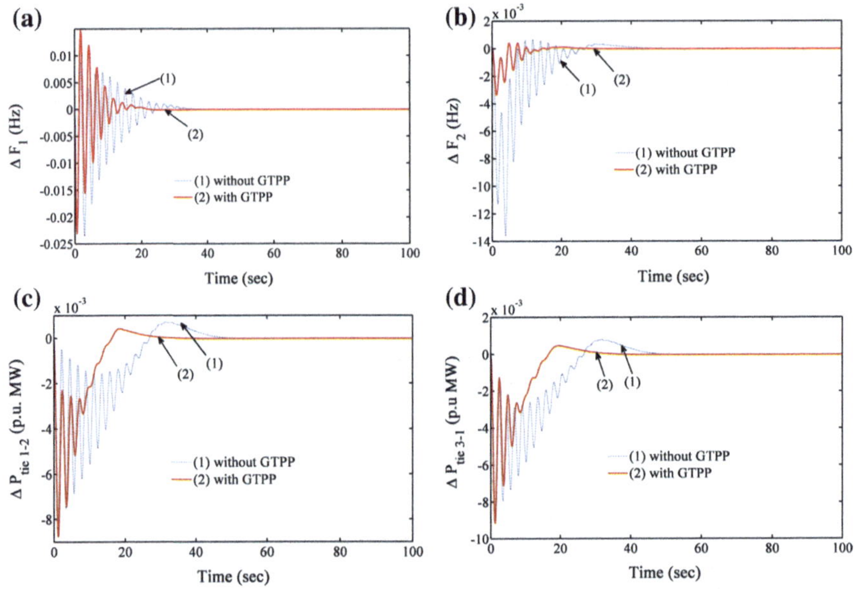

Fig. 4 System response comparison with inclusion of GTPP versus time, **a** ΔF_1, **b** ΔF_2, **c** $\Delta P_{\text{tie 1–2}}$ and $\Delta P_{\text{tie 3–1}}$

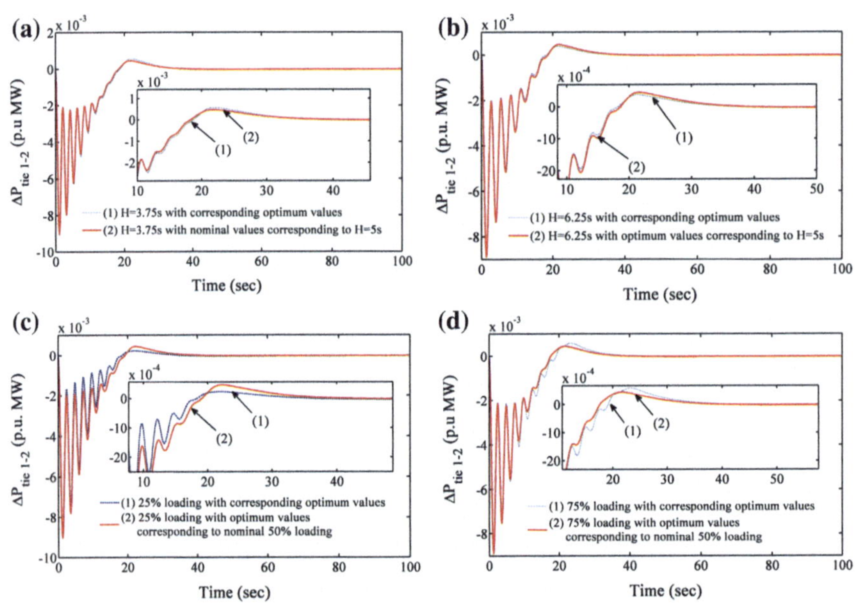

Fig. 5 Comparison of frequency deviation in Area-1 versus time for sensitivity analysis, **a** ΔF_1 for $H = 3.75$ s, **b** ΔF_2 for $H = 6.25$ s, **c** ΔF_1 for 25% loading, **d** ΔF_1 for 75% loading

Appendix

Nominal parameters are $F = 60$ Hz; $K_{ri} = 0.5$; $T_{gi} = 0.08$ s; $T_{pi} = 20$ s; $T_{gs} = 1.0$ s; $T_{ti} = 0.3$ s; $T_{ts} = 3.0$ s; $K_s = 1.8$; $\mathrm{T}_{ri} = 10$ s; $T_{ggeo} = 0.0500$ s; $T_{tgeo} = 0.100$ s; $T_s = 1.8$ s; $H_i = 5$ s; $D_i = 8.33 \times 10^{-3}$ pu MW/Hz; $K_{pi} = 120$ Hz/pu MW; $B_i = \beta_i = 0.425$ pu MW/Hz; $R_i = 2.4$ Hz/pu MW; $T_{12} = T_{13} = T_{23} = 0.086$ pu MW/rad; loading = 50%; SLP = 1% in Area-1; Area-1: Area-2: Area-3 = 1000 MW: 2000 MW: 4000 MW

References

1. Elgerd OI, Fosha CE (1970) Optimum megawatt-frequency control of multiarea electric energy systems. IEEE Trans Power Appar Syst 89(4):556–563
2. Sharma Y, Saikia LC (2015) Automatic generation control of a multi-area ST–Thermal power system using Grey Wolf Optimizer algorithm based classical controllers. Int J Electr Power Energy Syst 73:853–862
3. Tasnin W, Saikia LC (2018) Maiden application of an sine–cosine algorithm optimised FO cascade controller in automatic generation control of multi-area thermal system incorporating dish-Stirling solar and geothermal power plants. IET Renew Power Gener 12(5):585–597
4. Jagatheesan K, Anand B, Samanta S, Dey N, Ashour AS, Balas VE (2019) Design of a proportional-integral-derivative controller for an automatic generation control of multi-area power thermal systems using firefly algorithm. IEEE/CAA J Autom Sin 6(2):503–515
5. Sharma Y, Saikia LC (2015) Automatic generation control of a multi-area ST–Thermal power system using Grey Wolf Optimizer algorithm based classical controllers. Electr Power Energy Syst 73:853–862
6. Rathore NS, Singh VP (2018) Design of optimal PID controller for the reverse osmosis using teacher-learner-based-optimization. Membr Water Treat Int J 9(2):129–136
7. Lal DK, Barisal AK, Tripathy M (2016) Grey wolf optimizer algorithm based Fuzzy PID controller for AGC of multi-area power system with TCPS. Procedia Comput Sci 92:99–105
8. Das DC, Roy AK, Sinha N (2012) GA based frequency controller for solar thermal–diesel–wind hybrid energy generation/energy storage system. Int J Electr Power Energy Syst 43(1):262–279
9. Rambabu N, Saikia LC (2019) Automatic generation control of a solar thermal and dish-stirling solar thermal system integrated multi-area system incorporating accurate HVDC link model using crow search algorithm optimised FOPI Minus FODF controller. IET Renew Power Gener. https://doi.org/10.1049/iet-rpg.2018.6089
10. Dutta A, Debbarma S (2018) Frequency regulation in deregulated market using vehicle-to-grid services in residential distribution network. IEEE Syst J 12(3):2812–2820
11. Singh SP, Prakash T, Singh VP (2019) Coordinated tuning of controller-parameters using symbiotic organisms search algorithm for frequency regulation of multi-area wind integrated power system. Eng Sci Technol Int J. https://doi.org/10.1016/j.jestch.2019.03.007
12. Prakash T, Singh SP, Singh VP (2019) Analytic hierarchy process based model reduction of higher order continuous systems using sine cosine algorithm. Int J Syst Control Commun
13. Prakash T, Singh VP, Mohanty SR, Singh SP (2017) Binary Jaya algorithm based optimal placement of phasor measurement units for power system observabilty. Int J Control Theory Appl 10(5):515–523
14. Pierezan J, Dos SCL (2018) Coyote optimization algorithm: a new metaheuristic for global optimization problems. In: 2018 IEEE congress on evolutionary computation (CEC), Rio de Janeiro, pp 1–8

15. Rahman A, Saikia LC, Sinha N (2015) Load Frequency control of a hydro-thermal system under deregulated environment using biogeography-based optimised three-degree-of freedom integral-derivative controller. IET Gener Transm Distrib 9(15):2284–2293
16. Raju M, Saikia LC, Sinha N, Saha D (2017) Application of antlion optimizer technique in restructured automatic generation control of two-area hydro-thermal system considering governor dead band. In: Innovations in power and advanced computing technologies, Vellore, pp. 1–6

Chapter 30
Interval Modeling of Riverol-Pilipovik Water Treatment Plant and Its Model Order Reduction

M. M. Chodavarapu, V. P. Singh and Ramesh Devarapalli

1 Introduction

System modeling is a key branch of control systems. The modeling of any system should be performed in such a way so that the mathematical model obtained should describe all essential dynamics of the system. Generally, a nominal model is obtained for any physical system. Nominal model is generally preferred due to simplicity in analysis, simpler control design, etc. However, the nominal model does not perform satisfactorily when the parameter variations occur. To incorporate the variations of parameters, an interval model of such a system can be derived. Recently, some systems [1] like cold rolling milk, DC shunt motor, oblique wing aircraft, etc., are modeled as interval systems.

Recently, some models are proposed for desalination of sea and brackish water. In desalination, the salty seawater is converted into freshwater. Several methods for desalination are available in the literature. However, reverse osmosis (RO) is preferred over others due to its advantages like lower power consumption, etc. The some important RO-based desalination system is Doha water treatment plant [2, 3], Chaabene water treatment plant [4], and Riverol-Pilipovik (RP) water treatment plant [5, 6].

The desalination models proposed in the literature are generally interacting multi-input-multi-output (MIMO) in nature. The Doha water treatment plant has two

M. M. Chodavarapu
Department of Electrical Engineering, NIT Raipur, Raipur, India

V. P. Singh
Department of Electrical Engineering, MNIT Jaipur, Jaipur, India

R. Devarapalli (✉)
Department of Electrical Engineering, BIT Sindri, Dhanbad, India
e-mail: ramesh.2015dr1103@ee.ism.ac.in

V. K. Giri et al. (eds.), *Computing Algorithms with Applications in Engineering*,
Algorithms for Intelligent Systems, https://doi.org/10.1007/978-981-15-2369-4_30

manipulated variables (pressure and pH) and two controlled variables (flux and conductivity). Chaabene model [4] also has two manipulated variables, namely the angular speed of pump and reject valve aperture, and two controlled variables, namely flux and conductivity. Similarly, RP water treatment plant has two manipulated variables (pressure and pH) and two controlled variables (flux and conductivity).

In this contribution, an interval model is developed for Riverol-Pilipovik (RP) water treatment plant. The interval model is derived by considering a certain amount of uncertainty in all parameters of different coefficients of transfer function of RP water treatment plant. Once the interval model is derived, it is also reduced using Jaya algorithm. The results obtained prove that the model effectively resembles the interval modeled RP water treatment plant.

The remaining of the work is organized as follows. Section 2 presents the Riverol-Pilipovik (RP) water treatment plant. However, its interval model is derived in Sect. 3. Section 4 presents the reduction procedure for interval model. The Jaya algorithm is discussed in Sect. 5. In Sect. 6, the results and discussion are provided. Finally, the contributions are summarized in Sect. 7.

2 The Riverol-Pilipovik Water Treatment Plant

The generalized block diagram of Riverol-Pilipovik (RP) water treatment plant is given in Fig. 1. This block diagram can be divided into four main stages, namely, pretreatment, high-pressure pump, membrane module, and post-treatment.

Figure 1 contains all four segments, manipulated variables and control variables. Manipulated variables are P_f (pressure) and pH_f (pH value) at feed stream, respectively, and controlled variables are F_p (flux) and C_p (conductivity) at permeate stream, respectively.

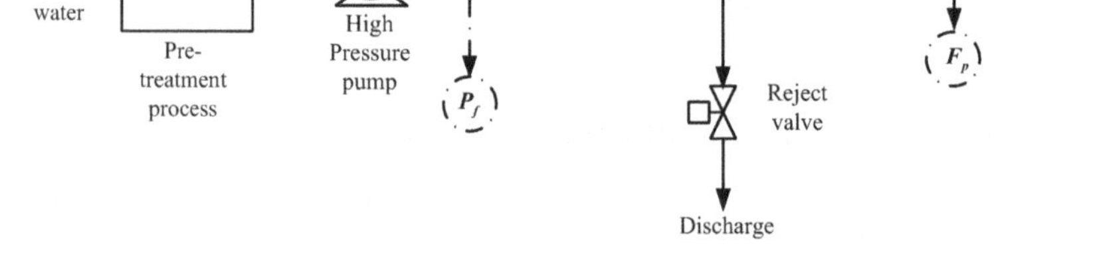

Fig. 1 Block diagram of RP RO system

The model of RO is generally obtained with the help of mass transfer equations. Finite-difference methods are generally applied to approximate these transfer equations into algebraic equations. In this investigation, multi-input-multi-output (MIMO) model developed by Riverol and Pilipovik [6] is considered for simulation. In this model, there are two loops, one for pressure and another for pH. Pressure and pH both affect the flow and conductivity at output side.

The transfer function model for RP water treatment plant is given as

$$\begin{bmatrix} F_p \\ C_p \end{bmatrix} = \begin{bmatrix} \rho_{11} & \rho_{12} \\ \rho_{21} & \rho_{22} \end{bmatrix} \begin{bmatrix} P_f \\ \mathrm{pH}_f \end{bmatrix} \quad (1)$$

where F_p is flux rate in m²/d and C_p is conductivity in μS/cm at input side, however, P_f in kPa, and pH_f are pressure and pH at output side.

The different transfer functions ρ_{11}, ρ_{12}, ρ_{21}, and ρ_{22} considered in (1) are given as

$$\rho_{11} = \frac{0.0045(0.104s + 1)}{0.012\, s^2 + s + 1} \quad (2)$$

$$\rho_{12} = 0 \quad (3)$$

$$\rho_{21} = \frac{-0.12\, s + 0.22}{0.1\, s^2 + 0.3\, s + 1} \quad (4)$$

$$\rho_{22} = \frac{10\,(-3s + 1)}{s^2 + 5\, s + 1} \quad (5)$$

3 Interval Modeling of RP Water Treatment Plant

The model of RP water treatment plant given in (1)–(5) is converted into interval model by considering 10% deviation from the nominal parameters. The transfer functions are given in (2)–(5) by considering 10% deviations become

$$\rho_{11} = \frac{[0.0004212, 0.0005148]s + [0.00405, 0.00495]}{[0.0108, 0.0132]\, s^2 + [0.90, 1.10]\, s + [0.90, 1.10]} \quad (6)$$

$$\rho_{12} = 0 \quad (7)$$

$$\rho_{21} = \frac{[-0.132, -0.108]\, s + [0.198, 0.242]}{[0.09, 0.11]\, s^2 + [0.27, 0.33]\, s + [0.90, 1.10]} \quad (8)$$

$$\rho_{22} = \frac{[-33, -27]s + [9, 11]}{[0.90, 1.10]\, s^2 + [4.5, 5.5]\, s + [0.90, 1.10]} \quad (9)$$

The interval model obtained in this section is reduced by using the Jaya algorithm. The description of Jaya algorithm is provided in Sect. 4.

4 Jaya Algorithm

Jaya algorithm [7] is one of the population-based optimization algorithms [8–17]. The advantage of Jaya algorithm is its simplicity in implementation. Further, the Jaya algorithm does not possess any algorithm-specific parameter. For obtaining the solution, there are only two common-control parameters, namely population size and number of iterations, are required. Due to these advantages, this algorithm is utilized for optimization in several engineering problems [18–20].

The key philosophy of algorithm is based on moving solutions toward best solution and moving away from the worst solution. It is assumed that the population size is I. It means that there are total I solutions in whole population. The dimension of each considered solution is J. All solutions can be represented by $X_{i,j}$ where $i = 1, 2, \ldots, I$ and $j = 1, 2, \ldots, J$. The best and worst solution of whole population can be denoted as $X_{\text{best},j}$ and $X_{\text{worst},j}$, respectively.

The solution in each iteration is updated as

$$X^n_{i,j} = X_{i,j} + a\left(X_{\text{best},j} - \left|X_{i,j}\right|\right) - b\left(X_{\text{worst},j} - \left|X_{i,j}\right|\right) \tag{10}$$

where $X^n_{i,j}$ is new solution for $X_{i,j}$. However, variables a and b denote two random numbers generated in range [0, 1]. The term $\left|X_{i,j}\right|$, shows absolute value of $X_{i,j}$. The movement toward best solution is denoted by the term $\left(X_{\text{best},j} - \left|X_{i,j}\right|\right)$. While a is responsible for the degree of movement toward the best solution. Similarly, the movement away from the worst solution is given as $-\left(X_{\text{worst},j} - \left|X_{i,j}\right|\right)$. However, b is responsible for the degree of movement toward the worst solution.

For next iteration, $X^n_{i,j}$ is considered if it has better objective function value otherwise $X_{i,j}$ is retained in the population. This process is continued for certain time until satisfactory solution is obtained.

5 Model Order Reduction

For obtaining the reduced model of transfer functions presented in (6)–(9), the technique presented in [21], is used. The details of technique are provided in [21]. In [21], first time moments of system and reduced model are matched for obtaining the reduced model along with minimization of error in between some time moments and Markov parameters.

6 Results and Discussion

For obtaining the reduced model, the lower bound of transfer function given in (9) is considered. The transfer function for lower bound of transfer function given in (9) becomes

$$G(s) = \frac{-33s + 9}{0.90s^2 + 4.5s + 0.90} \tag{11}$$

Suppose, a first-order model given by

$$H(s) = \frac{a}{bs + c} \tag{12}$$

is desired.

By following the technique presented in [21], the function given by

$$F = \left(1 + \frac{3a}{110b}\right)^2 \tag{13}$$

is to be minimized subject to

$$a = 10c \tag{14}$$

when matching of first time moments is considered along with minimization of error in between first Markov parameters. The first-order model obtained after minimizing (13) using Jaya algorithm is given as

$$H(s) = \frac{939.43}{99.99\,s + 93.94} \tag{15}$$

Figure 2 shows the step responses of first-order model and second-order system given in (15) and (11), respectively. From Fig. 2, it is clear that model is approximating the system.

7 Conclusion

In this contribution, first interval model is derived for Riverol-Pilipovik (RP) water treatment plant. Thus, the obtained interval model is obtained by considering a certain amount of uncertainty in all parameters of different coefficients of transfer function of RP water treatment plant. After deriving the interval model, a reduced-order model is obtained using Jaya algorithm. The results obtained show that model obtained is effectively resembling the considered system. The future line of research for the

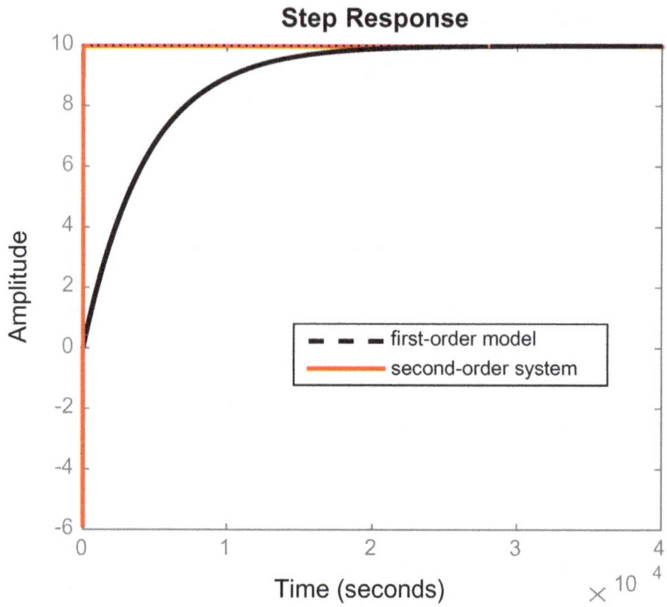

Fig. 2 Step responses of system and model

proposed work is to obtain the interval models for interval system using techniques presented for model reduction of interval systems [22–25].

Acknowledgements The work is sponsored under "TEQIP Collaborative Research Scheme" by NPIU, a unit of MHRD, Government of India (CRS application ID: 1-5766329561 and Institute PID: 1-466196671).

References

1. Choudhary AK, Nagar SK (2018) Order reduction in z-domain for interval system using an arithmetic operator. Circuits Syst Signal Process 1–16
2. Rathore NS, Singh V, Kumar B (2018) Controller design for Doha water treatment plant using grey wolf optimization. J Intell Fuzzy Syst Prepr 1–8
3. Rathore N, Chauhan D, Singh V (2015) Luus-jaakola optimization procedure for PID controller tuning in reverse osmosis system. In: International conference on electrical, electronics, and robotics (IRAJ-IACEER 2015)
4. Chaabene AB, Sellami A (2013) A novel control of a reverse osmosis desalination system powered by photovoltaic generator. In: 2013 International conference on electrical engineering and software applications (ICEESA). IEEE, pp 1–6
5. Rathore NS, Singh V, Phuc BDH (2019) A modified controller design based on symbiotic organisms search optimization for desalination system. J Water Supply: Res Technol-Aqua

6. Riverol C, Pilipovik V (2005) Mathematical modeling of perfect decoupled control system and its application: a reverse osmosis desalination industrial-scale unit. J Anal Methods Chem 2005(2):50–54
7. Rao R (2016) Jaya: A simple and new optimization algorithm for solving constrained and unconstrained optimization problems. Int J Ind Eng Comput 7(1):19–34
8. Singh SP, Prakash T, Singh VP (2019) Coordinated tuning of controller-parameters using symbiotic organisms search algorithm for frequency regulation of multi-area wind integrated power system. Eng Sci Technol Int J
9. Prakash T, Singh VP, Mohanty SR (2019) A synchrophasor measurement based wide-area power system stabilizer design for inter-area oscillation damping considering variable time-delays. Int J Electr Power Energy Syst 105:131–141
10. Rathore NS, Singh V (2019) Whale optimisation algorithm-based controller design for reverse osmosis desalination plants. Int J Intell Eng Inform 7(1):77–88
11. Prakash T, Singh V, Patnana N (2019) Gray wolf optimization-based controller design for two-tank system. In: Applications of artificial intelligence techniques in engineering. Springer, Berlin, pp 501–507
12. Prakash T, Singh V, Singh SP, Mohanty S (2018) Economic load dispatch problem: quasi-oppositional self-learning TLBO algorithm. Energy Syst 9(2):415–438
13. Rathore NS, Singh V (2018) Design of optimal PID controller for the reverse osmosis using teacher-learner-based-optimization. Membr Water Treat 9(2):129–136
14. Singh V (2017) Sine cosine algorithm based reduction of higher order continuous systems. In: 2017 International conference on intelligent sustainable systems (ICISS). IEEE, pp 649–653
15. Shrivastava S, Singh VP, Dohare R, Singh SP, Chauhan DPS (2016) PID tuning for position control of dc servo-motor using TLBO. In: National conference on process, automation and control. National Institute of Technology, Jaipur
16. Singh V, Prakash T, Rathore NS, Singh Chauhan DP, Singh SP (2016) Multilevel thresholding with membrane computing inspired tlbo. Int J Artif Intell Tools 25(06):1650030
17. Prakash T, Singh VP, Mohanty SR (2018) A novel binary whale optimization algorithm-based optimal placement of phasor measurement units. In: Handbook of research on power and energy system optimization. IGI Global, Hershey, pp 115–138
18. Singh SP, Prakash T, Singh V, Babu MG (2017) Analytic hierarchy process based automatic generation control of multi-area interconnected power system using Jaya algorithm. Eng Appl Artif Intell 60:35–44
19. Prakash T, Singh V, Singh S, Mohanty S (2017) Binary Jaya algorithm based optimal placement of phasor measurement units for power system observability. Energy Convers Manag 140:34–35
20. Prakash T, Singh VP (2018) A novel membrane computing inspired Jaya algorithm based automatic generation control of multi-area interconnected power system. Hybrid Metaheuristics: Res Appl 84:89
21. Singh S, Singh V, Singh V (2019) Analytic hierarchy process based approximation of high-order continuous systems using TLBO algorithm. Int J Dyn Control 7(1):53–60
22. Bokam J, Singh V, Raw S (2017) Comments on large scale interval system modelling using Routh approximants
23. Singh V, Chauhan D, Singh S, Prakash T (2017) On time moments and markov parameters of continuous interval systems. J Circuits Syst Comput 26(03):1750038
24. Singh V, Chandra D (2012) Model reduction of discrete interval system using clustering of poles. Int J Model Ident Control 17(2):116–123
25. Singh VP, Chandra D (2012) Reduction of discrete interval systems based on pole clustering and improved Padé approximation: a computer-aided approach. Adv Model Optim 14(1):45–56

Chapter 31
An Effective and Secure Key Management Protocol for Access Control in Pay-TV Broadcasting Systems Using Theory of Numbers

Vinod Kumar, Rajendra Kumar and S. K. Pandey

1 Introduction

With the rapid development of the Internet, Digital Pay-TV system has become more and more popular application, which facilitates the transmission of video signals from a Service Provider (SP) to authorized subscribers on a payment basis [1]. The Conditional Access System (CAS) is an essential component of any Pay-TV system to provide the access control on media delivery contents. CAS only allows legitimate subscribers to access broadcast video signals. At one time, a subscriber can subscribe to one or more channels according to his/her choice. A subscriber can subscribe any channel or combination of channels at any time and he/she can also unsubscribe the channels at any time [2]. According to forward secrecy, the leaving subscribers cannot access the future broadcast video signals. Similarly, according to backward secrecy, the newly joined subscribers cannot access the past broadcasted video signals. For access control on media contents, the CAS based Pay-TV systems requires key management protocol with minimum computational, communication and storage complexity. The main role of SP is to generate and distribute the scramble/descramble key to legitimate subscribers and maintain the forward and backward secrecy. The SP and legitimate subscribers use the key to scrambling/descrambling the video singles.

V. Kumar (✉)
Department of Electronics and Communication (Computer Science and Engineering), University of Allahabad, Allahabad, UP, India
e-mail: vinod@allduniv.ac.in

R. Kumar
Department of Computer Science, Jamia Millia Islamia, New Delhi, India

S. K. Pandey
Department of Electronics and Information Technology, Ministry of Communication and Information Technology, New Delhi, India

V. K. Giri et al. (eds.), *Computing Algorithms with Applications in Engineering*, Algorithms for Intelligent Systems, https://doi.org/10.1007/978-981-15-2369-4_31

This scramble/descramble key is need to be update frequently after every 5–20 s. The key is also refreshed after every change in group membership. This is called rekeying.

The main challenging issues of key management in centralized environment like Pay-TV broadcasting systems are efficiency, security and scalability. Many key distribution protocols [1–15] exist in the literature. Most of these protocols suffer from the problems of security, scalability and higher computational, communication, storage complexities. Thus, to design an effective and highly secure key management protocol is necessary for access control in Pay-TV broadcasting systems. In this paper, we propose a new key management protocol to solve above-mentioned issues. The main contributions are as follows: (1) design a new protocol for conditional access control system in Pay-TV systems; (2) the computation complexity at service provider side and member side reduced significantly; (3) the communication complexity and storage overhead is also minimized; (4) the proposed protocol is extended to handle multiple access control; (5) the proposed protocol provides key authentication, resistance to passive attack, resistance to collision attack and resistance to reply attack along with forward and backward secrecy. From both theoretical and experimental analysis, it is observed that our protocol is far superior in comparison with other related protocols.

The remainder of this paper is organized as follows. In Sect. 2, we present the review of related work. In Sect. 3, we describe our proposed protocol in detail. In Sect. 4, we extend our protocol for multiple access control. Section 5 provides security and performance analysis. In Sect. 6, we discuss experimental results of our protocol and related protocols. We conclude the work in Sect. 7.

2 Related Work

Many key management protocols [1–15] existing in the literature. Wong et al. [3] presented a new key management protocol for the solution of scalability problems using key graphs. They state that the protocol provides superior results when the degree of key graphs is 4. Kumar et al. [6] introduced a new key distribution protocol for secure multicast communication using RSA cryptosystem. Naranjo et al. [7] proposed a key distribution protocol using extended Euclid algorithm which supports frequently update of group key. However, the protocol is susceptible to an impersonation attack. Chou et al. [1] proposed a key management scheme for Pay-TV system using Logical Key Hierarchy (LKH). The protocol minimizes the communication complexity at member join. However, the computation complexity of the protocol is very high at member leave/join. Pal et al. [2] proposed an efficient key distribution scheme for Conditional Access System (CAS) of pay-TV systems in which channel package allotment is handled by Optimal Binary Search Tree (OBST). The protocol reduces the communication complexity and storage complexity. However, the computational complexity is high. Moreover, the protocol is susceptible to impersonation attack.

Joshi et al. [11] proposed a key transport protocol for Pay-TV system using Chinese remainder theorem which requires high computation cost at member leave/join. Lin et al. [12] introduced a key management scheme for multicast communication using RSA. In this protocol, the rekeying is not required for any change in group membership. However, scalability is the main issue in this protocol. Kumar et al. [14] introduced a new polynomial based non-interactive key computation protocol for secure communication in dynamic groups. The protocol has drastically reduced computation complexity. Kumar et al. [16] proposed an enhanced and secure RSA cryptosystem using Chinese remainder theorem. The protocol is resistant to factorization attack and timing attack. Kumar et al. [17] proposed a new fully homomorphic encryption cryptosystem using Euler's theorem over integers.

3 Proposed ESKDP Protocol

3.1 Initialization

Initially, the SP chooses the two large prime values $\mathcal{p}$ and $\mathcal{q}$ for defining multiplicative group $\mathbb{Z}_{\mathcal{p}}^*$ and threshold value δ, where $\boldsymbol{\delta} = \boldsymbol{\alpha} + \mathcal{q}$. The value of α is randomly selected from $\mathbb{Z}_{\mathcal{p}}^*$ and hence the value of δ increases when the value of α increases. The threshold δ is used to select the private keys for the members from $\mathbb{Z}_{\mathcal{p}}^*$ and random element α is used to compute the shared encryption key γ. The SP selects prime numbers as the private keys $\mathcal{PK}_i$ for the members of the group from $\mathbb{Z}_{\mathcal{p}}^*$. The selected private keys should be unique for each group member. For selection of private keys, the essential condition is that all private keys $\mathcal{PK}_i$ should be greater than δ. $\mathcal{PK}_i > \delta$ for every $i = 1, \ldots, n$). In case the condition is not fulfilled, the value of α must be modified so that it is possible to choose $\mathcal{PK}_i$. The SP sends $\mathcal{PK}_i$ to the members using secure unicast channel. These private keys are known only to members and SP.

3.2 Key Distribution

The KS executes the following steps to generate and distribute the group key among the members of the group.

First, SP chooses an element β randomly from $\mathbb{Z}_{\mathcal{p}}^*$ for all the members $\mathcal{M}_i$

Next, the encryption key to be distributed is computed by SP as

$$\gamma = \beta^{\alpha} \bmod \mathcal{p} \tag{1}$$

The SP computes $\mathcal{M} = \prod_{i=1}^{n} \mathcal{PK}_i$ and kept secret in its own database.

SP also selects two elements $\mathcal{K}$ and $\mathcal{SK}$ such that $\mathcal{K} > SK$ and $gcd(\mathcal{K}, \mathcal{SK}) = 1$.

Next, the SP encrypts δ and generates the cipher text $\mathcal{CT}$ as

$$\mathcal{CT} = ((\mathcal{K} \times \mathcal{M} + 1) \times \delta) \bmod (\mathcal{M} \times \mathcal{SK}) \tag{2}$$

The SP picks the current time-stamp $\mathcal{TS}_1$ and computes h_s using one-way hash function as

$$h_s = h(\mathcal{TS}_1 \parallel \delta \parallel \beta \parallel p \parallel q) \tag{3}$$

Finally, SP multicast (makes public) $\{\mathcal{CT}, \beta, p, q, h_s, \mathcal{TS}_1\}$ to group members.

3.3 Key Recovery

Upon receiving the multicast message $\{\mathcal{CT}, \beta, p, q, h_s, \mathcal{TS}_1\}$ from SP, the legitimate subscribers of the group execute the following steps to compute and authenticate the encryption key.

First, the subscriber S_i computes δ as

$$\delta = \mathcal{CT} \bmod \mathcal{PK}_i \tag{4}$$

After that, the subscriber S_i checks the freshness of the time-stamp $\mathcal{TS}_1$ i.e., $\mathcal{TS}_2 - \mathcal{TS}_1 = \Delta\mathcal{TS}$, where $\mathcal{TS}_2$ is current time-stamp and $\Delta\mathcal{TS}$ is maximum transmission delay. Therefore, the subscriber discards the received message due to the invalid transmission delay.

Next, the subscriber S_i computes h'_s as $h'_s = h(\mathcal{TS}_1 \parallel \delta \parallel \beta \parallel p \parallel q)$

If $h'_s \neq h_s$ then authentication is failed and he/she discards the received message.

If $h'_s = h_s$ authentication is successful and he/she computes the encryption as

$$\gamma = \beta^{(\delta - q)} \bmod p \tag{5}$$

The γ computed by the subscriber S_i must be equal to the γ computed by SP in step 2.

3.4 Scrambling and Descrambling of Video Signal

In the proposed protocol the standard symmetric key encryption algorithms such as DES, AES [18, 19] are used for scrambling/descrambling the video signal vs. The following steps describe the scrambling/descrambling procedure for video signal vs.

The SP scrambles the video signal vs, using symmetric key encryption algorithm and scramble key γ as $E_\gamma\{vs\}$

Upon receiving the scrambled video signal from SP, the subscriber $\mathcal{S}_i$ can descramble the signal using symmetric key decryption algorithm and descramble key γ as $vs = D_\gamma\{E_\gamma\{vs\}\}$.

There is need to refresh the scrambling/descrambling key γ if any new subscriber wants to subscribe to some set of channels and if the subscription of any member ends and he/she doesn't wish to continue.

3.5 Member Join

When a new member wishes to take the subscription of some set of channels, the SP generates the private key $\mathcal{PK}_{new}$ and sends it to newly joined members using secure unicast channel. To distribute the new scrambling key γ to newly joined members, the SP executes the following steps.

Computes $\mathcal{M}' = \mathcal{M} \times \mathcal{PK}_{new}$
Generates new cipher text $\mathcal{CT}'$ as $\mathcal{CT}' = \left((\mathcal{K} \times \mathcal{M}' + 1) \times \delta\right) mod \left(\mathcal{M}' \times \mathcal{SK}\right)$
SP computes $h'_s = h(\mathcal{TS}'_1 \parallel \delta \parallel \beta \parallel p \parallel q)$
SP unicast $\{\mathcal{CT}', \beta, p, q, h'_s, \mathcal{TS}'_1\}$ to new member $\mathcal{M}_{new}$.

Upon receiving the multicast message $\{\mathcal{CT}', \beta, p, q, h'_s, \mathcal{TS}'_1\}$ from SP, the newly joined subscriber executes the steps described in key recovery phase to compute and authenticate the scramble/descramble key.

3.6 Member Leave

When the subscription of any subscriber $\mathcal{S}_l$ ends and he/she doesn't wish to continue then its private key $\mathcal{PK}_l$ is deleted from the database. The SP selects a new random element α from $\mathbb{Z}_p^*$ and re-computes δ'. The SP also generates new scrambling key γ'. To distribute the new scrambling key γ' to remaining members, the SP executes the following steps.

Computes new δ' and γ'
Computes $\mathcal{M}' = \mathcal{M}/\mathcal{PK}_l$
Selects $\mathcal{K}'$ and $\mathcal{SK}'$ such that $\mathcal{K}' > SK'$ and $gcd\left(\mathcal{K}', \mathcal{SK}'\right) = 1$
Generates new cipher text $\mathcal{CT}'$ as $\mathcal{CT}' = \left((\mathcal{K}' \times \mathcal{M}' + 1) \times \delta'\right) \bmod \left(\mathcal{M}' \times \mathcal{SK}'\right)$
SP computes $h'_s = h(\mathcal{TS}'_1 \parallel \delta' \parallel \beta' \parallel p \parallel q)$
SP multicast $\{\mathcal{CT}', \beta', h'_s, \mathcal{TS}'_1\}$ to remaining new members.

Upon receiving the multicast message $\{\mathcal{CT}', \beta', h'_s, \mathcal{TS}'_1\}$ from SP, the remaining subscribers execute the steps described in key recovery phase to compute and authenticate the scramble/descramble key.

4 ESKDP Protocol for Multiple Access Control

Multiple access control requires separate communications with different groups of members such as different channel packages require different scramble/descramble key in the same TV platform. For handling such kind of situation, the SP only needs to maintain a different key pair for each channel package (group of members). We assume that a service provider provides the services of n number of channels called as a set of channels $\mathcal{CH} = \{ch_1, ch_2, ch_3 \ldots\ldots\ldots ch_n\}$ and these channels are grouped into m subgroups which are called as channel packages $\mathcal{CHP}_1, \mathcal{CHP}_2, \mathcal{CHP}_3, \ldots, \mathcal{CHP}_m$, for example, $\mathcal{CHP}_1 = \{ch_1, ch_3\}$, $\mathcal{CHP}_2 = \{ch_2, ch_5, \ldots ch_k\}$, etc. A member may subscribe to one or more channel packages at a time. The subscriber may change the channel package at any time.

1. First, the SP selects common p_j, q_j, α_j and β_j and computes δ_j and γ_j for every $j = 1, 2, 3 \ldots, m$, using the procedure described in Sects. 3.1 and 3.2
2. Next, the SP computes various chipper texts $\mathcal{CT}_j = ((\mathcal{K} \times \mathcal{M}_j + 1) \times \delta_j) \bmod (\mathcal{M}_j \times \mathcal{SK})$ for every $j = 1, 2, 3 \ldots, m$
3. The SP also computes $h_{s_j} = h\left(\mathcal{TS}_{1_j} \parallel \delta_j \parallel \beta_j \parallel p \parallel q\right)$ for every $j = 1, 2, 3 \ldots, m$
4. Finally, the SP multicast the message $\left\{\mathcal{CT}_j, \beta_j, p_j, q_j, h_{s_j}, \mathcal{TS}_{1_j}\right\}$ to the members of respective channel packages.

Every leave or join operation require to update the keys of affected channel package only. In case of Pay-TV system, it can happen that a subscriber can subscribe two or more channel package at the same time, i.e., at the same time a member may be enlisted in two or more groups. Therefore, it is clear that when that member performs join or leave operation, the refreshment of the keys are required in every channel package in which he/she belongs.

5 Security and Performance Analysis

In this section, we analyze the security and performance of our proposed protocol and provide the comparison for the same with surviving protocols like Chou et al. [1], Pal et al. [2] and Manisha et al. [11].

5.1 Security Analysis

Security analysis of proposed protocol is presented in this section. Our protocol is resistant to passive attack, Collusion attack, replay attack and maintained forward and backward secrecy.

Passive Attack: If an attacker is outsider and he/she does not have any knowledge about the group key and member's private keys, then it is not possible to perform passive attack on our protocol. Therefore, the proposed protocol is resistant to passive attack and no outsider can access the communication.

Forward Secrecy: When the subscription of any subscriber is ended then his/her private key is deleted from the database and the same will not be used for distributing the new scrambling/descrambling keys. Therefore, forward secrecy is maintained in the proposed EDKMS protocol.

Backward Secrecy: When any new subscriber wishes to access any new channel package then the required keys are frequently updated so that newly joined subscriber is not able to access old keys. Therefore, backward secrecy is maintained in our protocol.

Collision Attack: Two or more subscribers who have left the group may collaboratively compute the scrambling/descrambling key by sharing the keying information they know. In our protocol, on every change in the membership of the group, the scrambling/descrambling key and other required keys are updated. Therefore, our protocol is resistant to a collision attack.

Reply Attack: In our protocol, every multicast message is associated with current time stamp; therefore, an adversary cannot redistribute and outdated rekeying messages to subscribers. Thus, reply attack is not possible in our proposed protocol.

5.2 *Performance Analysis*

This section presents performance analysis of our protocol in terms of computation complexity, storage complexity and communication complexity and provides the comparison for the same with surviving protocols like Chou et al. [1], Pal et al. [2] and Manisha et al. [11]. The notations used for comparisons are defined as follows: n is the number of subscribers and other notations T_{md}, T_{hmg}, T_{mp}, T_{eed}, T_{ed}, T_{as}, and T_{mi} represent the computation time taken to perform multiplication/division, hash/mod/gcd, modpow, extended Euclid algorithm, symmetric encryption/decryption, addition/subtraction, and multiplicative inverse operations respectively.

Computation Complexity: The computation complexity is computed in terms of time taken to perform various mathematical operations by service providers and subscribers during every change in group membership. Table 1 describes the computation complexity of various protocols. From Table 1, it is clear that our protocol significantly reduces the computation complexity at member join/leave as compared with other surviving protocols.

Storage Complexity: The storage complexity is computed in terms of number of keys stored by service provider and a subscriber in its storage area. From Table 2 it is clearly observed that our proposed protocol requires less number of keys to store at service provider side and subscriber side in comparison with other related protocols.

Table 1 The computation complexity of various protocols

Protocols	Service provider (SP)	Member
Chou et al. [1]	$(\log_2 n - 2)T_{ed} + T_{hmg} + T_{ed}$	$(\log_2 n)T_{ed}$
Pal et al. [2]	$nT_{md} + T_{hmg} + T_{mp} + T_{eed}$	$T_{as} + T_{hmg} + T_{mp} + T_{mi} + T_{md}$
Manisha et al. [11]	$nT_{md} + T_{hmg} + T_{mp} + T_{eed}$	$T_{hmg} + 2T_{ed}$
Proposed	$4T_{md} + 2T_{hmg} + T_{mp}$	$T_{as} + 2T_{hmg} + T_{mp}$

Table 2 The storage complexity of various protocols

Protocols	Service provider (SP)	Member
Chou et al. [1]	$2n - 1$	$\log_2 n + 2n - 1$
Pal et al. [2]	$n + 9$	5
Manisha et al. [11]	$5n + 3$	4
Proposed	$n + 5$	3

Table 3 The communication complexity of various protocols

Protocols	Member join	Member leave
Chou et al. [1]	2 unicast	$(2\log_2 n - 3)$ multicast
Pal et al. [2]	1 unicast + 1 multicast	1 multicast
Manisha et al. [11]	1 unicast + 1 multicast	1 multicast
Proposed	1 unicast	1 multicast

Communication Complexity: The Communication complexity is computed in terms of number of unicast and multicast messages transmitted during every change in group membership. From Table 3, it is clear that our protocol requires to transmit only one unicast message at the join and it required one multicast message at leave. Therefore, our protocol is required less number of unicast and multicast messages during every join/leave as compared with other protocols.

6 Experimental Results

Our protocol is implemented in JAVA and tested on centralized environment based on star tree structure. For implementation, the BigInteger class of JAVA is used to perform various mathematical functions like modulo exponentiation, multiply, add and subtract, etc. For comparative study, the results are computed for the time taken to update scrambling/descrambling key at every join/leave. For performance analysis, the computation time is computed in millisecond for different groups of size varying from 500 to 8000 subscribers. The experimental results represented in Figs. 1 and

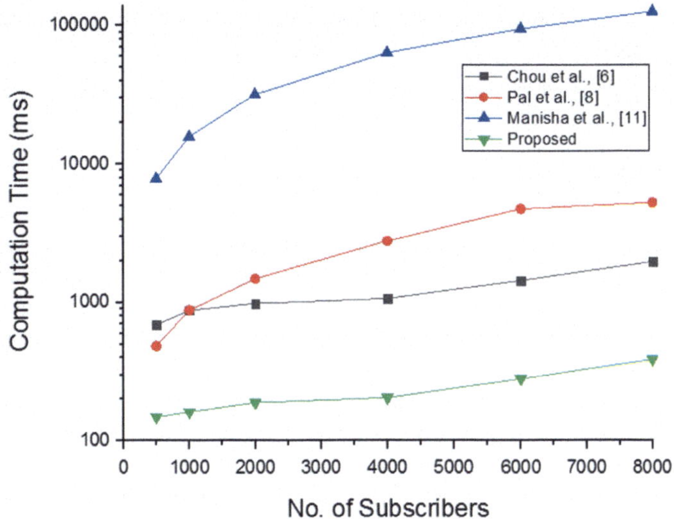

Fig. 1 Key updating time at service provider side

2 describe the computation time needed to refresh scrambling/descrambling key at service provider side and subscriber side. Figure 1 illustrates that the computation time taken by service provider in our protocol is very less in comparison with other surviving protocols. Figure 2 clearly illustrates that the computation time taken by subscribers in our protocol is less as compared to related protocols. Thus, our protocol

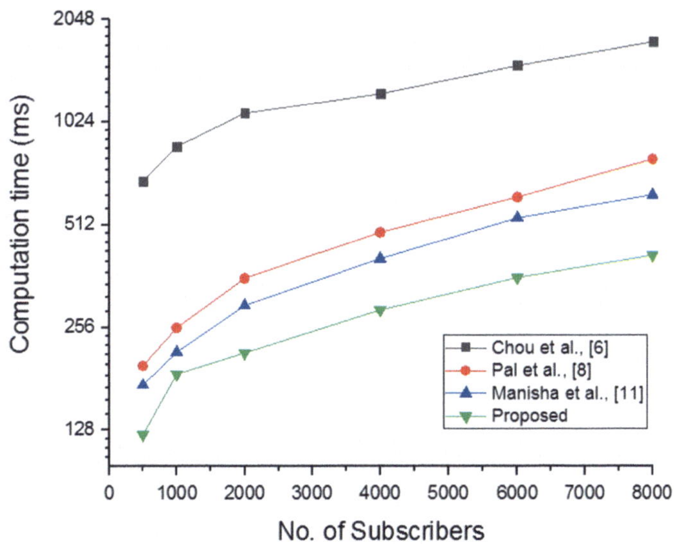

Fig. 2 Key recovery time at subscriber side

drastically reduced computation time. Therefore, our protocol is far better than other reference protocols.

7 Conclusion

This paper proposes a new key management protocol to minimize the computation, communication and storage complexity while providing access control in digital Pay-TV systems. The computational cost of our protocol is greatly reduced because it requires to perform only four basic operations such as addition, multiplication, hash function and modular exponentiation for updating the required keys when any subscriber leave/join the system. The communication complexity of the protocol is also reduced since it requires one unicast at member join and one multicast at member leave. The storage complexity of our protocol is $n + 5$ and 3 at SP side and subscriber side respectively, which is less as compared with other related protocols. The key authentication is also provided by the protocol. The protocol is also extended to handle multiple access control. The proposed protocol is secure against the various attacks and preserves the forward and backward secrecy. The proposed protocol is well-suited for the digital Pay-TV standard. The theoretical and empirical analysis clearly shows that our protocol is far better than other related protocols.

References

1. Chou KY, Chen Y-R, Tzeng W-G (2013) An efficient and secure group key management scheme supporting frequent key updates on Pay-TV systems. In: 3th Asia-Pacific network operations and management symposium (APNOMS), Taipei, Taiwan, pp 1–8
2. Pal O, Alam B (2019) Efficient and secure conditional access system for pay-TV systems. Multimed Tools Appl. https://doi.org/10.1007/s11042-019-7257-5
3. Wong CK, Gouda M, Lam SS (2000) Secure group communications using key graphs. IEEE/ACM Trans Networking 8(1):16–30
4. McGrew DA, Sherman AT (2003) Key establishment in large dynamic groups using one-way function trees. IEEE Trans Softw Eng 29(5):444–458
5. Xu L, Huang C (2008) Computation-efficient multicast key distribution. IEEE Trans Parallel Distrib Syst 19(5):577–587. https://doi.org/10.1109/TPDS.2007.70759
6. Kumar V, Kumar R, Pandey SK (2018) A computationally efficient centralized group key distribution protocol for secure multicast communications based upon RSA public key cryptosystem. J. King Saud Univ Comput Inf Sci. https://doi.org/10.1016/j.jksuci.2017.12.014
7. Naranjo JAM, Lopez-Ramos JA, Casado LG (2010) Applications of the extended Euclidean algorithm to privacy and secure communications. In: Proceedings of the 10th international conference on computational and mathematical methods in science and engineering, CMMSE
8. Tang S, Xu L, Liu N, Huang X, Ding J, Yang Z (2014) Provably secure group key management approach based upon hyper-sphere. IEEE Trans Parallel Distrib Syst 25(12):3253–3263. https://doi.org/10.1109/TPDS.2013.2297917
9. Goshi J, Ladner RE (2003) Algorithms for dynamic multicast key distribution trees. In: Proceedings ACM symposium principles of distributed computing, pp 243–251

10. Zheng XL, Huang CT, Matthews M (2007) Chinese remainder theorem based group key management. In: Proceedings of 45th ACMSE, Winston-Salem, NC, USA, pp 266–271
11. Joshi MY, Bichkar RS (2013) Scalable key transport protocol using Chinese remainder theorem. In: International symposium on security in computing and communication. SSCC 2013. Communications in computer and information science (CCIS), vol 377. https://doi.org/10.1007/978-3-642-40576-1_39
12. Lin I-C, Tang S-S, Wang C-M (2010) Multicast key management without rekeying processes. Comput J 53(7):939–950
13. VijayaKumar P, Bose S, Kannan A (2013) Centralized key distribution protocol using the greatest common divisor method. Comput Math Appl 65(9):1360–1368. https://doi.org/10.1016/j.camwa.2012.01.038
14. Kumar V, Kumar R, Pandey SK (2018) Polynomial based non-interactive session key computation protocol for secure communication in dynamic groups. Int J Inf Technol, pp 1–6. https://doi.org/10.1007/s41870-018-0140-1
15. VijayaKumar P, Bose S, Kannan A (2014) Chinese remainder theorem based centralized group key management for secure multicast communication. IET Inf Secur 8(3):179–187. https://doi.org/10.1049/iet-ifs.2012.0352
16. Kumar V, Kumar R, Pandey SK (2017) An enhanced and secured RSA public key cryptosystem algorithm using Chinese remainder theorem. In Third international conference, NGCT 2017, smart and innovative trends in next generation computing technologies, communications in computer and information science(CCIS), pp 1–12. https://doi.org/10.1007/978-981-10-8660-1_42
17. Kumar V, Kumar R, Pandey SK, Alam M (2018) Fully homomorphic encryption scheme with probabilistic encryption based on Euler's theorem and application in cloud computing. In Aggarwal VB, Bhatnagar V, Mishra DK (eds) Big data analytics. AISC, vol 654. Springer, Singapore, pp 605–611. https://doi.org/10.1007/978-981-10-6620-7_58
18. Schneier B (1996) Applied cryptography. Wiley, New York
19. Daemen J, Rijmen V (2001) Rijndael: the advanced encryption standard. Dr. Dobb's Journal

Chapter 32
System Reduced by Using Residue of Pole in Pole Clustering Technique and Differential Method

Maneesh Kumar Gupta and Rajnish Bhasker

1 Introduction

Higher order models are complicated to use in real-time system. The higher order model is complex and difficult to handle because of computational problems. The reduced order model makes simpler for controlling the system, reduces the complexity and gives the best result. The authors reduces the higher order of transfer fuction with several different techniques [1–7]. Here, we have taken input–output relationship of the system in the form of transfer function.

The proposed method is a mixed method of the residue of poles in modified pole clustering and differential method. In literature [8], differential method used with the residue of pole in pole clustering method [9] in six orders. Now, we are checking in higher order system's example. We are also used to reduce the denominator by residue of pole in pole clustering method and numerator reduced by differential method for checking performance with original system with preserving stability.

2 Problem Formulation

Let the single-input single-output (SISO) higher order transfer function of the system is:

$$G(s) = \frac{N(s)}{D(s)} = \frac{a_o + a_1 s + a_2 s^2 + \cdots + a_m s^{n-1}}{b_0 + b_1 s + b_2 s^2 + \cdots + b_n s^n}$$

where a and b are scalar constants.

Let the corresponding reduce rth order model is

M. K. Gupta (✉) · R. Bhasker
Electrical Engineering Department, UNSIET, VBS Purvanchal University, Jaunpur, UP, India

V. K. Giri et al. (eds.), *Computing Algorithms with Applications in Engineering*, Algorithms for Intelligent Systems, https://doi.org/10.1007/978-981-15-2369-4_32

$$G_r(s) = \frac{\hat{N}(s)}{\hat{D}(s)} = \frac{d_o + d_1 s + d_2 s^2 + \cdots + d_{r-1} s^{r-1}}{e_0 + e_1 s + e_2 s^2 + \cdots + e_r s^r}$$

where d and e are scalar constants.

2.1 Pole Clustering Techniques

Sinha and Pal [10] used the inverse distance measure (IDM) criterion for solving the pole clusters from the poles of the original system. The poles of the higher order system may be all real, all imaginary or a combination of complex poles. It solved individually as equation. Cluster center is obtained as:

$$p_c = \left[\left(\sum_{i=1}^{k} -1/|p_i|\right) \div k\right]^{-1}$$

where k is the number of pole, p is value of pole and p_c is pole cluster center. Pole is arranged in ascending order as

$$|p_1| < |p_2| < |p_3| < \cdots |p_k|.$$

2.2 Modified Pole Clustering

In modified pole clustering [11], some procedure is added in pole clustering algorithm for increasing stability and accuracy. Follow this step of modified pole clustering techniques:

Step 1: First, solve the equation
Step 2: Set $c = c + 1$
Step 3: Find a modified cluster center from

$$p_c = \left[\left(\frac{-1}{|p_1|} + \frac{-1}{p_{c-1}}\right) \div 2\right]^{-1}$$

Step 4: Is $k = c$? If no, and then go to Step 4, otherwise go to Step 5.
Step 5: Modified cluster center of the kth cluster as $p_k = p_c$.

2.3 Pole Clustering of Residue Method

The denominator polynomial is obtained by the dominant pole-based pole clustering approach [9]. The higher model converts in the residue form as:

$$G(s) = \sum_{k=1}^{n} \frac{R_k}{s - p_k}$$

Then corresponding to every pole p_k is the ratio of residue to pole intended as ratio of residue to pole

$$Q_k = \frac{|R_k|}{|\text{real}(p_k)|}$$

The pole p_k arranges in descending value of Q_k and then poles $p_1, p_2, p_3, \ldots, p_k$. arranged in group as most dominant pole. After then, this dominant pole is reduced by using the same above method of modified pole clustering.

2.4 Differentiation Method

The proposed method is based on differentiation method [8]. The numerator polynomial reduces by the simple differentiation method. Following steps in this algorithm:

Step 1: The reciprocal of the numerator of equation is taken as.

$$N(s) = a_m + a_{m-1}s + \cdots + a_1 s^{n-2} + a_0 s^{n-1}$$

Step 2: The reciprocated numerator is differentiated successively many (*n*-*r*) time to get desired order.
Step 3: The reduced numerator function is again reciprocated.
Step 4: Defined the correction factor (ψ) for steady-state correction as:

$$\psi = \frac{a_0/b_0}{d_0/e_0} = \frac{\text{Steady state factor of original system}}{\text{Steady state factor of reduced order system}}$$

Step 5: Finally, the steady-state correction is applied to the reduced order numerator system by multiplying the correction factor (ψ) as:

$$\check{N}(s) = \psi^* N_r(s)$$

Straight forward differentiation is discarded because the zero with large modules tends to be approximated than those with small modules.

3 Numerical Example

With residue:

Consider a six-order model [12].

$$G(s) = \frac{s^5 + 1014s^4 + 14{,}069s^3 + 69{,}140s^2 + 140{,}100s + 100{,}000}{s^6 + 222s^5 + 14{,}541s^4 + 248{,}420s^3 + 1{,}454{,}100s^2 + 2{,}220{,}000s + 1{,}000{,}000}$$

The poles of higher order system are;
$-1, -1, -10, -10, -100, -100$
Denominator reduces by pole clustering of residue method expressed as:

$$G(s) = \frac{2.0385}{s+100} + \frac{982.622}{s+100} + \frac{-1.084}{s+10} + \frac{2.4143}{s+10} + \frac{0.0455}{s+1} + \frac{0.0201}{s+1}$$

Ratio of residue to pole

$$|Q_k| = 0.0204, 9.826, 0.108, 0.2414, 0.0455, 0.0201$$

Arrange in descending order as

$$|Q_k| = 9.826 > 0.2414 > 0.108 > 0.0455 > 0.0204 > 0.0201$$

Pole is also set according to descending order of Q_k.

$- P_i = 100, \quad 10, \quad 10, \quad 1, \quad 100, \quad 1$

Cluster—1: Poles $(-1, -100, -1)$
Cluster—2: Poles $(-10, -100, -10)$.
Now, pole reduces by improved pole clustering method
$P_1 = -1.1$
$P_2 = -10.81$
The reduce order denominator

$\check{D}(s) = s^2 + 12.9626s + 14.0898$

Numerator reduces by differentiation method by above reduced denominator. The numerator polynomial is differentiated successively four times for converting six-order system to two-order system. Then

$$N_r(s) = 1.2\text{e}7s + 33.624\text{e}5$$

Then correction factor

$$\Psi = \frac{10\text{e}5/10\text{e}6}{3.624\text{e}5/11.89} = 3.536\text{e} - 7$$

$$\check{N}(s) = \psi^* N_r(s) = 5.0284s + 1.412$$

Then reduce transfer function is (Figs. 1, 2 and 3)

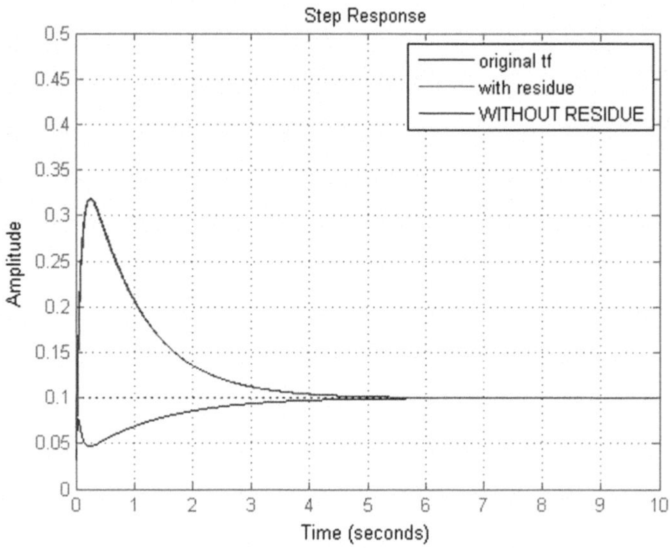

Fig. 1 Comparison of step responses of system

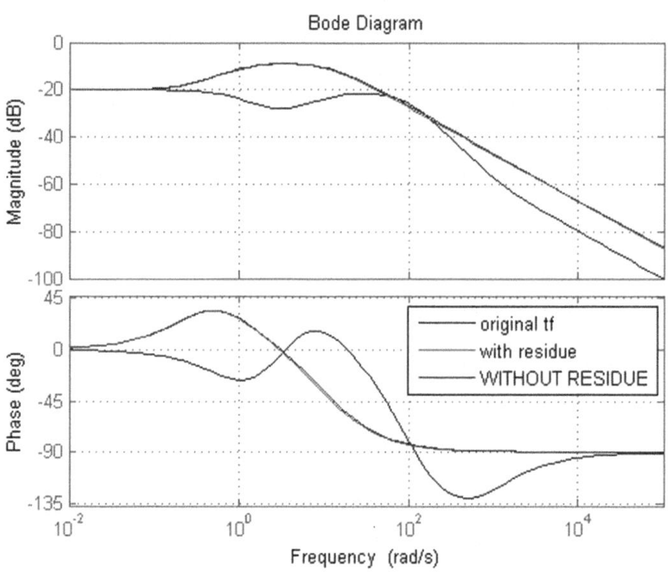

Fig. 2 Comparison of Bode plot of system

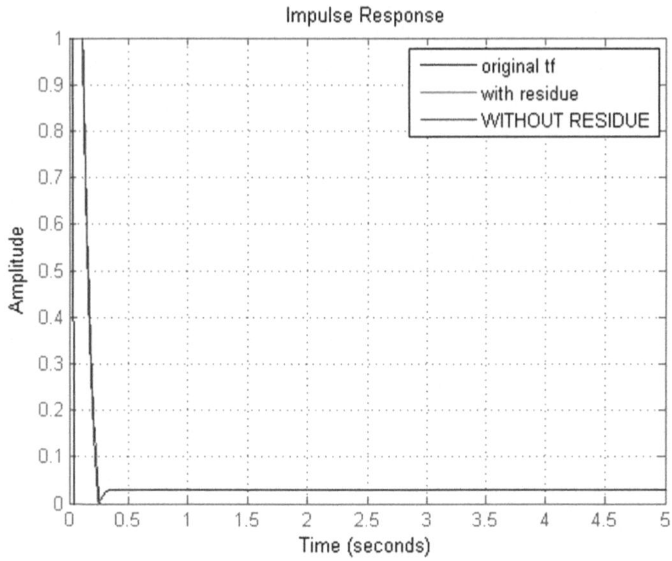

Fig. 3 Comparison of Bode plot of system

$$G_r(s) = \frac{4.2434s + 1.19}{s^2 + 11.91s + 11.891}$$

Without residue:

Pole is also set as increasing order without residue.

$$-P_i = 1, \quad 1, \quad 10, \quad 10, \quad 100, \quad 100$$

Cluster—1: Poles $(-1, -1, -10)$
Cluster—2: Poles $(-10, -100, -100)$
Now, pole reduces by improved pole clustering method
$P_1 = -1.081$
$P_2 = -11.7647$

The reduce order denominator

$$\check{D}(s) = s^2 + 12.8457s + 12.7176$$

Numerator reduces by differentiation method by above reduced denominator. The numerator polynomial is differentiated successively four times for converting six-order system to two-order system. Then

$$N_r(s) = 1.2\text{e}7s + 33.624\text{e}5$$

Then correction factor

Table 1 ISE comparison

Method of order reduction	Reduced model	ISE
Pole clustering with residue	$G_r(s) = \frac{4.2434s+1.19}{s^2+11.91s+11.891}$	0.0526
Pole clustering without residue	$G_r(s) = \frac{4.5387s+1.27176}{s^2+128457s+127176}$	0.0539

$$\Psi = \frac{10e5/10e6}{33.624e5/12.7176} = 3.7823e - 7$$

$$\check{N}(s) = \psi^{*}\mathrm{N_r(s)} = 4.53876\mathrm{s} + 1.27176$$

Then reduce transfer function is

$$G_r(s) = \frac{4.53876s + 1.27176}{s^2 + 12.8457s + 12.7176}$$

4 Comparison of Methods

The performance comparison of both the proposed algorithm is given in Table 1. The step responses of both reduce order models are compared with the original model, which is shown in the figure. In the table, comparing the integral square error of transient part of step response is in between original and reduced order models. The ISE is

$$\mathrm{ISE} = \int_0^\infty [y_o(t) - y_r(t)]^2 \mathrm{d}t$$

where y_o and y_r are unit step response of original reduce order system, respectively.

5 Conclusion

An algorithm which combined the advantage of two systems with differential method, one is residue of pole in pole clustering and other is without residue of pole in pole clustering means modified pole clustering. This method checked by an example, which reduced the numerator by differential method and denominator reduced by with residue of poles and without residue of pole in pole clustering. Now, the second method applies on same example. The result of step response, impulse response and Bode plot is shown in Fig. 1 and ISE error show in Table 1. ISE of residue of pole in pole clustering method is minimum. Reduced order model obtained is stable, mathematically simple, as well as in quality.

References

1. Shamash Y (1974) Stable reduced order model using padè type approximations. IEEE Trans Autom Control 19:615–616
2. Parthasarathy R, Jayasimhaj KN (1982) System reduction using stability-equation method and modified cauer continued fraction. Proc IEEE 70(10)
3. Gupta MK, Kumar A (2016) Model order reduction using Chebyshev polynomial, stability equation and Fuzzy C-means clustering. i-manager's J Instrum Control Eng 4(2)
4. Gupta MK (2017) Performance analysis and Com. of reduced order systems using Chebyshev polynomial, improved pole clustering and FCM clustering techniques. Int J Sci Res 6(5)
5. Bistritz Y, Langholz G (1979) Model reduction by Chebyshev polynomial techniques. IEEE Trans Autom Control 24(5):741–747
6. Singh D, Gujela OP (2015) Performance analysis of time moments, Markov's Parameters and Eigen Spectrum using matching moments. Int J Innov Res Comput Commun Eng 3(3)
7. Singh V, Chandra D, Kar H (2004) Improved routh pade approximants: a computer aided approach. IEEE Trans Autom Control 49(2):292–296
8. Sambariya DK, Manohar H (2016) Preservation of stability for reduced order model of large scale system using differentiation method. Br J Math Comput Sci 13(6). ISSN 2231–0851
9. Kumar A, Chandra D (2013) Improved padè pole clustering approximant. In: International conference on computer science and electronics engineering
10. Sinha K, Pal J (1990) Simulation based reduced order modeling using a clustering techniques. Comput Electr Engg 16(3):159–169
11. Sai Dinesh N, Siva Kumar M, Srinivasa Rao D (2013) Order reduction of discrete time system using modified pole clustering technique. Int J Eng Res Appl 3:565–569
12. Kumar V, Tiwari JP (2012) Order reducing of linear system using clustering method factor division algorithm. IJAIS 3(5)

Chapter 33
Machine Learning: An Overview of Classification Techniques

Anshita Malviya

1 Introduction

Experience refers to an event or happening which leaves an impression on us , and it helps us to learn different things which human beings face in their day-to-day life. Starting from the birth, humans learn many things which involve classifying between good or bad, they learn to speak and many more. In the same way, if they ponder about the working of computer, it follows the command given by them and can perform millions of instructions within seconds by giving the output. If he/she makes these computers learn from the experiences as they do, it refers to machine learning.

The term machine learning was coined by Arthur Samuel in 1959. The future of this world is machine learning. If today human or human beings do not focus on machine learning, then in the coming years, they will find themselves left behind. It is used to solve various real-time problems. According to Ryschard Michalski, machine learning is used to construct or modify representation of experiences [8]. Machine learning, a subset of artificial intelligence, is a discipline of making the machine learns new behaviors without giving complete instructions to machines.

Artificial intelligence is a branch of computer science which is used to incorporate human-like intelligence into machines through learning, reasoning and self-correction. Within few years, machine learning has emerged as one of the modern technology, and there are numerous applications in this field ranging from biometric recognition, to handwriting recognition, to medical diagnosis, to alignment of biological sequences, to detect fraudulent credit card transactions, to fault diagnostics, to natural language processing, to speech recognition and many more.

A. Malviya (✉)
Rajkiya Engineering College, Sonbhadra, UP, India

V. K. Giri et al. (eds.), *Computing Algorithms with Applications in Engineering*,
Algorithms for Intelligent Systems, https://doi.org/10.1007/978-981-15-2369-4_33

The remaining part of the paper consists of the following sections. Section 2 describes machine learning approaches. Classification techniques of machine learning are explained in Sect. 3. Section 4 is about the related work in this field. Methodology to achieve the objective is presented in Sect. 5. Further, Sect. 6 comprises the experimental work and result analysis. Section 7 concludes.

2 Machine Learning Approaches

There are mainly four types of machine learning approaches, which are as follows.

2.1 Supervised Learning

Supervised learning is the learning of the new data with the help of labeled examples to forecast the future events. In other words, it is the learning of the model where the analysis of a known training dataset is done which produces an inferred function to tell in advance about the output values/variables. The procedure of learning from the training dataset is known as supervised learning because it can be a perception of as a teacher who is guiding or can say supervising the whole learning process. Thus, this learning approach makes forecast on the training data which is improved by the "teacher" in the form of comparing its output with the correct, intended output and also finds error to modify the model and the learning is completed when the approach attains the desired level of performance. In supervised learning, there are notion of some attributes known as the dependent or output variable which are governed by the number of other attributes known as independent or input variables.

2.2 Unsupervised Learning

Unsupervised learning is the learning approach which does not consist of any classified or labeled examples to train the data. It cannot decide the correct output but can travel through the data in order to draw the conclusions from the datasets to explain the unseen forms from unlabeled data. In unsupervised learning, there is no notion on any particular attribute. There are no dependent attributes as all the attributes are independent variables. In this approach, the data is sorted into groups according to their similarity in characteristics.

2.3 Semi-supervised Learning

Semi-supervised machine learning contains the concept of both supervised and unsupervised learning approaches. It utilizes both the labeled and unlabeled data for the purpose of training. But the amount of labeled data is less in comparison with the unlabeled data. This approach is very helpful for the systems to improve their learning precision. It is selected only when the labeled data needs qualified and appropriate resources which would be used to train it or can learn from it.

2.4 Reinforcement Learning

It is a learning approach which is used to produce suitable actions and also discovers the errors or reward when it interacts with its environment. It uses trial and error interactions within a dynamic environment. It is forced to learn from its experiences. In this approach, the training data does not consist of answer key with it which could be used to train the model with the correct answer, so the reinforcement agent makes its choice to perform the given task.

3 Classification Techniques

Supervised learning approaches are used to solve the problems of classification. Classification technique consists of discrete values as its output rather than continuous values. In simple cases, there are only two feasible categories, so that is why it is called as binary classification. The different classification techniques are as follows:

3.1 K-Nearest Neighbor (KNN)

K-nearest neighbor algorithm is easy to understand and implement and is a simple, supervised machine learning algorithm which is used to solve the classification problem. KNN is based on learning by comparison, that is, it is used to compare a given test tuple with training tuples on the basis of similarity. In KNN, the training tuples are stored in n-dimensional space as they are described by n attributes, and each of the training tuple constitutes a point in the n-dimensional space. When an unknown tuple is given, then the KNN classifier finds the pattern space for the k training tuples which are nearest to the unknown tuple, and these are the k-nearest neighbors of that unknown tuple. The closeness is defined by the distance metric which is known as Euclidean Distance. The major disadvantage of KNN is that it becomes remarkably slower as the volume of the data (examples or predictors or independent variables) increases making it an impractical choice for the environment where forecast needs to be made faster.

3.2 Decision Tree Classifier (DTC)

Decision tree is a tree-based algorithm which is a powerful and popular tool to solve the classifications problems. In this algorithm, trees are constructed which are visually easy to understand and are used to give the production rules. These production rules are included into database access language like SQL so that the similar records can easily be retrieved. In this algorithm, the data needs not to be preprocessed, and there is an assumption on the distribution of the data. The colinearity is also handled efficiently by the decision tree approach. One of the major drawbacks of this algorithm is that it may grow to be very complex at the time of training the complicated datasets. Also, while handling the continuous variables, it loses its valuable information [9].

3.3 Linear Discriminant Analysis (LDA)

It is a dimensionality reduction technique which is used to solve the supervised classification problems. It is used for representing the differences in groups that is separation between multiple classes. It is used to estimate the attribute from higher dimension space into a lower dimension space. LDA is used as a preprocessing step for pattern classification. The applications of LDA are face recognition, medical and customer identification [8].

3.4 Support Vector Machine (SVM)

Support vector machines are used for learning and solving the classification problems. It is also an effective method for regression and general pattern recognition. SVM performs classification by constructing an n-dimensional hyperplane that optimally separates the data into two categories. In this algorithm, each data item of the dataset is plotted in n-dimensional space where n is the number of features in the dataset. After that, the classification is performed by finding the hyperplane that differentiates the two classes (hyper plane/line). Previously, the support vector machines were used only to solve the binary class pattern recognition problems, but nowadays, it is used for multiclass classification, regression estimation, feature selection and many more [3, 8].

3.5 Naive Bayes Classifier (NBC)

Naive Bayes Classifier is one of the simple probability-based classification technique. It uses the Bayes Theorem with assumption that the presence of a particular feature in a class is unrelated to the presence of any other feature. That's why it is called Naive which means immature. Despite of being a simple classification technique it generally gives good result in complex real-world situations.

3.6 Random Forest Classifier (RFC)

Random forest classifier is also a classification technique based on supervised learning approach. This classifier makes a set of decision tree. These decision trees are made from unplanned selected subset of training set. After that, it combines the polls from different decision trees to conclude the final class of test instance.

4 Related Work

This section presents the related work done in the field of classification techniques by various authors.

Jagtap et al. [1] discussed a comparative study of classification techniques—decision tree, rule-based algorithms, neural networks, support vector machines, Bayesian networks, genetic algorithms and fuzzy logic in terms of advantages and disadvantages.

Monika et al. [2] reported the result of eight research papers for comparison of different classification techniques. In addition to this, given the pros and cons of classification techniques like decision tree, Naive Bayes, rule-based, K-nearest neighbor (KNN), artificial neural network.

Tiwary [3] used Weka tool for comparison of decision tree, Naive Bayes, artificial neural network(ANN), support vector machine (SVM) on credit card approval data in terms of correctly/incorrectly classified instances, time taken, mean absolute error, root mean squared error, relative absolute error percentage and root relative squared error percentage.

Sanjay Kumar and Phaneendra [4] proposed SVM for classification of Berign and malignant brain tumor using magnetic resonance imaging (MRI) brain images.

Nikan [5] presented classification algorithms like C4.5, ID3, K-nearest neighbor classifier, Naive Bayes, SVM, ANN in terms of their features and limitations.

Rajput and Oza [6] explained classification techniques and reported the theoretical comparison between decision tree and KNN in terms of accuracy in general, speed of learning, speed of classification, tolerance to missing values, etc., of paper [7]. In addition to this, they also reported the simulation result in terms of correctly/incorrectly classified instances, time taken, Kappa Statistic of paper [3].

5 Methodology

To achieve the objective of this paper, the following methodology is used. The methodology consists of five major steps.

Step 1: Collect data from available resource
Step 2: Preprocessed the data if desired
Step 3: Train the models using SVM, KNN, DTC, LDA, NBC and RFC
Step 4: Test models
Step 5: Compare the accuracy of models.

6 Experimental Work and Result Analysis

6.1 Datasets

Two well-known datasets—fruits and iris flowers— are used in this study. Fruit dataset is taken from http://www.Kaggle.com/mjamilmoughal/fruits-colors-dataset, and iris flower dataset is taken from http://www.archieve.ics.uci.edu/ml/datasets/Iris. The fruits dataset consists of six attributes and fifty-nine instances in which fruit name is dependent variable, and fruit subtype, mass, width, height, color score are independent variables. The second dataset—iris flower consists of five attributes and one fifty instances. The dependent attribute is species, and independent variables are sepal length, sepal width, petal length and petal width.

6.2 Tool Used

Anaconda prepackaged distribution environment is used for the experimental work. Scikit-learn, Pandas and matplotlib are used in this experiment. These are prominent python libraries for machine learning.

6.3 Training the Models

KNN model, decision tree classifier model, liner discriminant analysis model support vector machine model, Naive Bayes classifier model and random forest classifier model are prepared by training them using both datasets separately. To evaluate the model performance, the dataset is splitted into two sets. One set of data is used to build machine learning model and is called a training set. The remaining data is used to access the performance of the model. This dataset is called the test data, test set or hold-out set. Seventy-five percentages (75%) of instances of both datasets are used to train the model, and remaining twenty-five percentages (25%) are used to test the accuracy of models.

6.4 Model Evaluation and Selection

Confusion Matrix

Given m classes, and entry in confusion matrix CM (i, j) indicates number of tuples in class i that were labeled by the classifier as class j. It is a table with rows representing the actual classes, and the columns represent the predicted classes. Confusion matrix is also called coincidence matrix [9]. Table 1 shows format of confusion matrix with two labeled classes.

Classifier Evaluation Metrics

Classifier accuracy, precision, recall and F measure are used to measure the performance of classifier.

- **Classifier Accuracy (Recognition rate)**: Percentage of test set tuples that are correctly classified

$$\text{Accuracy} = (\text{TP} + \text{TN})/\text{All}$$

- **Precision (Exactness)**: What % of tuples that the classifier labeled as positive are actually positive

$$\text{Precision} = \text{TP}/(\text{TP} + \text{FP})$$

Table 1 Confusion matrix format

Actual class\predicted class	C1	C2	Total
C1	True positive (TP)	False negative (FN)	
C2	False positive (FP)	True negative (TN)	
Total			

- **Recall (Completeness)**: What % of positive tuples did the classifier label as positive?

$$\text{Recall} = \text{TP}/(\text{TP} + \text{FN})$$

- **F measure (F_1 or F-score)**: Harmonic mean of precision and recall,

$$\text{F} = 2 \times \text{precision} \times \text{recall}(\text{precision} + \text{recall})$$

6.5 *Analysis*

Figure 1 depicts the relationship between no. of fruits and count. x-axis represents the fruit names, and y-axis represents count (instances of fruits). Figure 2 depicts the relationship between no. of iris flowers and count. x-axis represents the species of iris flowers, and y-axis represents count (instances of species).

Table 2 represents training and test sets of both dataset. Table 3 shows accuracy of models on training and test on both datasets.

Table 3 revels that for the training set of fruit dataset, DTC and RFC have given accuracy of 100%, whereas SVM has given minimum accuracy that is 61%. It also reveals that for the test set, KNN has given 100% accuracy, whereas SVM has given 33% of accuracy.

Table 3 depicts that for the training set of iris flower dataset, DTC and RFC have given accuracy of 100%, whereas SVM and NBC have given equal accuracy that is

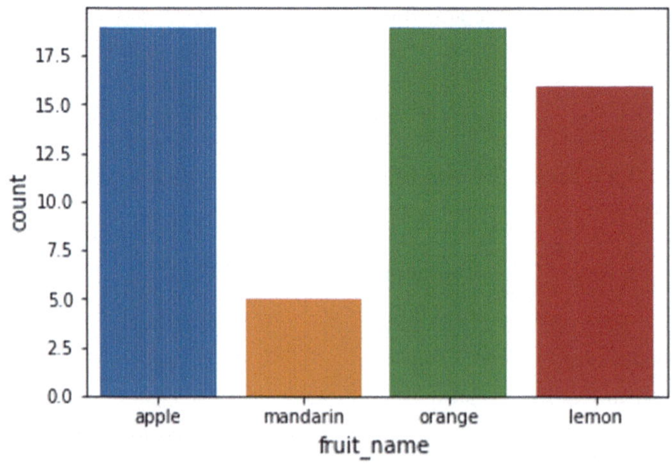

Fig. 1 Bar chart between no. of fruits and count

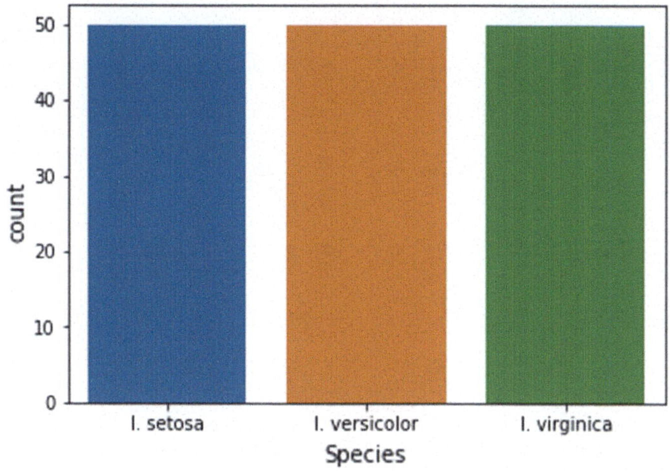

Fig. 2 Bar chart between species of iris and instances

Table 2 Training and test sets

Number of instances	Fruit dataset	Iris flower dataset
Training	44	112
Test	15	038
Total	59	150

Table 3 Comparison of classification models

Datasets		Fruit dataset		Iris flower dataset	
S. no.	Model name	Accuracy on training set	Accuracy on test set	Accuracy on training set	Accuracy on test set
1	KNN	0.95	1.00	0.96	0.97
2	DTC	1.00	0.67	1.00	0.97
3	LDA	0.86	0.67	0.98	0.97
4	SVM	0.61	0.33	0.95	0.97
5	NBC	0.86	0.67	0.95	1.00
6	RFC	1.00	0.80	1.00	0.97

95%. It also shows that for the test set, NBC has given 100% accuracy, whereas other classifiers have given equal accuracy of 97%.

Table 4 depicts precision, recall, F1-score, support of discussed classifiers on both the dataset. It concludes that KNN has maximum precision, recall, F1-score, whereas SVM has minimum on the fruit dataset. NBC has maximum precision, recall, F1-score, whereas DTC, LDA and RFC have minimum precision, SVM has minimum

Table 4 Classification report

S. no.	Classification techniques	Fruit dataset (Dataset 1)					Iris flower dataset (Dataset 2)				
		Fruit names	Precision	Recall	F1-score	Support	Species of iris flower	Precision	Recall	F1-score	Support
1	KNN	Apple	1.00	1.00	1.00	4	Setosa	1.00	1.00	1.00	13
		Mandarin	1.00	1.00	1.00	1	Versicolor	1.00	0.94	0.97	16
		Orange	1.00	1.00	1.00	8	Virginica	0.90	1.00	0.95	9
		Lemon	1.00	1.00	1.00	2					
2	DTC	Apple	0.67	1.00	0.80	4	Setosa	1.00	1.00	1.00	13
		Mandarin	1.00	1.00	1.00	1	Versicolor	1.00	0.94	0.97	16
		Orange	1.00	0.75	0.86	8	Virginica	0.90	1.00	0.95	9
		Lemon	1.00	1.00	1.00	2					
3	LDA	Apple	0.33	0.25	0.29	4	Setosa	1.00	1.00	1.00	13
		Mandarin	1.00	1.00	1.00	1	Versicolor	1.00	0.94	0.97	16
		Orange	0.67	0.75	0.71	8	Virginica	0.90	1.00	0.95	9
		Lemon	1.00	1.00	1.00	2					
4	SVM	Apple	0.29	1.00	0.44	4	Setosa	1.00	1.00	1.00	13
		Mandarin	0.00	0.00	0.00	1	Versicolor	0.94	1.00	0.97	16
		Orange	0.00	0.00	0.00	8	Virginica	1.00	0.89	0.94	9
		Lemon	1.00	0.50	0.67	2					

(continued)

Table 4 (continued)

S. no.	Classification techniques	Fruit dataset (Dataset 1)					Iris flower dataset (Dataset 2)				
		Fruit names	Precision	Recall	F1-score	Support	Species of iris flower	Precision	Recall	F1-score	Support
5	NBC	Apple	0.50	1.00	0.67	4	Setosa	1.00	1.00	1.00	13
		Mandarin	1.00	1.00	1.00	1	Versicolor	1.00	1.00	1.00	16
		Orange	1.00	0.38	0.55	8	Virginica	1.00	1.00	1.00	9
		Lemon	0.67	1.00	0.80	2					
6	RFC	Apple	0.80	1.00	0.89	4	Setosa	1.00	1.00	1.00	13
		Mandarin	1.00	1.00	1.00	1	Versicolor	1.00	0.94	0.97	16
		Orange	1.00	0.75	0.86	8	Virginica	0.90	1.00	0.95	9
		Lemon	0.67	1.00	0.80	2					

Table 5 Confusion matrix

S. no.	Classification techniques	Dataset 1		Dataset 2				
1	KNN	4	0	0	0	13	0	0
		0	1	0	0	0	15	1
		0	0	8	0	0	0	9
		0	0	0	2			
2	DTC	4	0	0	0	13	0	0
		0	1	0	0	0	15	1
		2	0	6	0	0	0	9
		0	0	0	2			
3	LDA	1	0	3	0	13	0	0
		0	1	0	0	0	15	1
		2	0	6	0	0	0	9
		0	0	0	2			
4	SVM	4	0	0	0	13	0	0
		1	0	0	0	0	16	0
		8	0	0	0	0	1	8
		1	0	0	1			
5	NBC	4	0	0	0	13	0	0
		0	1	0	0	0	16	0
		4	0	3	1	0	0	9
		0	0	0	2			
6	RFC	4	0	0	0	13	0	0
		0	1	0	0	0	15	1
		0	0	6	2	0	0	9
		0	0	0	2			

recall, except NBC all classifier has same F1-score on iris flower dataset. Table 5 shows the confusion matrix of all the classifiers discussed in this paper for both datasets.

7 Conclusions

This paper presented a comparison of some supervised learning techniques on two datasets. DTC and RFC have high accuracy on training set of both dataset. SVM model gives the least accuracy on training sets on both datasets. On other hand, KNN has high accuracy on test set of first dataset, and NBC has high accuracy on test set of second dataset. This is not the generalized result. More datasets are required to find out the strength and weakness of different machine learning classification techniques.

References

1. Jagtap N, Shevatekar PP, Mustary N (2017) A comparative study of classification techniques in data mining algorithms. Int J Mod Trends Eng Res 4(7):58–63
2. Monika E, Kaur A (2018) A comparative study of classification techniques for fraud detection. Int J Future Revolution Comput Sci Commun Eng 4(5):19–23
3. Tiwary DK (2014) A comparative study of classification algorithms for credit card approval Using weka. Int Interdisc Res J 2(3):165–174
4. Sanjay Kumar CK, Phaneendra HD (2018) Brain tumor classification with RBF kernel SVM. Int J Comput Math Sci 7(3):256–262
5. Nikam SS (2015) A comparative study of classification techniques in data mining algorithms. Int Open Free Access Peer Reviewed Res J 8(1):13–19
6. Rajput R, Oza BA (2017) A comparative study of classification techniques in data mining. Int J Creative Res Thoughts 5(3)
7. Ardhapure O, Patil G, Udani D, Jetha K (2016) Comparative study of classification algorithm for text based categorization. Int J Res Eng Technol 5(2):217–220
8. Kaushik S, Tiwari S (2018) Soft computing-fundamentals, techniques and applications, 1st edn. McGraw Hill Education(India) Private Limited, India
9. Ali ABMS, Wasimi SA (2009) Data mining-methods and techniques, 2nd edn. Cengage Learning India Private Limited, India

Chapter 34
Upcoming Power Crisis in India—Increasing Electricity Demand

Sushil Kumar, Kamlesh Kr. Bharati and Aman Shukla

1 Introduction

Industrial, service and agriculture sectors are the basis of economy in India. These sectors are majorly driven by electrical power. For a developing country like India, the need of electricity is increasing. This development includes development of these sectors internally. The various product and services industries are improving their technology to ensure better performance and features. But development is must be sustainable and non-emissive. In product industry the automobile sector plays vital role in economy. The automobile industry includes three types of vehicle production are diesel/gasoline engine vehicle, electric vehicle and hybrid vehicle (integration of engine and electric motor). In automobile sector, India officially committed to replace all the fuel engine car into electric vehicles till 2030. At present, we can see that E-Rickshaws are replacing the conventional auto rickshaws very rapidly. The mobile manufacturer companies are trying to implement wireless charging technologies to their upcoming phones, which will need more power to operate. Various home appliance manufacturers are integrating artificial intelligence (AI) technologies in devices like air conditioners, refrigerators, washing machines and smart homes. These systems demand more power for their functioning. In above advancements in technology, there will be huge increment of power demand. To meet them, India will have to install new power plants as planning commission forecast. But the plant type to be installed is thermal, i.e., coal- or gasoline-based. This is due to coal availability in India is very large. Mining of coal results in biodiversity and vegetation loss and burning of coal leads to emission of life threatening CO, CO_2, SO_x and NO_x gases also results in problems like acid rain and global warming. So this method of

S. Kumar · A. Shukla
B.N. College of Engineering and Technology, Lucknow, India

K. Kr. Bharati (✉)
Rajkiya Engineering College Kannauj, Kannauj, India

V. K. Giri et al. (eds.), *Computing Algorithms with Applications in Engineering*, Algorithms for Intelligent Systems, https://doi.org/10.1007/978-981-15-2369-4_34

Table 1 Source: Central electricity authority (CEA)

Fuel	MW	% of total
Total thermal	221,803	64.3
Coal	196,098	56.9
Gas	24,867	7.2
Oil	838	0.2
Hydro (Renewable)	45,457	13.2
Nuclear	6780	2.0
RES[a,b] (MNRE)	70,649	20.5
Total	344,689	

[a]Installed capacity in respect of RES as on 30.06.2018
[b]RES (Renewable Energy Sources) include Small Hydro Project, Biomass Power, Urban and Industrial Waste Power, Solar and Wind Energy

increasing supply of electricity results in atmospheric pollution and is not a sustainable or persistent. India have to choose alternative methods, i.e., renewable energy resources can take place of conventional for long time span. So India must have to re-evaluate their planning to meet the upcoming power demand in an effective way [1]. International agencies and unions also warned India for limitation of pollution caused by burning of coal can also be noticed from space in form of white clouds. On August 2018, India has a total installed capacity of 344,689 MW [2].

1.1 Case Study of Energy from Various Sources

From Table 1, we can conclude that the major of the portion of power is produced by thermal, i.e., coal which leads to more emission But this source is non-renewable, i.e., it will end after few years. According to a research, oil will run out in 53 years, natural gas in 54 years, and coal in 110 years.

1.2 Case Study of Performance of Conventional Generation

Table 2 shows a gradual increase in growth rate of conventional power generation from 2010–11 to 2014–15 and after its growth rate decreases by taking pollution in consideration. In present session 2018–19, the percentage of growth is 3.66%. This signs good but not actually good. India has committed to shut down all of their thermal power plants but in current scenario power demand increases drastically. If we do not choose other eco-friendly alternative energy source, we should not stop use of thermal power plants completely.

Table 2 Generation and growth in conventional generation in county during 2009–10 to 2018–19

Year	Energy generation from conventional sources (Billion Units)	% of growth
2010–11	811.143	5.56
2011–12	876.887	8.11
2012–13	912.056	4.01
2013–14	967.150	6.04
2014–15	1048.673	8.43
2015–16	1107.822	5.64
2016–17	1160.141	4.72
2017–18	1205.921	3.95
2018–19[a]	527.388	3.66

[a]Till August 2018 (Provisional)
Source CEA

2 Case Study of Energy Demand and Supply

Data analysis of energy demand and supply in India (Fig. 1).

The above data clearly shows the energy demand and supply per year. From the table, we can see that India is energy-deficit country from past years till now. But in session 2018–19 energy deficit is only 0.6% but still not zero or generating surplus energy [3].

After analysis of data in above tables, we can predict three main conclusions:

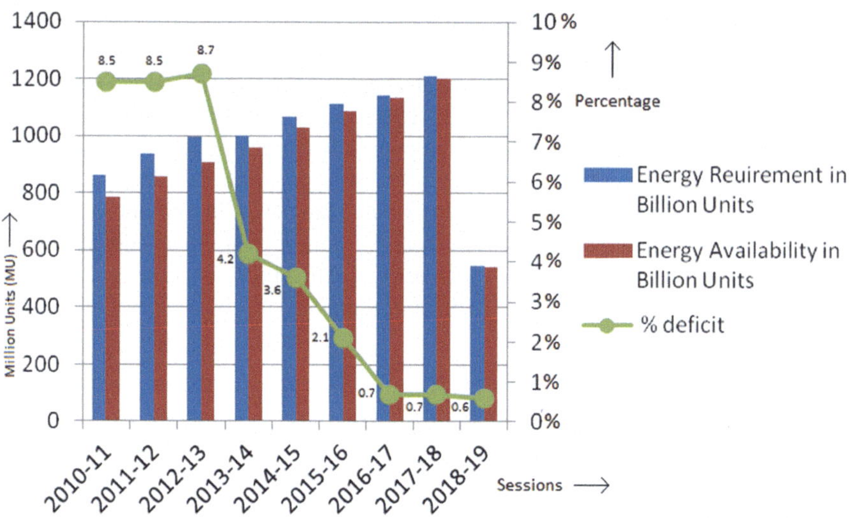

Fig. 1 Demand and supply curve. *Till August 2018 (Provisional), Source: CEA, **Graphical representation of energy demand, supply and energy deficit from year 2010–11 till now

- India's power sector is totally based on coal (fossil fuels).
- India is still growing its thermal (conventional) power plants.
- India is trying to be a power surplus country.

At present, India is nearest to meet the power demand. If we neglect various losses, i.e., generation, transmission, distribution and user end losses (losses in various equipments) India is producing surplus power. Most of the power loss is in agricultural and industrial sectors in terms of power theft. Due to vast geographical area and distributed (scattered) consumers power theft is difficult to detect and rectify. But due to recent advancements in power industry, i.e., with the introduction of smart grid technology, we are now able to rectify these problems. Above statement will contradict the power-saving systems sometimes and requires more power.

This paper presents the analysis of increasing power demand in India, factors affecting it and suggests the methods to reduce, i.e., limits power demand. This paper also recommends the ways to increase power capacity. It is also helpful for review of future power planning that will meet the upcoming power demand.

3 Comparative Analysis of Various Systems in Terms of Power Usage

For comparative analysis, we have to know all of the factors affecting the power demand.

Main factors affecting the power demand are: (i) automobile sector; (ii) electronic sectors; (iii) home appliance sectors.

(i) Automobile Sectors: Globally automobile companies started to manufacture electric vehicles rapidly. The total number of electric car reached 3 million units globally. Approximately, one million new electric cars were sold in 2017. The electric vehicle market is also expanding in India. There are 25,000 units of electric vehicle were sold at the end of 2016–17. Various automobile companies are going to launch their electric vehicles in 2019–20 [4]. Besides these preparation of launches firstly we have to analyze their impact on Utility Grid. Many charging points were made up for charging of batteries of electric vehicles. So these charging points also draw electricity from electricity grid. They will result in extra burden over the grid. Finally, we have to review our planning for such a huge change.

3.1 Comparison of Internal Combustion (IC) Engine Car with Electric Vehicles

From Table 3, it shows the advantage of electric vehicle over conventional IC engine vehicles. So in few years people will likely prefer electric vehicles. We have also analyzed its impact on grid. Averagely, an electric car consumes 4.5–5.5 km/kwh

Table 3 Comparison of IC engine with electric car

Parameters	IC engine car	Electric car
Vehicle cost	Moderate	High
Size	Engine and fuel assemblies acquire most of car space	Motor and battery require less space
Emissions	CO_2, SO_x and NO_x	Almost no emissions
Durability	Moderate	High
Maintenance	High	Low
Riding cost	Low	High
Riding range	Moderate	Lower than IC engine car
Refueling time	Lowest	Very high
Energy efficiency	17–21%	85–90%

energy which is equivalent to adding an additional house load burden on the grid [5]. Its charging assembly also introduces power loss in terms of AC-DC conversions in conventional chargers and AC-DC-AC-DC in modern advance semiconductor charging equipment. In that condition, national grid may able to deliver such additional energy demand but local infrastructure is not capable to support such huge power transmission and delivery. In hybrid and DC microgrid systems such devices offer more losses due to use of cycloconverters and chopper-controlled DC converters. Chargers like switched-mode power supply (SMPS) perform well, i.e., also in load and variable voltage outputs but they also suffers AC-DC-AC-DC losses we will discuss more about them in second factor.

There are various types of losses in different types of electric drive application in electric vehicles: (a) series PHEV, (b) parallel PHEV, (c) blended PHEV

*PHEV (Plug-in Hybrid Electric Vehicle).

In case of hybrid electric vehicles, there will be electric motor drive is only for no fuel and emergency conditions. This will not be responsible for huge increment of burden on grid [6]. But those vehicles which are only of electrical type can be responsible on greater impact on load increment on the grid. Because in this case, the motor battery system is only present which will draw more amperes from the main grid [7].

On the basis of motor and its supply technology, the losses result in additional load as follows:

a. AC supply-based induction or synchronous motor.
b. DC supply-based stepper or brushless DC (BLDC) motor.

In the case of AC supply-based motors, induction motors are generally used for more convenient operation than synchronous motor. So in induction motors, losses are generally due to its slip and losses due to nature of input supply. This will increase the apparent power intake of the motor from supply due to its low power factor [8]. But in case of DC supply-based motors, BLDC motors are generally used due to its smooth, simpler and reliable operation. BLDC motor runs at synchronous speed, i.e., almost zero slip and at a better power factor [9]. But DC systems and transient losses also come into play. This also results in a huge ohmic loss due to voltage drop across at least value of resistances also. Hence resulting output mechanical power is reduced as per electrical input power. As we know that a small quadcopter BLDC motor must draw 5–10 amperes initially so for BLDC motors having higher current drawing capacities, i.e., almost 5–8 times greater than the smaller one. This will result in fast battery drainage which is connected as a source. Again charging of this battery at lesser time intervals increased energy demand from the grid.

Finally, we have to analyze which type of electric drive system offers greater efficiency at various desired conditions for drive operation. We should have also analyzed the efficiencies of different converters (controllers) which are based on different technologies, which will help in planning of power system operation and control [10]. We have to accept the fact this time that no any electrical machine can give as much efficiency as static electrical machines hence always results in power loss and draw more input power to operate under rated parameters.

(ii) Electronic Sectors (wireless power transmission and battery charging): Electronic sectors here refer to semiconductor devices which convert the one form to another, i.e., alternating current to direct current and vice versa. Many of the companies involved in this sector they usually made wireless chargers for smartphones, laptops, electric cars and buses. Instead of wired charging methods, wireless chargers are less efficient, i.e., for same power output wire charging draws more power from input. It also requires a certain distance limit for working it efficiently and possesses primary winding and core losses when charging device is kept far from distance limit. For improving this, we have to introduce Tx and Rx, i.e., transmitter and receiver coils in which we increase the frequency of alternating current (AC) to reduce transmission losses [11]. But unfortunately while achieving this it results in increased harmonics and distortions. This affects both input and output side of the power transmission system, i.e., supply as well as phone end receiver's converter. This will lead to increased total harmonic distortion and decreases supply power quality [12]. This may be considered as extra burden on grid. In current scenario of electronics sector manly in mobile products like smartphones, they are continuously trying to give more and more features like latest telecommunication technology like 4G and 5G. Every generation is trying to reduce power consumption but due to competitive market and user demand for more bit rate which is also as a result of HD voice, video calling and real-time gaming applications. This indirectly demands more power for such high-speed communication system. Smartphone manufacturing companies are trying to improve and add the new features to their products. By doing a simple comparison, we can analyze the power difference among batteries which were used in past, i.e., 850 mAh were sufficient for one week use for mobile phones. But if we take example of present smartphones, by improving battery technologies we are able to install capacities of 5000–1000 mAh batteries but they not run a day completely after charging. This also makes a huge difference in power demand. These advancements are done to add on more features to smartphones like camera pixels and various sensors for enabling artificial intelligence features. Commonly used sensors like proximity, gyroscope, accelerometer, gravity, barometer, temperature sensor, orientation, rotational vectors, Magnetic compass and light sensor consume on average of 2–5 mA of current results in huge power loss when are continuously used [13].

Latest coming smartphones are equipped with artificial intelligence (AI) microprocessors and they are capable of multitasking and multithreading operations. These microprocessors consume less power in small-scale applications but in large-scale like industries they will demand extreme energies and were implemented by industries to reduce human hence regarding cost. These companies are also influenced by high production quality and accuracy. Nowadays electronic manufacturing companies producing smartphone, tablet are introducing wireless charging in their devices and also introduced reverse wireless charging. Like reverse wired charging, reverse wireless charging can either get charged and can also charge other devices wirelessly. In this feature, the device consists of high-frequency coil that facilitates both receiving and transmission coil. User can select mode of operation or automatically chosen by device.

But this will result in more power loss due to wireless technology will used consecutively two or more times hence power loss and consumption will be increased.

Comparison between Wired and Wireless Charging Technologies Process chart (Fig. 2).

From above chart, we can observe that overall charging efficiency of wired charging after calculation is 42.79% after using same conversion technology as wireless charging. But wireless charging has calculated overall efficiency of 34.57%. These both efficiencies are calculated by keeping other parameters same, i.e., conversion and energy storage efficiency. Energy storage efficiency are 99% by considering Lithium ion battery in both wired and wireless charging technologies [14].

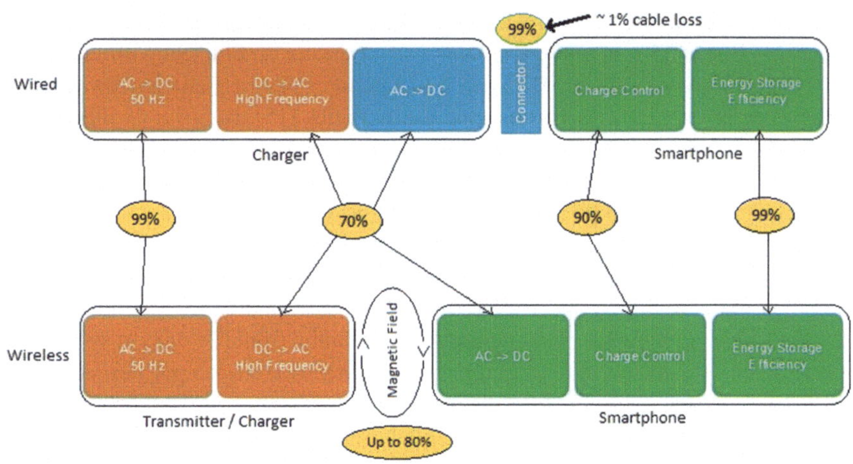

Fig. 2 Comparison between wired and wireless charging. *Wireless Charging of smartphone, **Source: MT5009—Analyzing Hi-Technology Opportunities—Group Project

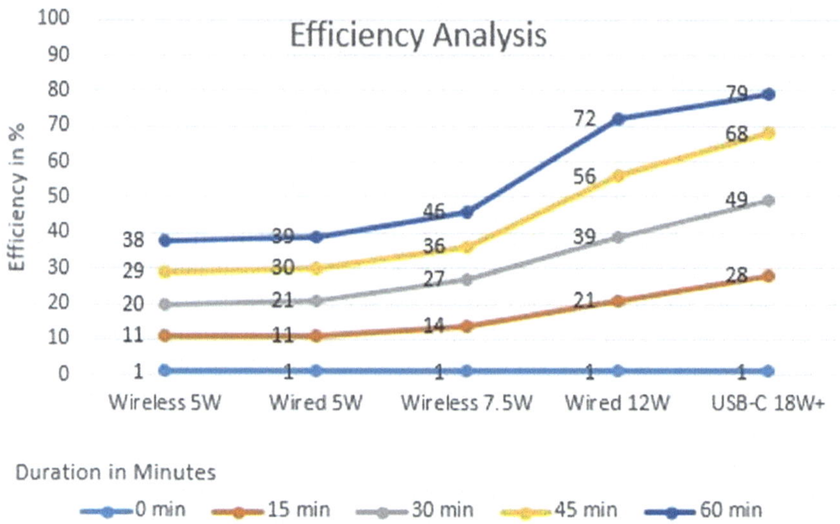

Fig. 3 Various charging techniques. *Analysis of wired/wireless USB-C (15 min intervals), **Source: MacRumors

The calculated efficiency difference is 8.22%, i.e., wired charging efficiency is still greater than wireless charging. This 8.22% more power is drawn by wireless charging to keep the output power same as wired charging.

Efficiency Analysis for various charging techniques is considering charging durations (Fig. 3).

From above graph, we can observe the real-time operation for same charging power and duration, i.e., 5 W wireless and wired charger have no efficiency difference at starting. But at 15 min and onwards we can measure an efficiency difference of 1%. Universal Serial Bus (USB) Type C has greater charging efficiencies when it is used with wired and wireless charging technologies [15]. But overall efficiencies are always differed by 1%. Due to this difference, individually they negligibly affect power demand. But when huge wireless chargers are connected on the grid they will make a huge incremental power demand on the grid.

(iii) Changing Lifestyle of People: In last two decades, technological revolution takes place in which human are not untouched. Technologies have greater impact on human living and their survival. Today we are surrounded by technologies and their associated equipment. These equipments are found from life support systems to entertainment for humans. Many equipments like air purifier, water purifier, smart air-conditioning systems, bio-medical diagnostics and treatment, smartphones, tablets, laptops, smart televisions, refrigerators, kitchen appliances like air fryer, electric chimneys, microwaves, induction cookware, and other smart appliances are making our life easier but making us more electrical energy dependent. These technologies are seemed to easier and flexible to use but they demand more electrical power. Our per-capita energy demand is increasing tremendously from recent few decades

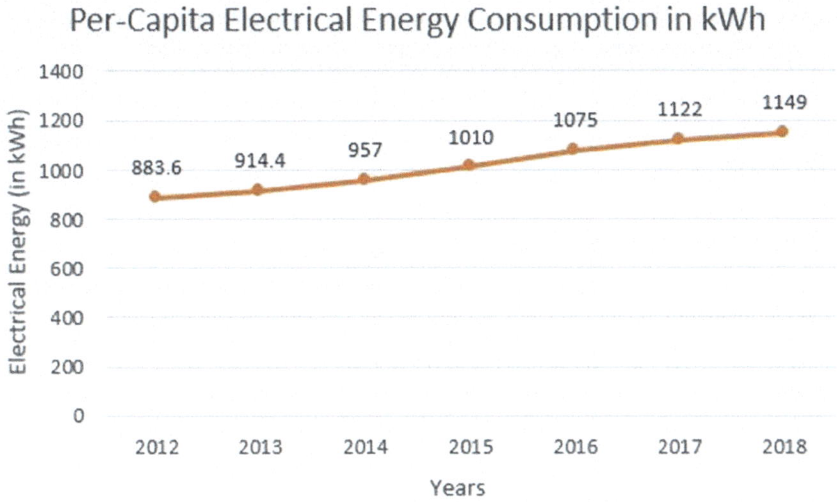

Fig. 4 Per capital electrical energy consumption. *Analysis of per-capita energy consumption in past years. **Source: Wikipedia

[16]. With the emergence of new technologies, our non-energy-consuming portion of lifestyles turned into energy-consuming portion. Technologies like smart grid are energy efficient but consume energy from main grid under no operating conditions like non-operating durations of renewable energy sources. Smart wireless communication system is getting integrated with conventional equipments which are beneficial to enhanced remote operation and control, but may draw more power than conventional in communication units.

Analysis of Per-Capita Electrical energy consumption (Fig. 4).

From above chart, the per-capita energy consumption is increased 265 kWh from year 2012 to 2018 shows tremendous increase in energy demand. This will result in future energy crisis when not compensated by renewable energy resources. Reason for this energy crisis is increased demand of facility and services which are usually for ease of human efforts.

4 Conclusions

This paper concludes that energy demand per person is huge increased in past decades and increasing till now. By collecting and analyzing the previous data, we have to increase the electricity production much more than forecasted. To make India power surplus country, we have to produce more electricity in eco-friendly way. We have to search new technologies to extract electrical energy more efficiently. India has to install more solar, wind, tidal, geothermal power plant than expected. India has to discover new and clean energy resources to overcome upcoming power crisis. We have also improved transmission efficiency and reduce the power theft.

References

1. http://cbs.teriin.org/pdf/EnCore/power_sector_report.pdf
2. Power sector in India—IMC (Indian Merchants' Chamber)
3. http://www.imcnet.org/cms/public/content/ertf_thoughtpaper/2.%20Power%20sector%20in%20India.pdf
4. Varghese G, Eapen LM (2016) Power sector in India—recent challenges and measures undertaken. Asian J Res Bus Econ Manage 6(1):33–46
5. Power Sector at a Glance All India—Ministry of Power, Govt. of India
6. https://powermin.nic.in/en/content/power-sector-glance-all-india
7. The Indian Automotive Industry—Evolving dynamics—KPMG
8. https://www.kpmg.de/docs/Auto_survey.pdf
9. Argueta R (2010) A technical research report: the electric vehicle. University of California Santa Barbara College of Engineering
10. Chan CC Electric vehicles. University of Hong Kong, Hong Kong, China
11. https://www.eolss.net/Sample-Chapters/C05/E6-39A-05-08.pdf
12. Aakib J, Sayyad N, Sarvade P (2014) Wireless power transmission for charging mobiles. Int J Eng Trends Technol 12(7):331–336
13. Rizaev A "Visual mining of smartphone sensor data". Thesis for the Master of Science in Computer Science, University of Fribourg
14. Wang Z, Wei X (2015) Design considerations for wireless charging systems with an analysis of batteries. Energies 8(10):10664–10683. https://doi.org/10.3390/en81010664
15. Enos N, Gosselin B "A primer on USB type-C and power delivery applications and requirements" in Texas Instruments
16. Growth of Electricity Consumption in India https://en.wikipedia.org/wiki/Electricity_sector_in_India
17. Power sector report final—Teri BCSD (Business Council for Sustainable Development) http://cbs.teriin.org/pdf/EnCore/power_sector_report.pdf

Chapter 35
Performance Analysis of AES, RSA and Hashing Algorithm Using Web Technology

Diksha Tiwari, Anand Singh and Abhishek Prabhakar

1 Introduction

There are various types of cryptographic algorithm which are defined as follows:

- Symmetric key cryptographic algorithm
- Asymmetric key cryptographic algorithm
- Hashing.

1.1 Symmetric Encryption

Also known as single-key encryption [1] was in use before the invention of public key encryption in the 1970s. The original message is called plaintext while the coded message is called cipher text. The conversion of plaintext to cipher text is called encryption and restoration of plaintext from cipher text is called decryption.

This method uses two different algorithms for the purpose of encryption and decryption and the same key is used by sender and receiver. The encryption algorithm is used by sender to encrypt the data and the decryption algorithm is used to decrypt the data. In this paper, we have analyzed the efficiency of a symmetric key [2] cryptography algorithm, for example, AES [3] and compare it with RSA and Hashing Algorithm using web technology.

D. Tiwari (✉)
Naraina College of Engineering and Technology, Kanpur, Uttar Pradesh 208020, India

A. Singh
Kantipur City College, Putalisadak, Kathmandu, Nepal 44600, India

A. Prabhakar
Ambedkar Institute of Technology for Handicapped Awadhpuri,
Kanpur Uttar Pradesh, 208024, India

V. K. Giri et al. (eds.), *Computing Algorithms with Applications in Engineering*,
Algorithms for Intelligent Systems, https://doi.org/10.1007/978-981-15-2369-4_35

1.2 Asymmetric Key Cryptography

This type of cryptographic [4] method uses a public key for encryption and a private key for decryption. The public key is used by sender to encrypt the message. The private key is used by receiver to decrypt the message. One of the widely used asymmetric key cryptographic algorithms is RSA [3]. In this paper, we have analyzed the efficiency of RSA, AES and hashing algorithm.

2 Hashing Cryptography

A hash function H is a algorithm that takes input a message (any length) M, and a fixed-length key K (Keyed hash function) and outputs a fixed-length message called D (message digest).

$$H(K, M) = D$$

where

D = Output message
K = key length (fixed)
M = message length.

3 Related Work

In previous researches hashing security can be improved by the use of message authentication code (MAC). The message authentication code is used between the sender and receiver to check transmitted data. This procedure is called as HMAC(Hash-based message authentication code) [5].

An improved wireless protocol has been described in [6]. HMAC with WTLS provides high-level security [7]. Base on security and architecture authors in [8] provide a comparison analysis of symmetric cryptography. In [9] encryption and decryption of image and text are done. In [10], author describes AES, DES, RSA encryption considering computational time and memory usage [11]. Cryptographic tool is used for analysis. In [12] use of cloud for file saving is mentioned by authors. In the present paper, we have used web technology for measuring performance of AES, RSA and hashing algorithm.

Table 1 Different algorithm with its size and time

S. no.	Algorithm	Key size (bit)	File size (kb)	Encryption time (in ms)
1	RSA	512	2	425
		512	4	800
		512	6	1.1 k
		512	8	1.4 k
		512	10	1.7 k
2	AES	256	5	
				1 ms (approx)
		256	10	1 ms (approx)
		256	15	1 ms (approx)
3	HASHING	SHA1		
		SHA 256		

4 Experimental Platform and Environment

Software specifications:—Xammp 1.8.3, PHP 5.5.15, PHP Myadmin 4.2.7.1, Apache web server 2.4.10 win(32)

Hardware: Algorithms are tested on AMD A4 -3530 MX APU with Radeon HD GRAPHICS 2.20Ghz, operating system windows 7 (64 bit)

Key management algorithms: RSA 512 and AES 256 and SHA1, SHA256 and MD5.

5 Experimental Result and Analysis

Experiments are shown as in Table 1 for RSA, AES and Hashing algorithms. Results are calculated with variable file sizes and time has been calculated for encryption. Key size has also been mentioned in the table. All the experiments are performed using the same processor in same system. Figure 1 shows RSA encryption and decryption. Figure 2 shows RSA performance which indicates as message size increases encryption time also increases. But in AES as message increases the encryption time remains nearly constant as shown in Fig. 4. Encryption for hashing algorithm is stated in Fig. 5 and performance of hashing algorithm is shown in Fig. 6 (Fig. 3).

6 Conclusion and Future Scope

We have analyzed the performance of RSA, AES, and Hashing algorithm using web technology. We have found that AES takes less encryption and decryption time as

Fig. 1 RSA encryption

Fig. 2 RSA performance

compared to RSA and hashing algorithms. Encryption and decryption time are taken in MS and message size in kb that too in one system. Since we have analyzed the text messages, we can also take audio and video messages.

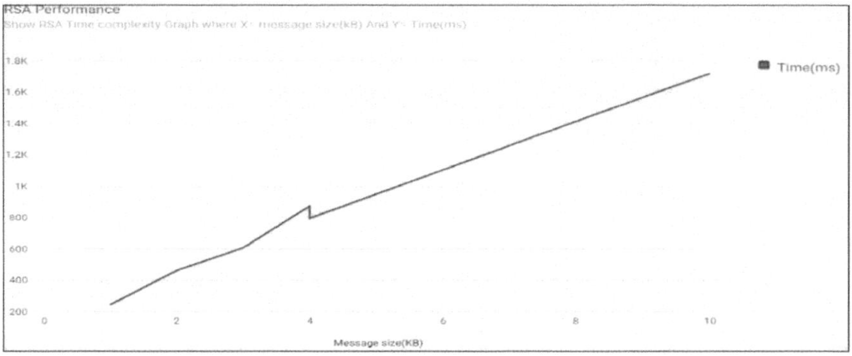

Fig. 3 RSA time complexity Graph

Fig. 4 Hashing algorithm

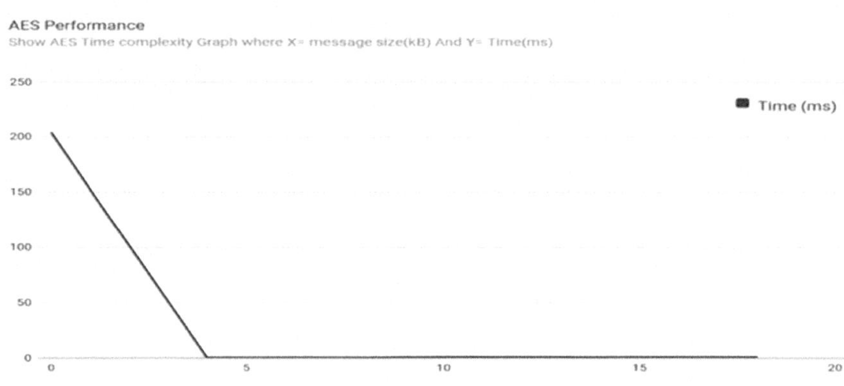

Fig. 5 AES performance

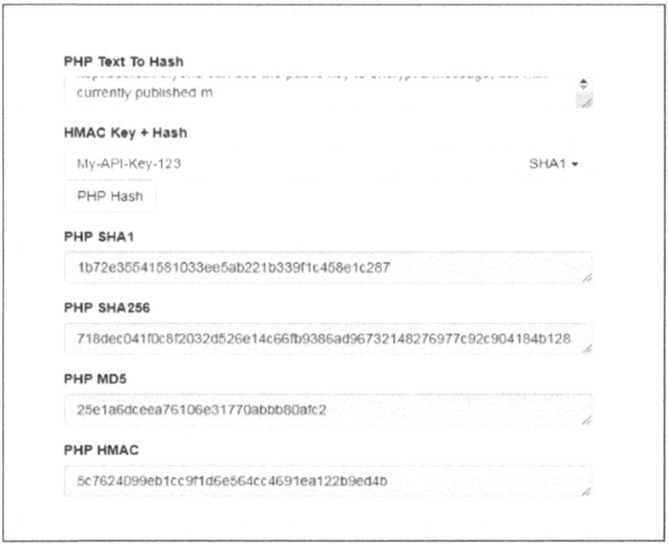

Fig. 6 Hashing algorithm using SHA1, SHA 256 and MD5 Key

References

1. Stallings W (2009) Cryptography and network security principles and practices, 4th edn. Pearson Education, Prentice Hall
2. Tseng Y-M (2007) An efficient two-party identity-based key exchange protocol. Informatica 18(1):125–136
3. Mare SF, Vladutiu M, Prodan L (2011) Secret data communication system using Steganography, AES and RSA. In: International symposium for design and technology in electronic packaging, vol 2, pp 339–344
4. Kahate A (2008) Cryptography and network security. Tata McGraw-Hill Companies
5. Jungles P, Simos M, Godard B, Bialek J, Bucher M, Waits C, Peteroy W, Garnier T (2014) Defending against Pass-the-Hash Attacks, mitigating Pass-the-Hash and other credential theft. The Microsoft Security Intelligence Report (SIR), Microsoft Corporation
6. Berry R, Berry K, Kumar A (2016) Review on network security and cryptography. Int J Innov Res Technol 3(7):44–53
7. Aguirre I, Alonso S (2012) Improving the automation of security information management: a collaborative approach. IEEE Secur Priv 10(1):55–59. https://doi.org/10.1109/msp.2011.153
8. Ebrahim M, Khan S, Bin Khali U (2013) Symmetric algorithm survey: a comparative analysis. Int J Comput Appl 61(20):12–19
9. Selimis G, Sklavos N, Koufopavlou O (2003) VLSI implementation of The Keyed-Hash message authentication code for the wireless application protocol. In: 10th IEEE international conference on electronics, circuits and systems, United Arab Emirates, 14–17 Dec 2003
10. Prajapati P, Patel N, Macwan R, Kachhiya N, Shah P (2014) Comparative analysis of DES, AES, RSA encryption algorithms. Int J Eng Manage Res 4(1):292–294
11. Kim HW, Lee S (2004) Design and implementation of a private and public key crypto processor and its application to a security system. IEEE Trans Consum Electron 50(1):214–224
12. Abutaha MS, Amro AA (2014) Using AES, RSA, SHA1 for securing cloud. In: International conference on communication, internet and information technology, Madrid, Spain

Chapter 36
A Unified Approach for Outage Analysis of Dual-Hop Decode and Forward Relay Network

Himanshu Katiyar, P. K. Verma, Arun Kumar Singh and Saurabh Dixit

1 Introduction

Establishment of highly reliable, spectrally efficient and high data rate communication link is a major thrust of research around the world [3]. Power, bandwidth restrictions and operation over a harsh multi-path fading channel are major bottleneck for high data rate wireless communication services. A new type of user cooperation between user terminals have been discussed in [16] and various issues of its implementation in real-world are analyzed in [17]. Various low complexity protocols for cooperative communication are discussed in [12]. Initially, space-time coding for distributed relay network is analyzed in [13] by Laneman et al. Outage performance for dissimilar Rayleigh fading is analyzed in [1]. Approximate outage performance for cooperative relay network with best relay selection criteria and arbitrary channel distribution is analyzed in [2]. Outage analysis of relay-based system in which end-to-end link is established with the help of multiple relays is discussed in [20]. Expressions in closed form for the average received SNR, channel capacity, average error rate and outage probability is established in [7, 11], here multi-antenna base station is receiving data from single antenna source and relay. In [9], the outage is analyzed for the case when a direct link between source and destination is absent and communication is assisted by two infrastructure-based multi-antenna relays. For multi-antenna regenerative relay network, average capacity and average SNR are derived in [8]. In [10], outage and BER expressions in closed form are established

H. Katiyar (✉) · P. K. Verma
Department of Electronics Engineering, Rajkiya Engineering College, Sonbhadra, UP, India

A. K. Singh
Department of Electronics Engineering, Rajkiya Engineering College, Kannauj, India

S. Dixit
Department of Applied Science, CIPET, Lucknow, India

V. K. Giri et al. (eds.), *Computing Algorithms with Applications in Engineering*, Algorithms for Intelligent Systems, https://doi.org/10.1007/978-981-15-2369-4_36

for a scenario in which communication is assisted by an infrastructure-based multi-antenna relay node. For improving coverage and capacity in remote areas, deployment of satellite terrestrial cooperative network is beneficial which is explored in [5]. A Survey on simultaneous wireless information and power transfer with cooperative relay is dine [6]. Vehicle to vehicle communication is useful for extending the coverage in highways with the help of cooperative communication technique is explored in [21]. An expression for the average capacity of relay based system is derived in [14] for Rayleigh channel. Regenerative relaying for the application of internet of things is presented in [15]. This paper investigates the outage probability of three-node relay system. The direct path between transmitter and receiver is completely obstructed hence communication is possible only through relay link. Nakagami-m fading channel is assumed in this work because we can model a more versatile and realistic scenario. The contributions of this work are summarized below:

- Established a mathematical model of dual-hop decode and forward serial relay network operating in Nakagami-m fading channel.
- Development of simulation model for Nakagami-m fading channel.
- Derived the expressions for outage probability.

Rest of the paper is organized as follows: Sect. 2, briefly discusses the channel modeling of Nakagami-m. Closed-form expressions of outage probability for dual-hop decode and forward serial relay network have been derived in Sect. 3. Numerical results are discussed in Sect. 4. In Sect. 5, tracks for future work have been discussed. Finally, conclusions are drawn in Sect. 6.

2 Channel Modeling

A three-node serial relay system shown in Fig. 1. Here, communication link between source (s) and destination (d) is established with the supporting relay (r). In this system model, s transmits signal to r (i.e. $s \rightarrow r$) in first time-slot. Received signal is first decoded at r and forwarded to d (i.e. $r \rightarrow d$) in second time-slot. At r and d, short-term signal variation occurs due to mobility of various scatters in environment.

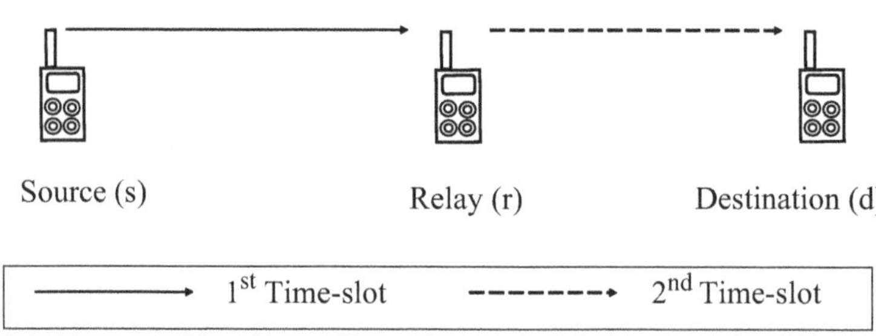

Fig. 1 A three-node relay network operating in fading channel

At any instant of time, signals transmitted from s or r may combine constructively or destructively at r or d, respectively. Such fading phenomena is relatively fast and is therefore responsible for the short-term signal variations. Random variation in signal strength can be statistically modeled by various fading models. The Nakagami-m fading model is versatile because it is suitable for indoor short distance communication, outdoor land mobile communication and scintillating ionospheric radio links. For a special case, we can easily model one-sided Gaussian distribution by taking $m = 1/2$ and the Rayleigh distribution by taking $m = 1$. The Nakagami-m fading channel converges to a non-fading AWGN channel by taking limit $m \rightarrow \infty$. The PDF of Nakagami-m fading amplitude α_{ij} can be written as [18, Eq. (2.20)]:

$$f_{\alpha_{ij}}(\alpha) = \frac{2m_{ij}^{m_{ij}}\alpha^{2m_{ij}-1}}{\Omega_{ij}^{m_{ij}}\Gamma(m_{ij})}\exp\left(-\frac{m_{ij}\alpha^2}{\Omega_{ij}}\right) \tag{1}$$

where $\Gamma(\cdot)$ is gamma function [4, Eq. (8.310.1)] and m_{ij} is the Nakagami-m fading parameter, which ranges from 1/2 to ∞. α_{ij} is a random variable (RV) which model the instantaneous fading envelop between node i and j, $i \in \{s, r\}$, $j \in \{r, d\}$, $\Omega_{ij} = E[\alpha_{ij}^2]$ and $E[\cdot]$ is expected value. The PDF of received SNR (γ_{ij}) of Nakagami-m fading model can be written as [19]:

$$f_{\gamma_{ij}}(\gamma) = \frac{m_{ij}^{m_{ij}}\gamma^{m_{ij}-1}}{\bar{\gamma}_{ij}^{m_{ij}}\Gamma(m_{ij})}\exp\left(-\frac{m_{ij}\gamma}{\bar{\gamma}_{ij}}\right) \tag{2}$$

Here, $\gamma = \alpha^2 E_s/N_0$ is the instantaneous signal to noise power ratio per symbol, $\bar{\gamma}_{ij} = \Omega_{ij}E_s/N_0$ the average signal to noise power ratio per symbol between node i and j, E_s is the energy per symbol and N_0 is one-sided power spectral density.

3 Modeling of Two-Hop Relay System

For the dual-hop decode and forward serial relay system which is shown in Fig. 1, the weakest link between $s \rightarrow r$, $r \rightarrow d$ will be effective. Hence, received SNR at d can be modeled by RV γ_{srd}:

$$\gamma_{srd} = \min(\gamma_{sr}, \gamma_{rd}) \tag{3}$$

The cumulative distribution function (CDF) for *RV* γ_{srd} can be written as:

$$\begin{aligned} F_{\gamma_{srd}}(\gamma) &= \Pr\{\gamma_{srd} \le \gamma\} \\ &= \Pr\{\min(\gamma_{sr}, \gamma_{rd}) \le \gamma\} \\ &= 1 - \Pr\{\min(\gamma_{sr}, \gamma_{rd}) > \gamma\} \end{aligned} \tag{4}$$

Here, $\Pr\{.\}$ is the probability of occurrence of an event. In (4), min $(\gamma_{sr}, \gamma_{rd}) > \gamma$ is true if and only if both $\gamma_{sr} > \gamma$ and $\gamma_{rd} > \gamma$ are true. So, (4) can be written as:

$$\begin{aligned} F_{\gamma_{srd}}(\gamma) &= 1 - \Pr\{\gamma_{sr} > \gamma, \gamma_{rd} > \gamma\} \\ &= 1 - \Pr\{\gamma_{sr} > \gamma\}\Pr\{\gamma_{rd} > \gamma\} \\ &= 1 - [1 - \Pr\{\gamma_{sr} \le \gamma\}][1 - \Pr\{\gamma_{rd} \le \gamma\}] \end{aligned} \tag{5}$$

In (5), γ_{sr}, γ_{rd} are independent *RV* because $s \rightarrow r, r \rightarrow d$ links observe independent fading phenomena. The CDF for *RV* γ_{srd} can be written as:

$$F_{\gamma_{srd}}(\gamma) = 1 - \left[\{1 - F_{\gamma_{sr}}(\gamma)\}\{1 - F_{\gamma_{rd}}(\gamma)\}\right] \tag{6}$$

Here, $F_{\gamma_{sr}}(\gamma) = \int_0^\gamma f_{\gamma_{sr}}(\gamma)d\gamma$ is CDF of *RV* γ_{sr}. Similarly, CDF of *RV* γ_{rd} can be calculated as $F_{\gamma_{rd}}(\gamma) = \int_0^\gamma f_{\gamma_{rd}}(\gamma)d\gamma$. Communication link in outage when strength of received SNR drops below a certain value. In this case, outage probability can be calculated for a given threshold χ [18, Eq. (1.4)]:

$$\begin{aligned} P_{\text{out}} &= \Pr\{\gamma_{srd} \le \chi\} = F_{\gamma_{srd}}(\gamma)\big|_{\gamma=\chi} = F_{\gamma_{srd}}(\chi) \\ &= 1 - \left[1 - \frac{\Gamma\left(m_{sr}, \frac{m_{sr}\chi}{\bar{\gamma}_{sr}}\right)}{\Gamma(m_{sr})}\right]\left[1 - \frac{\Gamma\left(m_{rd}, \frac{m_{rd}\chi}{\bar{\gamma}_{rd}}\right)}{\Gamma(m_{rd})}\right] \end{aligned} \tag{7}$$

here $\Gamma(\cdot, \cdot)$ is upper incomplete gamma function [4, Eq. (8.350.1)].

4 Numerical Results

This section presents the outage performance of dual-hop decode and forward serial relay network operating in Nakagami-*m* fading channel. Nakagami-*m* distributed *RV* are generated as given in APPENDIX which is mapped with (2) for $m = 1$, 2, 3 and plotted in Figs. 2, 3 and 4, respectively. Analytical expression of outage probability for such type of system has been derived in (7). For evaluating the outage performance in various SNR, we have fixed the value of outage threshold (χ) to unity. In Fig. 5, outage performance of dual-hop serial relay network have been plotted for the various value of *m*. Outage performance for various outage threshold is calculated by fixing the value of SNR to unity (i.e. 0 dB) and plotted in Fig. 6.

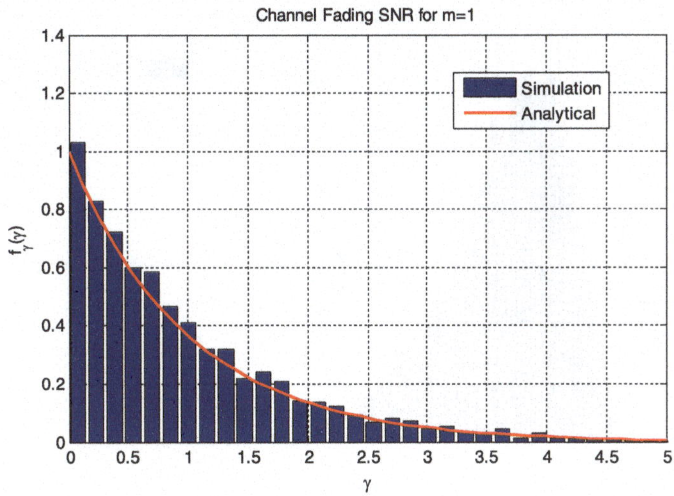

Fig. 2 Mapping of Nakagami-m distributed RV for $m = 1$

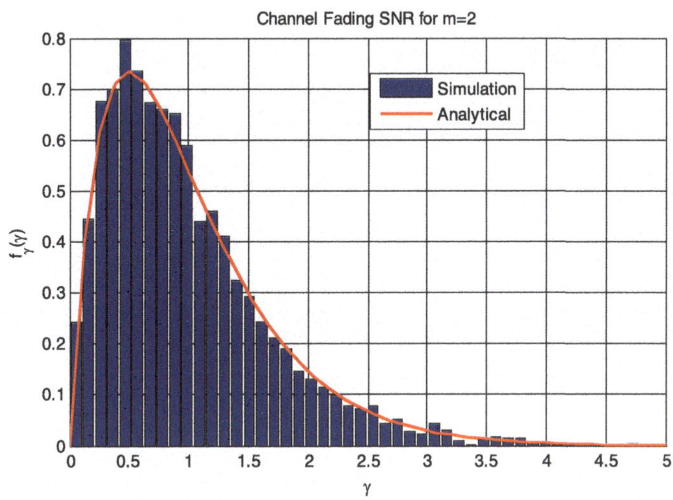

Fig. 3 Mapping of Nakagami-m distributed RV for $m = 2$

5 Tracks for Future Work

In this work, we have only analyzed the outage performance of dual-hop decode and forward serial relay network operating in Nakagami-m fading channel. Other performance parameters like average fade duration, amount of fading, average error probability, average link capacity, end-to-end average SNR, etc. yet to be analyzed.

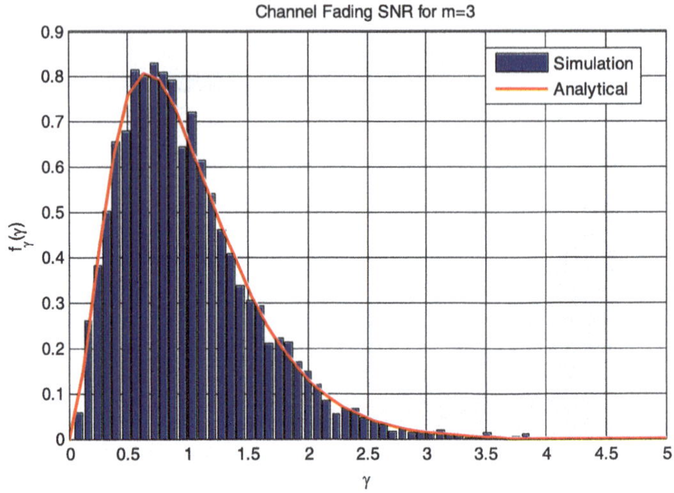

Fig. 4 Mapping of Nakagami-*m* distributed *RV* for $m = 3$

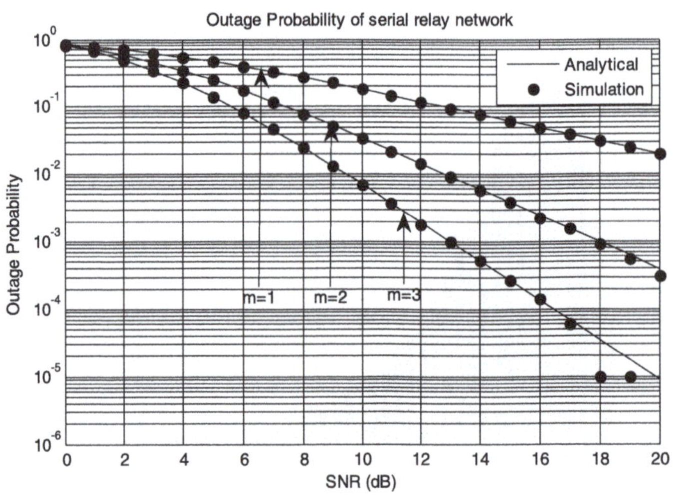

Fig. 5 Outage probability of dual-hop decode and forward relay network at various SNR

This work can also be further extended to other fading channels such as Ricean, Hoyt, Rice, κ-μ, α-μ, η-μ, etc.

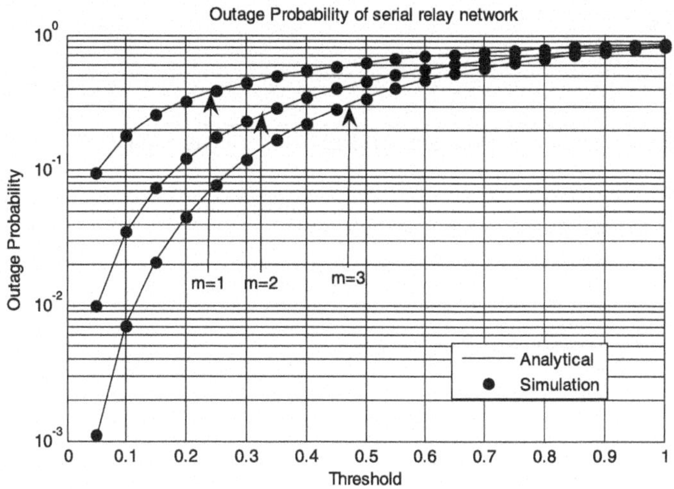

Fig. 6 Outage probability of dual-hop decode and forward relay network at various outage threshold

6 Conclusion

This paper investigates the outage performance of dual-hop decode and forward relay network in Nakagami-*m* fading channel. Validity of analytical results can be crosschecked with the help of Monte Carlo simulation (running simulator freely with 10^6 samples) and it is found that such results are perfectly matched with the simulation results. We found that outage probability decreases with increment of SNR and *m* which is fading parameter of Nakagami-*m* channel (i.e. line of site component increases). However, outage probability increases with increment of outage threshold.

Acknowledgements This work is supported by Collaborative Research and Innovation Program (CRIP) funding through TEQIP-III scheme of Dr. A.P.J. Abdul Kalam Technical University (AKTU), Lucknow. Authors are highly obliged for their support and gratefully acknowledge.

Appendix

Algorithm 1 Generation of Nakagami-m distributed random variable for integer value of m.

1: **procedure** Nakagami-m distributed RV

2: m←Parameter of Nakagami-m fading channel

3: μ←Mean of random variable - Mean is zero

4: σ^2←Variance of random variable - Variance is unity

5: n←Channel length

6: count←zero matrix of order 1×n

7: **for** $i \leftarrow 1, m$ **do**

8. $$count = count + \frac{1}{\sqrt{2}}\left|N\left(m,\sigma^2\right) + jN\left(m,\sigma^2\right)\right|^2$$

-A matrix of order 1×n containing random variables, here $N(\cdot,\ \cdot)$ represents Gaussian distribution.

9: **end for**

10: $H \leftarrow count/m$ - Here, H is Nakagami-m distributed RV

11: **end procedure**

References

1. Beaulieu NC, Hu J (2006) A closed-form expression for the outage probability of decode-and-forward relaying in dissimilar rayleigh fading channels. IEEE Commun Lett 10(12):813–815. https://doi.org/10.1109/LCOMM.2006.061048
2. Beres E, Adve R (2007) Outage probability of selection cooperation in the low to medium SNR regime. IEEE Commun Lett 11(7):589–597. https://doi.org/10.1109/LCOMM.2007.070097
3. Goldsmith A (2005) Wireless communication, 1st edn. Cambridge University Press, New York. www.cambridge.org/9780521837163
4. Gradshteyn IS, Ryzhik IM (2007) Table of integrals, series and products, 7th revised edn. Academic Press Inc, Cambridge
5. Hajipour P, Shahzadi A, Ghazi-Maghrebi S (2019) Improved performance for a heterogeneous satellite-cooperative network with best relay node selection. China Commun 16(5):93–105. https://doi.org/10.12676/j.cc.2019.05.008
6. Hossain MA, Md Noor R, Yau KA, Ahmedy I, Anjum SS (2019) A survey on simultaneous wireless information and power transfer with cooperative relay and future challenges. IEEE Access 7:19166–19198. https://doi.org/10.1109/ACCESS.2019.2895645
7. Katiyar H, Bhattacharjee, R (2009) Performance of regenerative relay network operating in uplink of multi-antenna base station under Rayleigh fading channel. In: Proceedings of the

TENCON 2009, IEEE Region 10 Conference, Singapore, pp 1–5. https://doi.org/10.1109/TENCON.2009.5395929
8. Katiyar H, Bhattacharjee R (2011) Average capacity and SNR analysis of multi-antenna regenerative cooperative relay in Rayleigh fading channel. IET Commun 5:1971–1977. https://doi.org/10.1049/iet-com.2010.0969
9. Katiyar H, Bhattacharjee R (2011) Outage performance of multi-antenna relay cooperation in the absence of direct link. IEEE Commun Lett 15(4):398–400. https://doi.org/10.1109/LCOMM.2011.020411.101863
10. Katiyar H, Bhattacharjee R (2011) Performance of two-hop infrastructure based multi-antenna regenerative relaying in Rayleigh fading channel. Phys Commun 4(3):190–195. https://doi.org/10.1016/j.aeue.2011.04.007 (Elsevier)
11. Katiyar H, Bhattacharjee R (2012) On the performance of decode-and-forward re-laying with multi-antenna destination. AEU Int J Electron Commun 66:1–6. https://doi.org/10.1016/j.aeue.2011.04.007
12. Laneman J, Tse D, Wornell G (2004) Cooperative diversity in wireless networks: efficient protocols and outage behavior. IEEE Trans Inf Theory 50(12):3062–3080. https://doi.org/10.1109/TIT.2004.838089
13. Laneman J, Wornell G (2003) Distributed space-time-coded protocols for exploiting cooperative diversity in wireless networks. IEEE Trans Inf Theory 49(10):2415–2425. https://doi.org/10.1109/TIT.2003.817829
14. Lee M, Oh S (2019) A closed-form ergodic capacity expression for a generic cooperative diversity network in rayleigh fading channels. J Commun Netw 21(3):307–318. https://doi.org/10.1109/JCN.2019.000029
15. Lyu B, Yang Z, Guo H, Tian F, Gui G (2019) Relay cooperation enhanced backscatter communication for internet-of-things. IEEE Internet Things J 6(2):2860–2871. https://doi.org/10.1109/JIOT.2018.2875719
16. Sendonaris A, Erkip E, Aazhang B (2003) User cooperation diversity. part I. system description. IEEE Trans Commun 51(11):1927–1938. https://doi.org/10.1109/TCOMM.2003.818096
17. Sendonaris A, Erkip E, Aazhang B (2003) User cooperation diversity part II implementation aspects and performance analysis. IEEE Trans Commun 51(11):1939–1948. https://doi.org/10.1109/TCOMM.2003.819238
18. Simon MK, Alouini MS (2005) Digital communication over fading channels, 2nd edn. Wiley, Hoboken, NJ. http://as.wiley.com/WileyCDA/WileyTitle/productCd-0471649538.html
19. Yang HC, Alouini MS (2005) MRC and GSC diversity combining with an outputthreshold. IEEE Trans Veh Technol 54(3):1081–1090. https://doi.org/10.1109/TVT.2005.844634
20. Zhao Y, Adve R, Lim TJ (2005) Outage probability at arbitrary SNR with co-operative diversity. IEEE Commun Lett 9(8):700–702. https://doi.org/10.1109/LCOMM.2005.1496587
21. Zhou J, Tian D, Wang Y, Sheng Z, Duan X, Leung VCM (2019) Reliability-optimal cooperative communication and computing in connected vehicle systems. IEEE Trans Mob Comput p 1–1. https://doi.org/10.1109/TMC.2019.2907491

Chapter 37
Load Frequency Control of Hybrid Power System Using Soft Computing Approach

Shashi Kant Pandey, Vikas Pandey, Sudheer Tiwari, S. R. Mohanty and V. P. Singh

1 Introduction

The main goal of the control approach in a power system is to generate and deliver electricity as economically and reliably as conceivable whereas preserving the power quality such as frequency and voltage profile within permissible limits. This paper presents the load frequency control (LFC) problem for the hybrid power system of Photovoltaic (PV) with the integration of the thermal power system. The detailed model of PV includes maximum power point tracking (MPPT), boost converter, voltage source inverter to get the available power at AC bus for the subsequent controller design. It is worthy to note that the controller design which simply relies on the system model in the form of matrix is not feasible with a detail model of PV. The frequency control is designed based on artificial intelligence methods such as an artificial neural network (ANN), fuzzy logic (FL), and a combination of both. In this context, ANFIS is designed for LFC problem. Currently, in power systems, PV is the fastest growing and most extensively employed renewable energy technology. The PV power generation has intermittent in nature, having varying and unpredictable characteristics. Thus, power generation through PV is not completely manageable and its availability is also depending on sunny patterns.

The LFC objective with integration of DG is a new thrust of research which is increasingly gaining momentum. A literature review on LFC for Conventional and Distribution Generation (DG) power systems are discussed in detail in [1]. The LFC

S. K. Pandey (✉) · V. P. Singh
Department of Electrical Engineering, REC, Sonbhadra, India

V. Pandey
Department of Electrical Engineering, NITTTR, Chandigarh, India

S. Tiwari
Department of Electrical Engineering, SIET, Allahabad, India

S. R. Mohanty
Department of Electrical Engineering, IIT, BHU, Varanasi, India

V. K. Giri et al. (eds.), *Computing Algorithms with Applications in Engineering*, Algorithms for Intelligent Systems, https://doi.org/10.1007/978-981-15-2369-4_37

of conventional power systems such as thermal and hydro with conventional and intelligent controllers were described in [1–6]. The PV-Diesel hybrid system and the impact of PV generation on load frequency objective is discussed in [7]. In [8], hybrid power system of wind-diesel is considered for LFC incorporation with a traditional PI controller whereas wind-diesel-micro-hydro considered for LFC as hybrid power system in [9, 10]. In [11], wind-diesel unit is considered as hybrid power system for LFC using fuzzy logic controller (FLC) with its parameter optimization by Particle Swarm Optimization (PSO). Further ANFIS based load frequency controllers are described for thermal and hydrothermal systems in [12, 13]. In this paper, ANFIS based load frequency controllers are proposed for thermal-PV power generation unit as hybrid power system. Simulations are executed with different load conditions and solar isolations with the proposed ANFIS based load frequency controllers and results are compared with conventional PI and FLC controllers.

2 System Description

Figure 1a exhibited the block diagram of proposed hybrid power system. The proposed hybrid power system comprises thermal power unit, a PV power unit armed with a maximum power point tracking (MPPT) control, PV inverter, and AC load. Figure 1b is explaining the detail of proposed thermal-PV hybrid power system used for simulation purposes. In which the solar insolation, open-circuit voltage and short-circuit current of the PV array are represented by S_i, V_{OC} and $I_{SC,}$ respectively. Whereas the command power of MPPT, output power of MPPT, PV inverter command power, inverter output power and generated thermal power are represented by $P^*_{\max}$, $P_{\max}$ and P^*_{inv}, P_{pv} and $P_{\text{TH,}}$ respectively. Other parameters expressed as droop characteristic by R, integral control gain by K_{i}, T_{g} is the governor time constant, T_{t} is turbine time constant, K_{P} is system gain and T_{P} is system time constant of the thermal-generator unit, Δf is the deviation in frequency, P_{L} is the AC load, and P_{sys} is the output power of thermal-PV hybrid power system. The values of parameters of proposed hybrid power system are taken from [7].

3 ANFIS Controller Design

3.1 Simple Fuzzy Logic Model

The application of fuzzy logic in various problems in power system has already gained its remarkable momentum. The common rule adopted is "if-then" and fuzzy set is made through membership function. The basic structure comprises of three blocks fuzzification, knowledge base, defuzzification as shown in Fig. 2. Real valued variable is transformed to fuzzy variable through fuzzification. The second block is

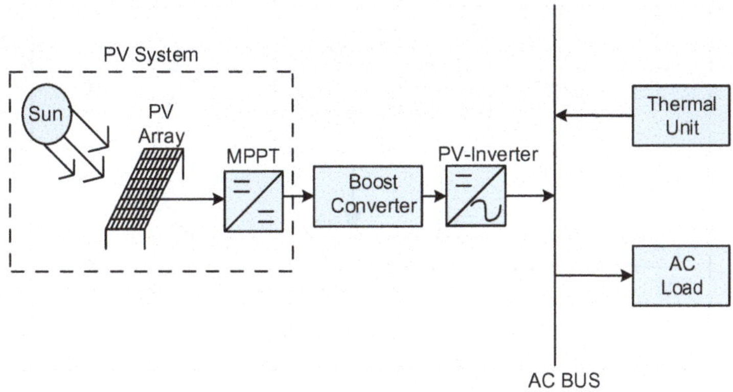

(a) Block model of Thermal-PV.

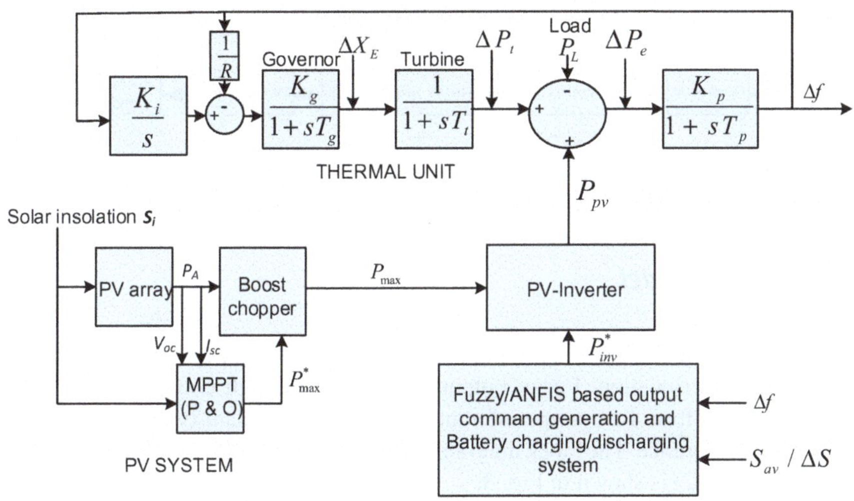

(b) Model of Thermal-PV as hybrid power system.

Fig. 1 Projected hybrid power system

knowledge based which is the main component of fuzzy system. The last block is defuzzification, which converts the output fuzzy variable to a crisp value, which can be used for control purposes. The rules made are used to govern the fuzzy controller action.

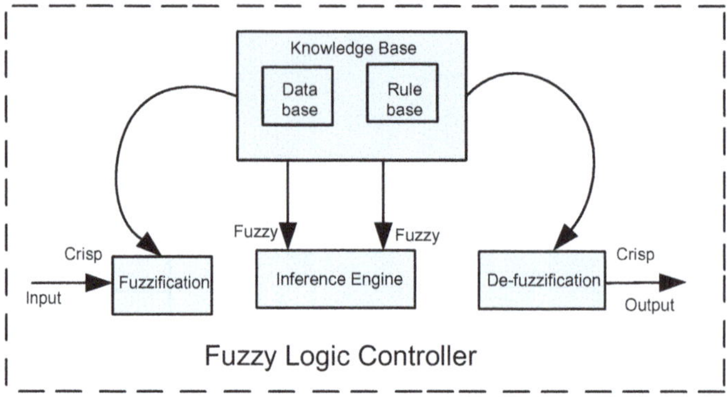

Fig. 2 Simple fuzzy logic controller

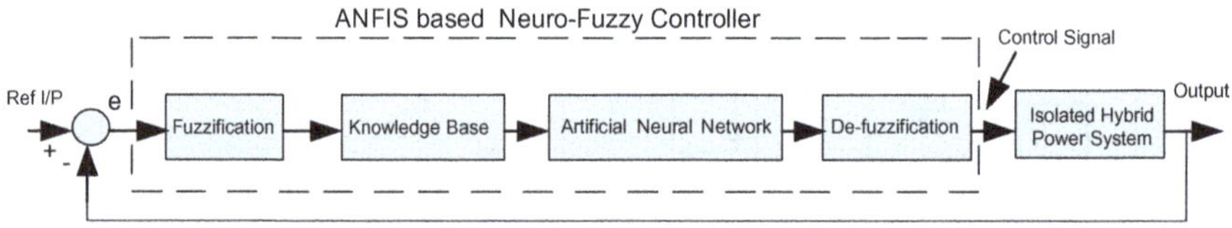

Fig. 3 Block diagram of ANFIS based ANFC

3.2 ANFIS Model

The common process for a fuzzy system modelling is Fuzzification, Fuzzy Inference, and Defuzzification. Fuzzy output is obtained through defuzzification. The utmost common form fuzzy-type inference comprises Mamdani-type inference and Sugeno-type inference. The block diagram of the projected ANFIS based neuro-fuzzy controller (ANFC) is shown in Fig. 3.

4 Controller Design

4.1 Fuzzy Logic Controller (FLC)

Figure 4 signifies the output power command generation system constructed on fuzzy logic in order to govern the PV output power. From this control method, we can realize that two controllers are used. Although fuzzy-I has two inputs as average insolation and frequency deviation though fuzzy-II has also two inputs as change of insolation ΔS and frequency deviation. The output power command P_{inv}^{*} is created as shown in Fig. 4.

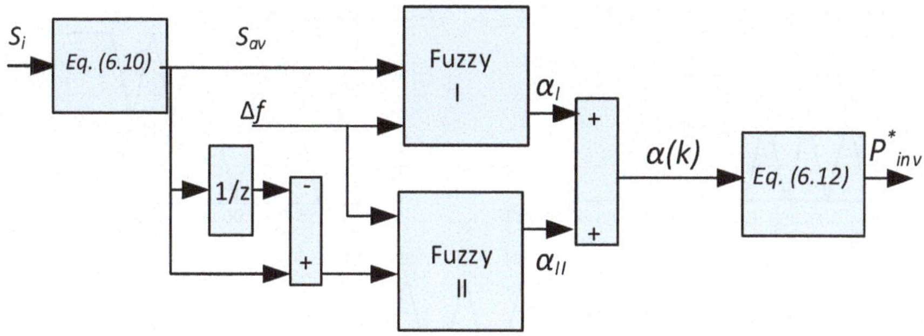

Fig. 4 Fuzzy logic controller

Fuzzy controller-I: In this controller design deviation in frequency Δf, and the average insolation S_{av} are selected as inputs and S_{av} is defined as:

$$S_{\mathrm{av}} = \frac{1}{T}\int_{t-T}^{t} S_i \mathrm{d}t \tag{1}$$

where T represents the integral interval, t is the present time, and S_i is the instantaneous insolation of PV system. Table 1 and Fig. 5 exhibited the fuzzy rules and membership functions of fuzzy controller-I, respectively.

Fuzzy controller-II: In this controller design, deviation in frequency Δf, and the change of insolation ΔS are selected as inputs and ΔS is defined as:

$$\Delta S = S_i(t-1) - S_i(t) \tag{2}$$

Table 2 and Fig. 6 exhibited fuzzy rules and membership functions of fuzzy controller-II, respectively.

Table 1 Fuzzy rules of fuzzy controller-I

		Δf						
		LN	MN	SN	ZE	SP	MP	LP
S_{av}	LN	LN	LN	MP	SP	ZE	MN	SN
	MN	ZE	LN	ZE	MN	ZE	SN	SP
	SN	LP	SP	MN	LN	SP	MP	LP
	ZE	LP	LN	MN	SN	SP	LN	ZE
	SP	MP	ZE	SP	ZE	SN	SP	MN
	MP	SP	MP	MN	MN	MN	SP	SP
	LP	SN	SP	LN	LP	LN	ZE	SP

LN large negative, *MN* medium negative, *SN* small negative, *LP* large positive, *MP* medium positive, *SP* small positive, *ZE* zero

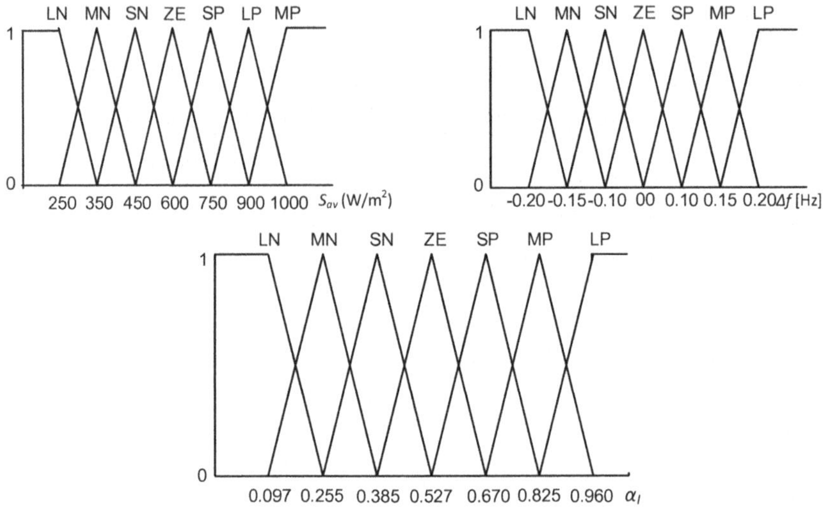

Fig. 5 Membership functions of fuzzy controller-I

Table 2 Fuzzy rules of fuzzy controller-II

		Δf						
		LN	MN	SN	ZE	SP	MP	LP
ΔS	LN	LP	MN	SP	MP	SN	MP	ZE
	MN	ZE	MN	MN	SP	MP	SP	MP
	SN	ZE	ZE	SN	SN	MP	ZE	LN
	ZE	LP	LN	MN	SP	MN	LP	MN
	SP	MN	SP	MP	ZE	ZE	MN	MP
	MP	SP	ZE	MP	MP	LN	MN	LN
	LP	SP	LN	LP	ZE	MN	LN	MN

The output power command P_{inv}^{*} is generated by the sum of outputs of both fuzzy controller and act as central power command which is described as:

$$P_{\text{inv}}^{*} = P_{\text{rated}}\left\{\alpha(k) + \frac{\alpha(k+1) - \alpha(k)}{T_s} f(t)\right\} \tag{3}$$

where, T_s represents the sampling time, $f(t)$ is a periodic function such that $f(t) = t$ $(0 < t < T_s)$ and P_{rated} is the rated power of the PV system.

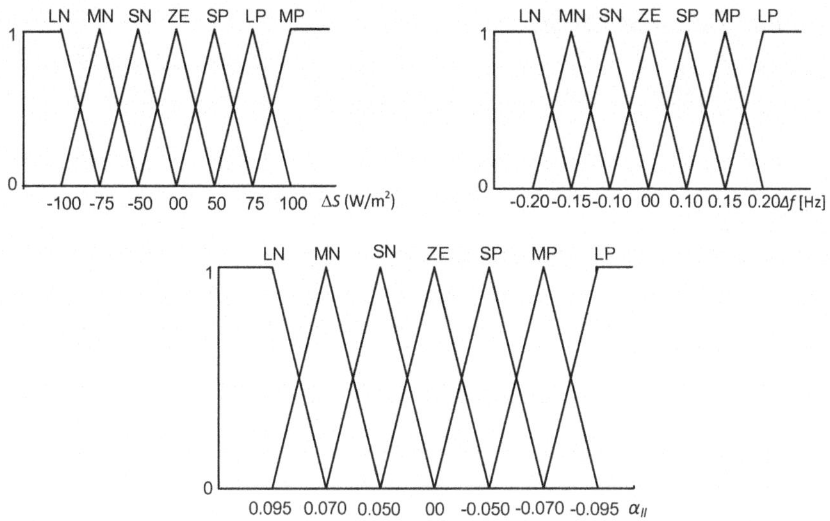

Fig. 6 Membership functions of fuzzy controller-II

4.2 Adaptive Neuro-Fuzzy Controller (ANFC)

ANFIS control structure-based control strategy is development to control the deviation in frequency of the proposed hybrid power system is offered here. There are two ANFC is development to control the deviation in frequency of the proposed system. Here, we considered that deviation in frequency Δf and average insolation S_{av} as inputs to the ANFC-1 and deviation in frequency Δf and the change of insolation ΔS as inputs to the ANFC-2 while the output is the corresponding signal to the inverter. Identical rules have been used as taken in the design of FLC. The proposed ANFC designed by using 7 gbell MFs and 49 rules and 10 epochs. The training of the ANFIS is the first step of ANFC design. Large amount of data is composed of in view of load fluctuations from 0 to 0.01 p.u. for training the ANFIS. The load fluctuations which have been measured for receiving the data are 0.002, 0.004, 0.008 and 0.01. Finally, the training set consists of 5000 elements.

5 Simulation Results and Discussion

Simulations were performed using the conventional PI controller, FLC and proposed ANFC for the proposed thermal-PV hybrid power system. This section deliberates the simulation analysis through different case studies done on above-designated controllers in thermal-PV hybrid power system.

5.1 Load Demand Change

The above-described controller response is verified for step load and random load change.

Step load change: In this case step load disturbance of 0.01 p.u. applied to the thermal-PV hybrid power system. The deviation in frequency profile for above-described controllers is shown in Fig. 7. The performance of hybrid power system with use of ANFC attains relatively better than those attained by PID and FLC.

Random step load change: In this case, step load disturbance varied as shown in Fig. 8a, considered applied to the proposed hybrid power system. The corre-

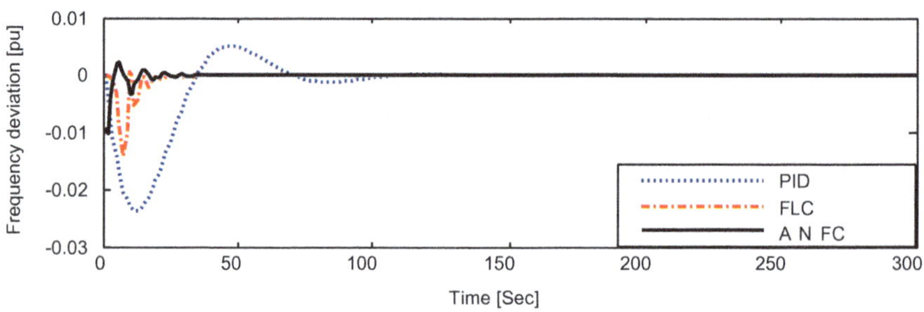

Fig. 7 Frequency deviation with step load change

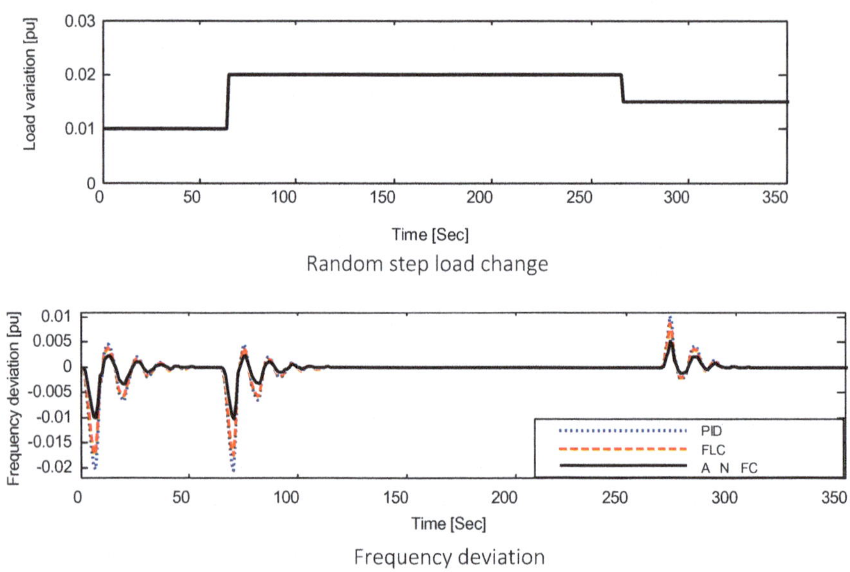

Fig. 8 Control response against arbitrary step load change

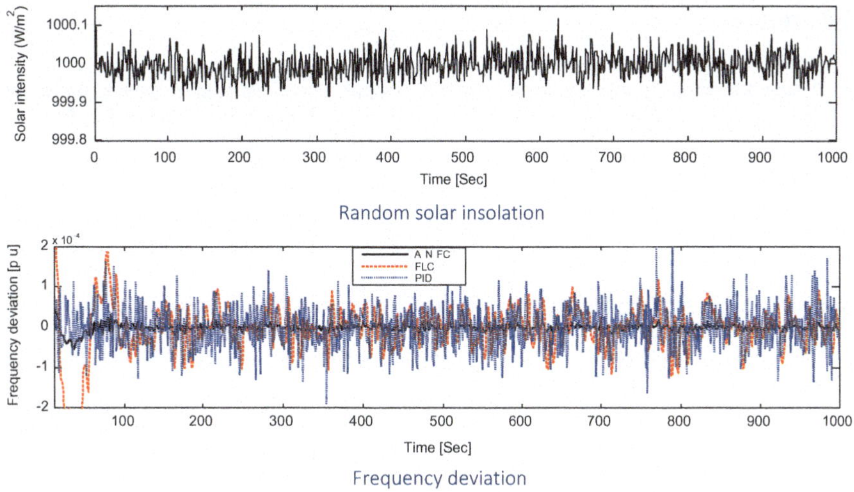

Fig. 9 Control response against random solar insolation

sponding deviation in frequency is shown in Fig. 8b. From the simulation result of Fig. 8b, observed that ANFC attains relatively better response for frequency deviation contour.

5.2 *Random Solar Insolation*

In this study, the random solar insolation as shown in Fig. 9a is applied to the considered hybrid power system. Figure 9b displays the corresponding deviation in frequency of the system. The performance of the system corresponding to ANFC remains extremely robust against arbitrary change in solar insolation as disturbance.

6 Conclusions

ANFIS-based load frequency controllers are projected for LFC of thermal-PV power generation entity as a hybrid power system. Simulations are performed with altered load settings and solar isolations with the projected ANFIS based load frequency controllers and results are compared with conventional PI and FLC controllers. The results accomplished with ANFC are superior to conventional PI controller, and FLC. It typically controls the frequency deviation of projected hybrid power system and thereby advances the dynamic performance. The performance of projected controller is superior to those achieved by conventional PI controller and FLC.

References

1. Pandey SK, Mohanty SR, Kishor N (2013) A literature survey on load frequency control for conventional and distribution generation power systems. Renew Sustain Energy Rev 25:318–334
2. Cam E, Kocaarslan I (2005) Load frequency control in two area power systems using fuzzy logic controller. Energy Convers Manag 46(2):233–243
3. Soundarrajan A et al (2003) Intelligent controllers for automatic generation control. In: Proceedings of the international conference on robotics, vision, information and signal processing, Malaysia, pp 307–311
4. Anand B, Ebenezer Jeyakumar A (2009) Load frequency control with fuzzy logic controller considering nonlinearities and boiler dynamics. J Autom Control Syst Eng 8(3)
5. Cam E (2007) Application of fuzzy logic for load frequency control of hydro electrical power plants. Energy Convers Manag 48:1281–1288
6. Soundarrajan A, Sumathi S (2009) Effect of non-linearities in fuzzy based load frequency control. Int J Electron Eng Res 1(1):37–51
7. Datta M, Senjyu T, Yona A, Funabashi T, Kim CH (2011) A frequency-control approach by photovoltaic generator in a PV-diesel hybrid power system. IEEE Trans Energy Convers 26(2):559–570
8. Bhatti TS, Al-Ademi AAF, Bansal NK (1997) Load frequency control of isolated wind diesel hybrid power system. Energy Convers Manag 38(9):829–837
9. Dhanalakshmi R, Palaniswami S (2011) Load frequency control of wind diesel hybrid power system using conventional PI controller. Eur J Sci Res 60(4):630–641
10. Bhatti TS, Al-Ademi AAF et al (1997) Load frequency control of isolated wind-diesel-micro hydro hybrid power systems. Energy Convers Manag 22(5):461–470
11. Chokpanyasuwan C, Pothiya S, Anantasate S, Pattaraprakorn W, Bhasaputra P (2008) Robust fuzzy logic-PID controller for wind-diesel power system using particle swarm optimization. In: GMSARN international conference on sustainable development: issue and prospects for the GMS, 12–14 Nov
12. Panda G, Panda S, Ardil C (2009) Hybrid neuro fuzzy approach for automatic generation control of two-area interconnected power system. Int J Comput Intell 5(1):80–84
13. Srinivasa Rao C (2010) Adaptive neuro-fuzzy based inference system for load frequency control of hydrothermal system under deregulated environment. Int J Eng Sci Technol 2(12):6954–6962

Chapter 38
Review and Analysis of Access Control Mechanism for Cloud Data Centres

Ajay Kumar Dubey and Vimal Mishra

1 Introduction

The exponential growth of cloud services has provided enormous opportunities but has also created serious security concern. Though companies offer regular security updates as a patchwork. Many security measures have been adopted regularly and new protocols have been developed to improve the security in the cloud environment. Even though to meet the security and performance of demanded cloud service is a tedious job due to the exponential increase in the demand. Many different types of users are accessing these open cloud services. It's very difficult to detect the behaviour of users and restrict the malicious behaviour. The cloud computing system should have provision for encouraging the good behaviour and restrict the malicious behaviour to prevent the security risk in cloud environment [1, 2]. The concept of crowdreviewing is also used in access control mechanism, which is based on the principles of crowdsourcing. "crowdsourcing is an act of outsourcing a task to large number of people in the form of open call" [3]. This crowdreviewing is useful in reputation computation of the users. The reputation and attribute-based access model [4, 5] is now very popular and useful for cloud-based applications.

2 Significance of Access Control Mechanism in Cloud Environment

In cloud computing environment the access control mechanism is required, when any user wants to access the resources from cloud server. Access to a system or a resource

A. K. Dubey (✉)
Dr. APJ Abdul Kalam Technical University, Lucknow 226021, Uttar Pradesh, India

V. Mishra
Institute of Engineering and Rural Technology, Prayagraj 211002, Uttar Pradesh, India

V. K. Giri et al. (eds.), *Computing Algorithms with Applications in Engineering*, Algorithms for Intelligent Systems, https://doi.org/10.1007/978-981-15-2369-4_38

to be controlled by some process and we say this process as access control mechanism. Access control mechanism defines the capability of the user to perform an authorized operations viz. read, write, modify, delete, execute, update, etc. The access control mechanism ensures confidentiality and security of the server resources. The important aspects of security are to protect the resources from unauthorized access, integrity of resources and ensuring availability of the resources for the authorized users. Access control mechanism shows the control flow of information between the different type of users and the resources. The components of access control mechanism communicate and interact with each other and work together to provide a secure environment. There are mainly the following components responsible for secure access control mechanism [6, 7].

1. **Identification**: The effective access control mechanism must have some kind of strong identification system to identify the entities. If the identification system is poor then there is the chance of malicious entity gets identified. This malicious entity got access permissions may cause severe damage to the system.
2. **Authorization**: The process of identification is based on authentication. On successful authentication of entity, the authorization process is performed. In this the access levels granted for that particular entity. Every user has different level of access permission. A database is maintained on cloud server for different access role. Every entity may have its associated access role. Authorization can be performed using any one of the following ways, the authenticated entity gets authorization and access control to perform the read, write, delete, update and execute operations.
3. **Accountability**: Accountability ensures that all actions of access control are attributed to an authenticated entity. Accountability is very important in access control mechanism in cloud environment; this is often achieved with the help of system logs and database maintenance. The system log stores all information like the successful and failed attempts toward the access of the resources. The auditing of database maintenance also ensures accountability; it is simply a tracking system for a particular application or services.

3 Access Control Framework for Cloud Environment

Access control mechanism explains the detail that which user can access what resources. The services or resources available in the cloud environment are elastic and loosely coupled which are highly prone to security attacks. So some defense mechanism is required for secure access to these services by users. The access control is the basic defense mechanism to restrict the access of the services or resources. The prerequisites for the access control framework are given below:

- There should be provision of access control in this mechanism, for sharing the services with a group of users.
- It should provide proper access to the user, i.e., the information is not shared without the permission of the data owner.
- The mechanism must be able to generate notifications about the usage of the services to the data owner.
- The cloud service provider should have proper access control method for both services and relationships between the user and services.
- The mechanism must be able to define access control constraints according to the defined rules.
- The mechanism must be able to define access control based on predefined constant as well as dynamic properties of the services.
- The mechanism provides the options that users can perform any specific operations on services, such as read/write/execute.
- The mechanism process must be transparent to its users.

The access control framework is developed to implement the security policy and to define how the users can access the resources. The access control mechanism is needed for the protection of the services from the access attacks. The access control mechanism decides whether the access permission of the services or resources to be granted or not to the requested user. There exist several models [8] for the development of access control mechanism. The access control framework for cloud environment consists of following components (i) Access Control Policies, (ii) Access Control Models and (iii) Access Control Mechanisms.

1. **Access Control Policies**: The access control policies define a set of protocols which are used while implementing access control mechanism. The request is processed by cloud server considering the services or resources available in the cloud environment. The cloud server determines whether to grant or deny the permission to user. The mechanism is implemented using some regulations established by the policy. Different mechanisms can be applied based on the permissions, by different means of security assurances.
2. **Access Control Models**: The access control model provides a formal presentation of the access control policy and its related functional working. It provides formal proof of security properties being designed for access control system. They define low-level and high-level functions that are defined by the policy.
3. **Access Control Mechanisms**: This defines the basic functions of model which implement the controls that are defined by the policy.

The mentioned concepts correspond to the conceptual separation between the different levels of the design and provide multilevel software development. The separation is between the policies and mechanisms applied to the services. It is possible to discuss the protection requirements based on the policies, compare different access policies and mechanisms and the enforcement of these policies on the entities involved in the communication. It should also provide tamperproof information, confined to a limited part of the system and apply rigorous verification methods. The various access control mechanisms are discussed in the following sections.

4 Types of Access Control Methods

There exist many types of access control method. These access control methods has been discussed and compared with reputation and attribute-based access control method as follows:

- **Identity-Based Access Control (IBAC)**:

In identity-based access control method, the authentication and their access right policies are to be publicized by the certificate authorities [9]. The access rights of a user to access any resources can be stored with the help of access control list (ACL), i.e., the entry in the ACL specifies whether a user will be permitted or denied to access any resources within the cloud network. The agents in the cloud network can store the list of authenticated users and their permissions in the databases. The requests of the entities can be fulfilled by other databases if it comes via an authenticated entity. The identity-based access control method is very straight forward but it may contain Information leakage problems.

- **Authorization-Based Access Control (ABAC)**:

The identity-based access control depends on the identification process of the entity and in this approach, the entities access control permissions are stored with the certificate authorities. While in authorization-based access control, if a service is requested by the entity then it is forwarded to the policy engine for determining the authorizations. The service is allocated to the entity on the basis of its authorization but not on the identity. The identity of the entity is forwarded to the policy engine and it returns the access permissions, which the entity is allowed to do. Entity requests the service by including the rights it wants to access. In this case, authorization gets priority over identity, which improves the security of the system [10]. Each domain or the system has a database of all the services available to its entities along with a set of access permissions and the policies. While IBAC has a list of entities associated with the service, ABAC has a list of access permissions for each entity. A policy language can be used to store which entity gets which rights. This has an advantage over IBAC as each database has the details about its own entities, no problem of distributed identity management. If one entity changes the duties to another, the access permissions are simply changed in that database and can revoke that are not appropriate. No need to inform other databases about the changes, as no other database is involved and no information leakage about the organization as the structure of the database is not revealed to another.

- **Rule-Based Access Control (R-BAC):**

In this system, the access to the services is based on the list of predefined rules for access to be granted. The rules are created by the owners and access controls are granted on the basis of the specified condition. In the ACL, the matrix can specify which entity is allowed to access what, whereas in rule-based system conditions are applied for the access of a particular application. Rule-based access control [11] is similar to discretionary access control. In this control system owner makes the rules according to its organization's need. Rules are enforced using a mediation mechanism which ensures the access by intercepting each and every request comparing with the access permissions of the entity.

- **Policy-Based Access Control (PBAC):**

In this case, all defined policies are stored in policy repository and system verifies whether an entity is authorized to access service according to the repository and agreed by the entity. The activities are controlled by the rules with diverse granularities on cloud server. The security policies are dependent on the role the entity plays within an organization. These policies are implemented by different levels of the owners of that particular department [12]. Since the different policies are described for different users, so it's very difficult to reach on common consensus for global view of the policies. Sometimes few policies may conflict and it is almost impossible to reach on common consensus. The solution of this problem is to enforce the predefined rules in a consistent way and keep track of the access rules, always keep them up to date. It should allow a wide variety of mechanisms to implement; the access status should be made clear and no direct access to the services. All these rules are to be considered and some meta-information to combine them for taking a final decision. A policy will specify how these rules are to be combined, composite aggregate rules using a composite pattern are considered. All the applicable rules are to be stored in the repository of the cloud server [13].

- **Discretionary Access Control (DAC):**

Discretionary access control is a very early form of authorization which are widely deployed in many architectures like UNIX, Linux etc., as it is flexible in nature [14]. These are implemented on the option of the user. Now-a-day's these are used to manage the owners' data in their own systems and the security of that files, OS as well as it is supported by Linux. The owner determines the privileges they can allocate for others. It is user-based and the owner has the right to allocate the access and to whom based on the requirements. The owner can allow, in general, unrestricted access or they can allow specific entity or set of entities to access the service. The owner of the service has full control regarding the access of the owned service by other entities and such controls are called discretionary. The matrix gets updated when the service and the entities are added or deleted. These are inadequate for enforcing the information policies. The policies control the access of the entities to the services on the basis of the entity's identity and specific access permissions, for individual entity or a class of entities which are allowed. If the access permission is related to the service, the access is allowed or denied. The problem with this approach is that there is no constraint on copying the information from owner service to other [15]. As well there is no actual assurance of information flow and does not impose any restriction once the entity receives the service. In mandatory systems, the diffusion of the service is controlled by restricting the information flow from high level to low-level entities.

- **Mandatory Access Control (MAC)**:

Mandatory access controls use data classification schemes, this mechanism gives limited access to the entities and the data owners. With DACs, the controls applied by the entity are free to their discretion but not based on the classification. Each set of service is given rates to specify the level of application that the entity can access. The ratings are collected and often referred to the sensitivity levels which indicate the confidentiality level of the service or the application. Each entity and linked service is assigned a security level. The security level linked with a service reflects the seriousness of the information contained in the service. MAC itself requires a system to manage the access control with the security policies. MAC is used for the systems and the services which are highly sensitive, where the owner does not want others to bypass the mandated access control. MAC provides a multi-layered service processing [16]. Based on the information provided by the owner like the level of control and the system makes the decision control, allows the access for a service. For this, the service needs to be labelled according to the classification, depending on the permissions, the entity or the group of entities are granted. The security level associated with an entity, also called clearance, reflects the entities trustworthiness not to disclose sensitive information to other entities that are not authorized.

- **Role-Based Access Control (RBAC)**:

The Role-based access control mechanism is most popular and widely used access control technique due to its low cost and optimum security. It was pioneered in the 1970s for the online systems which have begun with multi-user environment. The advantage of this technique is that it is not directly related to the policy but has been well known as a safety model and enforces the access control mechanism in an organizational way [17]. This has greatly simplified the management of permissions. The access control is allocated based on the roles the entities are assigned by the organization of a network. The owner governs what the roles can access and how they can access the services as in DAC or Policy-based or as with MACs. It regulates entity access on the basis of the entity activity execution in the system and its own access capabilities. Role-based policies depend on the identification of the roles in the system. A role is a set of activity, can be performed by the entity. Services are linked with a role that contains the privileges assigned to that role. Entities can be reassigned from one role to another. Entity supports the following three security concepts:

i. *Least privilege* ensures that only those permissions are assigned to the users, which are necessary for the completion of the tasks.
ii. *Separation of duties* is ensured by invoking the mutually exclusive roles to complete a particular service.

Table 1 Role access matrix

S. No.	Role of the user	Permission granted
1	Technical user	Read, write
2	Normal user	Read-only
3	Data owner	Read, write, append
4	Service provider	Read, write, execute

iii. *Data abstraction* issues abstracted permission to the entities rather than the read, write, execute which are allocated by the operating system.

The adoption of this model has several advantages like authorization management, roles hierarchy and separation of duties [18]. This allows better management by separating the entity assignment to the roles and the access control to the roles. It allows better static/dynamic constraint enforcement that restricts the number of roles allowed for a given privilege. This is true even for a group of entities and the roles the entities play in various groups [19]. When we look at the meaning of a group, it is a list of entities which have the same access permissions for a period of time using some defined procedures. It includes the capability to establish the relationships between the roles and the permissions as well as entities and roles. Roles can take the inheritance relations like the one role inherits the permissions assigned to a different role (Table 1).

5 Literature Survey

Mulimani and Rachh [20] reviewed existing access control mechanisms and analyses these schemes. According to them access control plays a major role in providing data security and privacy. It is the simplest way to protect cloud storage from unauthorized and malicious users. One of the major areas where access control is used is in medical health care, wherein sensitive information can be accessed only by authorized users such as doctors, researchers, and medical staff. It is concluded that Attribute-based Encryption (ABE) is the most suitable access control mechanism for the cloud computing environment. Attribute-based access control provides best access control by granting different access rights to a set of users and allows flexibility in specifying the access rights of individual users.

Aluvalu and Muddana [21] surveyed many access control models related to cloud computing. In this literature, they discussed merit and demerit of these models and possible solutions to overcome their limitations. In this paper they mainly analyzed the effective access model required for cloud system. Since cloud computing is a distributed and dynamic model, static policies are unsuitable for access control in cloud system. Hence, access control models that support dynamic policies and attribute-based access control models using encryption techniques are discussed and analyzed. In this paper access models using Extensible Access Control Markup

Language (XACML) are implemented for a comparative study. XACML is mainly an attribute-based access control system (ABAC). Their future work includes the implementation of a risk-aware role-based access control model integrated with a hierarchical attribute-based set.

Demchenko et al. [22] defined the basic model and architectural pattern for federated access control in heterogeneous multi-cloud and inter-cloud federation. There exist mainly two types of federations, i.e., client side federation and service provider side federation. The client side federation includes cloud-based services and client architecture while the service provider side federation includes the set of service providers and client. The main task of service provider side federation is to outsource their resources to the customer. The proposed access control method, in this paper uses the federated identity management (FIDM) model which rely on the trusted third party such as cloud service broker (CSB). This paper defines the Inter-Cloud Architecture Framework (ICAF) which resolves the problem of multi-domain heterogeneous cloud-based application integration and inter-platform interoperability. This research analyses different federated identity management scenarios and proposes solution for number of practical problems in multi-provider service integration and enterprise users.

Zhou et al. [23] proposed a trust model with cryptographic role-based access control (RBAC) for secure cloud data storage. The proposed system considers role inheritance and hierarchy to evaluate the trustworthiness of users and roles. The proposed system uses role-based encryption (RBE) to encrypt and decrypt user requests. In this paper, the concept of account role inheritance has been used to propose a trust model for role-based access control. This paper shows how the trust model can be integrated into a system that uses cryptographic RBAC scheme. The trust model is analyzed and evaluated with a real-time application. The results show that trust models can be used to reduce risks and enhance the decision-making skills of data owners and role managers in cloud storage. However, the performance of this system is poor due to the large size of the encryption and decryption key.

Zhou et al. [24] proposed a novel cryptographic administrator model for managing and evaluating a cryptographic role-based access control model (AdC-RBAC). Cryptographic techniques are used to verify that administrative tasks are performed only by authorized administrators. The proposed RBE implements access policies on encrypted data for improved data privacy in cloud storage. In the proposed technique the data owner can encrypt the data into specific role. So only authorize users of this role can decrypt it. The advantage of this technique is that if administrator wants to add or delete any existing user from this role, it only need to update their role parameter without affecting any user or other roles. The AdC-RBAC model can be used in an untrusted environment since its security is guaranteed by using cryptographic RBAC technique.

6 Reputation and Attribute-Based Access Control (RAAC)

The reputation and attribute-based access control is used to secure the resources in cloud environment. The ultimate aim of this technique is to encourage reputed users and discourage malicious users. In this technique, the unintended behaviours of malicious users are detected and prevented their further action for privacy protection [25] in cloud environment. There exist various credit and reputation model [26, 27]. In recent years many researches are going on in this direction to develop an online application system based on credit and reputation model. Nowadays credit and reputation based applications are very popular in cloud environment. This application is also helpful in online financial transaction and online e-commerce marketing. Here, we discuss the conventional credit and reputation model and suggest a few modifications based on crowd reviewing.

6.1 Basic Design of Reputation and Attribute-Based Access Control Model

Here, we explain the basic principles of reputation and attribute-based access control (RAAC) model [28].

Encourages Reputed Users. The reputation and attribute-based access control model is designed to encourage the good users. In this model, a credit value is assigned to user on successful completion of every task. The users, who properly utilize the resources of cloud computing environment and does not perform any malicious activity gets the higher credit value. If the user continuously performs as good user then value of credit score will increase.

Discourages Malicious Users. The reputation and attribute-based access control model is designed to restrict the malicious behaviours of wicked users. This model safeguards the honest users and cloud resources. To discourage malicious user this model immediately reduces the credit value, if any user performs the unintended activity and if the user's credit value is below the threshold value then user won't be able to access any resources from cloud environment.

Credit Computation. The credit computation is the most important feature of reputation and attribute-based access control model to ensure the security in cloud environment. This system assigns a positive or negative credit value to users on each and every activity of the users. When user performs any upright activity then system appends positive credit value to its initial credit value and on every malicious action it appends negative credit value. The reputation and attribute-based access control model utilize this credit value to restrict the malicious user from accessing the cloud resources.

6.2 Modified Reputation and Attribute-Based Access Control System (M-RAACS)

This modified reputation and attribute-based access control system (M-RAACS) is based on crowd feedback mechanism of credit value for securing the cloud resources by authenticating users. In this technique after any transaction, every user gives the feedback to reputation centre via cloud service provider, in term of credit value. As shown in Fig. 1 the modified reputation and attribute-based access control system contains mainly three subsystems: Attribute Management Centre (AMC), Resource Management Centre (RMC) and Reputation Centre (RC).

Attribute Management Centre (AMC). The attribute management centre is very important subsystem of the modified reputation and attribute-based access control system. This is responsible for maintaining all attributes of users. When any user wants to access any resources in the cloud environment, the attribute management centre grants the permission on the basis of its attributes. So any users have to must register with the attribute management centre as a legitimate user. The attribute management centre works on the basis of user's attribute. It provides the identification and authentication services to users. The attribute management centre translates the

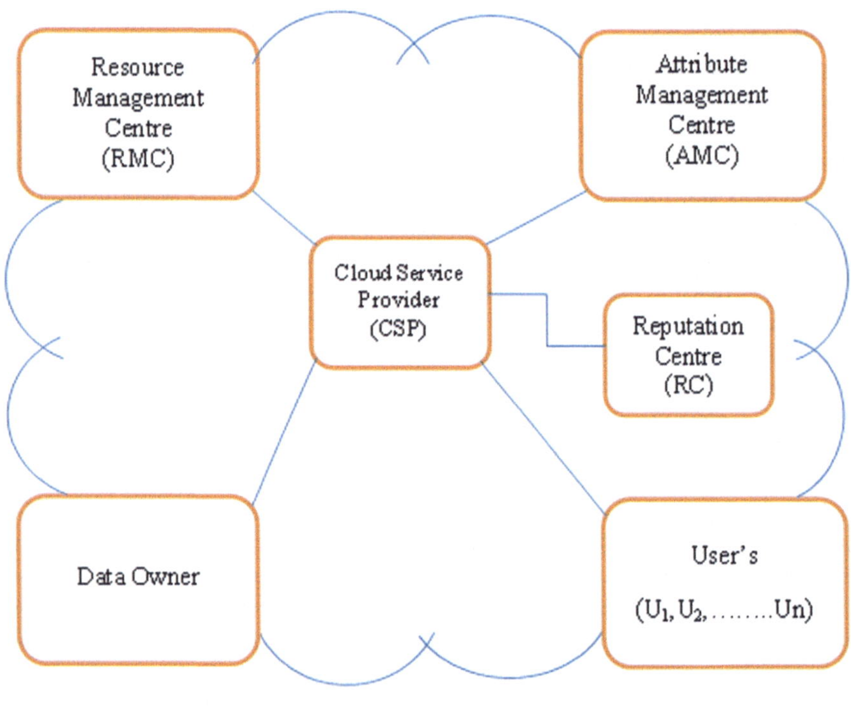

Fig. 1 Reputation and attribute-based access control system

real identity of a user to the confidential identity and links them by the certificate management.

Resource Management Centre (RMC). The first step is every user has to register with the attribute management centre. After successful registration, the reputation centre will assign the initial credit value to the user. Now user can login to the system to access the resources with the resource management centre. When the user login to the system, it is the responsibility of the resource management centre to check the user's credit value and verified it from reputation centre, if the credit value is lower than the threshold value then user will not get the access permission. If the user's credit value is above the threshold value and successfully authenticated then user can now access the resources. The resource management centre continuously checks the user's credit value and keeps control of the user's rights for accessing the resource in the system.

Reputation Centre (RC). The reputation centre is the most important part of the modified reputation and attribute-based access control system. The reputation centre is responsible for monitoring all the resources access by the user and calculation of credit value of the user, currently accessing the resources. The reputation centre receives the feedback credit value from all connected users and calculates the latest credit value for that user. The reputation centre also updates the credit value to the attribute management centre for further usage. If the user's current credit value is lower than the threshold value then reputation centre will restrict the user's resource access privilege (Fig. 2).

7 Conclusion

The security and integrity of the cloud data is the major concerns among cloud users. To ensure the security of the data, only authorized users are allowed to access the data from cloud server. The access control mechanism is one of the most popular ways to protect the illegal access of cloud server from malicious users. In this paper, a comprehensive review of the various access control methods has been presented, from which, it can be stated that reputation and attribute-based access control method is most suitable in a cloud environment. The reputation and attribute-based access control system provides the security to consistent users in cloud environment. This system encourages the consistent or honest user by increasing their credit value and discourages the inconsistent or dishonest user by decreasing their credit value. In future, the credit value computation mechanism can be improved by using more attributes and the crowdsourcing concepts.

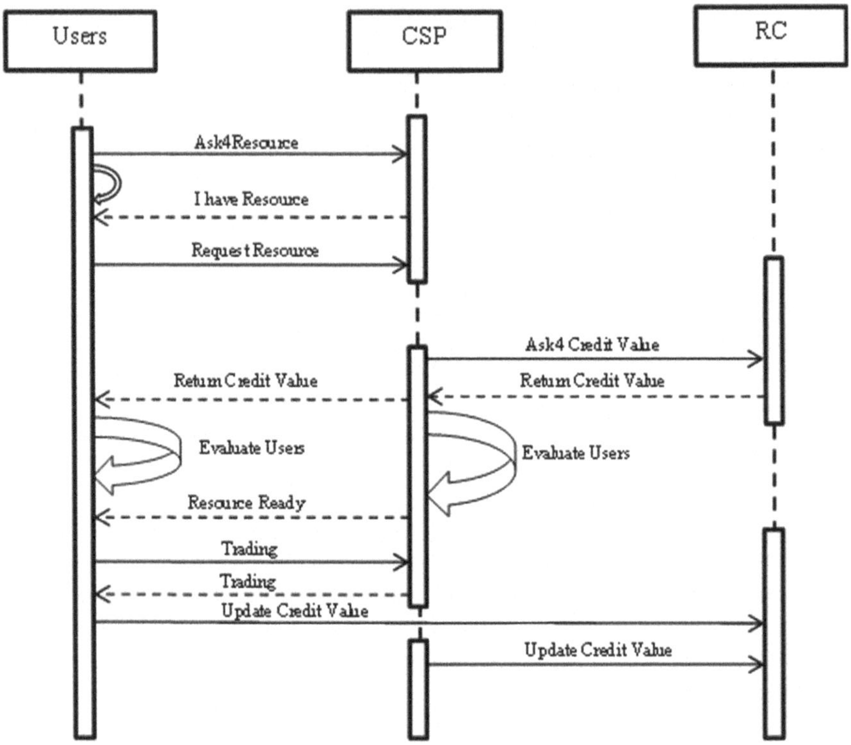

Fig. 2 Sequence diagram for transaction between users and CSP

References

1. Song R, Korba L, Yee G (2006) Pseudonym technology for e-services. In: Privacy protection for E-services. IGI Global, pp 141–171
2. Miconi A, Rainie L, Wellman B (2013) Networked: the new social operating system. Int J Commun 7:6
3. Abhishek T, Tahmasbi N, Khazanchi D, Najjar L (2014) Crowdsourcing typology: a review of is research and organizations. In: Proceedings of the Midwest Association for Information Systems (MWAIS)
4. Wang B, Li B, Li H (2014) Oruta: privacy-preserving public auditing for shared data in the cloud. IEEE Trans Cloud Comput 2(1):43–56
5. Ruohomaa S, Kutvonen L, Koutrouli E (2007) Reputation management survey. In: The second international conference on availability, reliability and security (ARES'07). IEEE, pp 103–111
6. Samarati P, de Vimercati SC (2000) Access control: policies, models, and mechanisms. In: International school on foundations of security analysis and design. Springer, Heidelberg, pp 137–196
7. Sandhu R, Samarati P (1996) Authentication, access control, and audit. ACM Comput Surv (CSUR) 28(1):241–243
8. Leune CJ et al (2007) Access control and service-oriented architectures. Technical report, Tilburg University, School of Economics and Management

9. Chen X, Berry D, Grimson W (2009) Identity management to support access control in e-health systems. In: 4th European conference of the international federation for medical and biological engineering. Springer, Heidelberg, pp 880–886
10. Karp AH (2006) Authorization-based access control for the services oriented architecture. In: Fourth international conference on creating, connecting and collaborating through computing (C5'06). IEEE, pp 160–167
11. Carminati B, Ferrari E, Perego A (2006) Rule-based access control for social networks. In: OTM confederated international conferences on the move to meaningful internet systems. Springer, Heidelberg, pp 1734–1744
12. Demchenko Y, Gommans L, Tokmakoff A, van Buuren R (2006) Policy based access control in dynamic grid-based collaborative environment. In: International symposium on collaborative technologies and systems (CTS'06). IEEE, pp 64–73
13. Reddivari P, Finin T, Joshi A et al Policy-based access control for an RDF store. In: Proceedings of the IJCAI-07 workshop on semantic web for collaborative knowledge acquisition, 2007
14. Sandhu R, Munawer Q (1998) How to do discretionary access control using roles. In: Proceedings of the third ACM workshop on role-based access control. ACM, New York, pp 47–54
15. Downs DD, Rub JR, Kung KC, Jordan CS (1985) Issues in discretionary access control. In: 1985 IEEE symposium on security and privacy. IEEE, pp 208–208
16. McCune JM, Jaeger T, Berger S, Caceres R, Sailer R (2006) Shamon: a system for distributed mandatory access control. In: 2006 22nd annual computer security applications conference (ACSAC'06). IEEE, pp 23–32
17. Park JS, Sandhu R, Ahn G-J (2001) Role-based access control on the web. ACM Trans Inf Syst Secur (TISSEC) 4(1):37–71
18. Mitseva A, Imine M, Prasad NR (2006) Context-aware privacy protection with profile management. In: Proceedings of the 4th international workshop on wireless mobile applications and services on WLAN hotspots. ACM, New York, pp 53–62
19. Boutaba R, Aib I (2007) Policy-based management: a historical perspective. J Netw Syst Manage 15(4):447–480
20. Mulimani M, Rachh R (2016) Analysis of access control methods in cloud computing
21. Aluvalu RK, Muddana L (2015) A survey on access control models in cloud computing. In: Emerging ICT for bridging the future-proceedings of the 49th annual convention of the computer society of India (CSI), vol 1. Springer, Heidelberg, pp 653–664
22. Demchenko Y, Ngo C, de Laat C, Lee C (2014) Federated access control in heterogeneous intercloud environment: basic models and architecture patterns. In: 2014 IEEE international conference on cloud engineering. IEEE, pp 439–445
23. Zhou L, Varadharajan V, Hitchens M (2015) Trust enhanced cryptographic role-based access control for secure cloud data storage. IEEE Trans Inf Forensics Secur 10(11):2381–2395
24. Zhou L, Varadharajan V, Hitchens M (2014) Secure administration of cryptographic role-based access control for large-scale cloud storage systems. J Comput Syst Sci 80(8):1518–1533
25. Sutcu Y, Li Q, Memon N (2007) Protecting biometric templates with sketch: theory and practice. IEEE Trans Inf Forensics Secur 2(3):503–512
26. Papaioannou TG, Stamoulis GD (2004) Effective use of reputation in peer to-peer environments. In: IEEE international symposium on cluster computing and the grid, 2004, CCGrid 2004. IEEE, pp 259–268
27. Resnick P, Zeckhauser R, Swanson J, Lockwood K (2006) The value of reputation on ebay: a controlled experiment. Exp Econ 9(2):79–101
28. Donghong S, Wu L, Ping R, Ke L (2016) Reputation and attribute based dynamic access control framework in cloud computing environment for privacy protection. In: 2016 12th international conference on natural computation, fuzzy systems and knowledge discovery (ICNC-FSKD). IEEE, pp 1239–1245

Chapter 39
Design and Analysis of Low-Noise Amplifier for Ku-Band Applications

Gaurav Maithani, Gaurav Upadhyay and Arvind Kumar

1 Introduction

Wireless communications applications have increased in popularity over the last decade. Nowadays, satellites provide worldwide TV channels, global messaging services, positioning information, communications, etc. Allocated satellite communication frequency bands include spectrum from as low as 2.5 GHz to almost 50 GHz. Ku-band is primarily used for satellite communications. LNA with a NF of 2.1 dB and forward gain of 12.5 dB at 15 GHz was reported by Aspemyr et al. [1] where a 90 nm radio-frequency complementary metal-oxide semiconductor technology and Cascode topology is implemented in its design. LNA at 17.5 GHz has demonstrated a NF of 2.12 dB and forward gain of 17.62 dB by Ehrampoosh and Hakimi [2], LNA with a NF of 2.29 dB and forward gain 9.11 dB at 10 GHz with Gallium Arsenide (GaAs) high-frequency transistor and common source topology has been implemented for designing circuit by Challal et al. [3]. Distributed and radial stub elements are used for designing input and output matching as well as biasing circuits. LNA with a NF of 2.72 dB and forward gain of 11.040 dB at 11 GHz was designed by Afshar and Niknejad [4] where a 180 nm RF-CMOS technology and two-stage Cascode topology is used for designing the circuit. The Cascode LNA is widely used architecture among the LNA designs. This Cascode architecture satisfies all the requirements for designing the LNA but Cascode LNA still has a problem like trade-off between input matching and NF, gain NF, load tuning, and output matching. The parasitic capacitances around the common source-common gate (CS-CG) stages degrade the noise performance. A narrowband 3–5 GHz LNA design was made by Kung et al. [5] but this paper does not contain proper input and output matching techniques. A very simple approach to achieved proper load matching and a low NF to overcome these problems can be designed by a series resonate inductor is explained in [6]. In this technique, the two stages CS-CG to suppress the parasitic capacitance

G. Maithani · G. Upadhyay (✉) · A. Kumar
Institute of Technology Gopeshwar, Gopeshwar, Uttarakhand 246424, India

V. K. Giri et al. (eds.), *Computing Algorithms with Applications in Engineering*, Algorithms for Intelligent Systems, https://doi.org/10.1007/978-981-15-2369-4_39

effects and a buffer stage is placed after output stage to get the trade-off between tuning the load and matching the output. This is a simple step which gives us proper output matching technique, better gain and NF.

2 LNA Parameters

2.1 Noise Figure

NF is the characteristic of the system. It is measured by degradation of signal to noise ratio (SNR) between input and output of a system. In noisy condition of network, the output SNR is decreased. Once the desired signal and noise are applied to the input of any noiseless network, both the signal and noise can be amplified or attenuated by the same amount. Therefore, SNR will not be changed. The following mathematical equation is used to calculate NF:

$$\mathrm{NF} = \frac{\frac{\mathrm{S_i}}{\mathrm{N_i}}}{\frac{\mathrm{S_O}}{\mathrm{N_O}}} = \frac{\mathrm{SNR_i}}{\mathrm{SNR_o}} \geq 1 \tag{1}$$

$$\mathrm{NF(dB)} = 10\log_{10}\mathrm{NF} \tag{2}$$

where $\mathrm{S_i}$, $\mathrm{N_i}$ and $\mathrm{S_O}$, $\mathrm{N_O}$ are input signal and noise power and output signal and noise power, respectively.

Using Friis' equation, NF of LNA of a radio receiver can be obtained by:

$$\begin{aligned}\mathrm{NF_{Total}} &= 1 + (\mathrm{NF_1} - 1) + \frac{(\mathrm{NF_2} - 1)}{G_1} + \frac{(\mathrm{NF_3} - 1)}{G_1 G_2} \\ &+ \cdots + \frac{(\mathrm{NF}_K - 1)}{G_1 G_2 \cdots G_{K-1}}\end{aligned} \tag{3}$$

where, G_K is gain of each stage and NF_K is NF of each stage. From Friis' equation, it can be understand that the total NF ($\mathrm{NF_{Total}}$) is totally dominated by the NF of first stage ($\mathrm{NF_1}$) which is NF of the LNA. In the same way, the gain of the first stage G_1 reduces the noise in the following stages [7].

2.2 Harmonic Distortion and Intermodulation

The term linearity defines that the highest acceptable signal level at the input of any system [8]. Nowadays, designs of LNA generally produce some amount of non-linearity. The distortion in the signal is a result of the non-linearity. The methods used to find non-linearity of any system are third-order intercept point and 1-dB compression point [8].

2.3 RF Transistor

[(*Field Effect Transistor (FET)* i.e. *Pseudomorphic High Electron Mobility Transistor (PHEMT)*)]

FETs are monopolar devices and bipolar junction transistors (BJTs) are bipolar devices, it means that FETs are single carrier devices that carry the current through the channel. In this design, Hetro-FET called PHEMT is used. The use of hetero-FET, to abrupt transitions between different layers of semiconductor materials, i.e., gallium aluminum arsenide (GaAlAs) to gallium arsenide (GaAs) or gallium indium arsenide (GalnAs) to gallium aluminum arsenide (GaAlAs) interfaces. High electron mobility transistor (HEMT) uses the difference in the bandgap energy between the two dissimilar semiconductor materials such as GalnAs and GaAlAs in order to exceed the upper-frequency limit of the metal-semiconductor field-effect transistor (MESFET) and at the same time keeping low-noise performance. The high-frequency behavior of HEMT is because of the separation of the carrier electrons from the donor sites at the boundary of the doped GaAlAs and n-doped GaAs called the quantum wall where they are limited to very thin layer in which motion is possible parallel to the interface.

2.4 Transistor Biasing

The biasing of transistor can be separated into two parts: first is to select the quiescent and second is to design the biasing network. The above two parts of transistor biasing are important for the operation of amplifier. The bias selection for HEMT/FET, i.e., drain to source voltage (V_{DS}), drain to source current (I_{DS}) and gate to source voltage (V_{GS}) depend on the type of application. For high-power applications, higher values of V_{DS} and I_{DS} are needed. The value of maximum V_{DS} is generally fixed for most of the transistors. The selection of V_{GS} is made upon the gain, class of operation and power-aided efficiency (PAE) requirements. In passive bias network, there is resistive network which provides suitable currents and voltages for the RF transistor. These types of networks are thermally unstable. Therefore, active bias networks are used for compensating this weakness.

2.5 Stability Analysis

Stability test is an important task to verify the stability of transistor. The stability of a transistor is verified by Rollet's conditions, i.e., [9]:

$$K = \frac{1 - |S_{11}|^2 - |S_{22}|^2 + |\Delta|^2}{2|S_{21}||S_{12}|} > 1 \quad (4)$$

$$\Delta = |S_{11}||S_{22}| - |S_{12}||S_{21}| \quad (5)$$

where K is Rollet's factor. If $|\Delta| < 1$ and $K > 1$, then the amplifier is said to be stable in the selected frequency range and bias conditions. By connecting a series resistance or a shunt conductance at output port or input port, an amplifier can be stabilized. After stabilization through adding resistors, there is a trade-off in low gain and high NF [9].

3 Design Methodology of LNA

The design methodology of LNA is described step by step in the following steps. The flow chart is shown in Fig. 1. The operating frequency of the design is 15 GHz which is designed and simulated by using ADS.

The flow chart describes in many steps. The selection of transistor is the very important for designing of the LNA. The properties of transistor within a frequency interval determine the overall power gain and NF performance of the LNA. Here

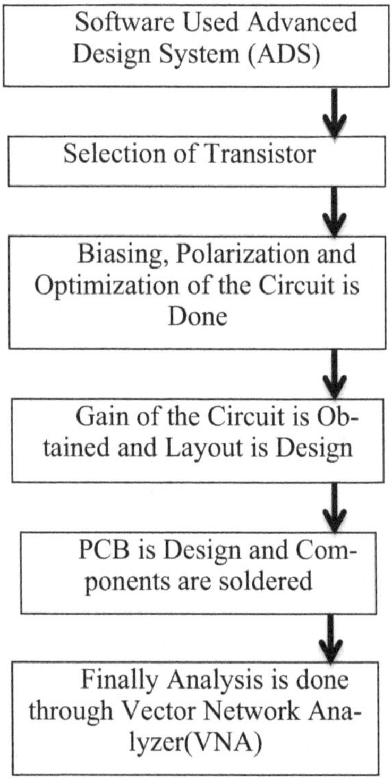

Fig. 1 Flow chart of LNA design

we have studied different transistors and we select ATF36163 transistor from Avago Technologies [10]. The transistor is a PHEMT for LNA application in a surface mount plastic package.

3.1 DC Bias Circuit Design

Using large signal model of the transistor simulate in ADS, transistor's output characteristics can be measured. Then through the data sheet of transistor, the biasing point and corresponding V_{GS} determined. By using these information, the biasing network can be designed to give the drain and gate voltages and the desired drain current in the simulation test bench which is shown in Fig. 2. The transistor's output characteristics is simulated by sweeping the values of V_{GS} and V_{DS}. The two voltage supplies are used as transistor operates in depletion mode, the V_{DS} has positive voltage and V_{GS} has negative voltage.

The following circuit (Fig. 2) is set up for I-V characteristics which is simulated in ADS. Figure 3 shows different I-V curves with different V_{GS} (Table 1; Fig. 4).

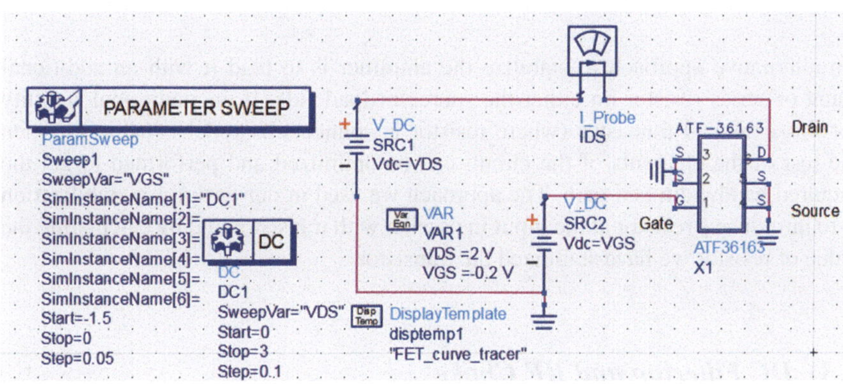

Fig. 2 Transistor simulation for V_{DS}/I_{DS} characteristic using "FET Curve Tracer"

Fig. 3 FET V_{DS}/I_{DS} output characteristic

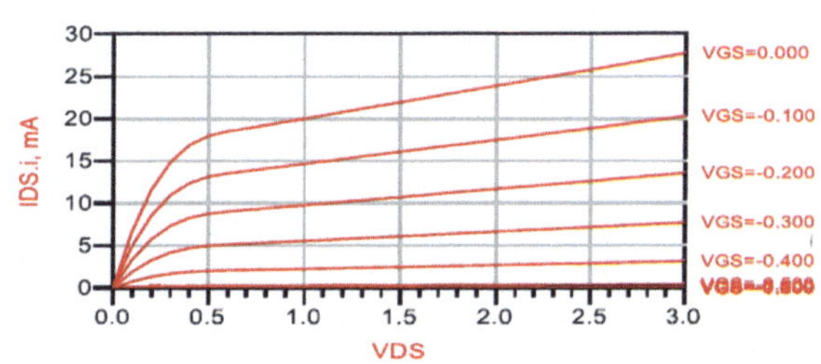

Table 1 Biasing points

Symbols	Parameters	Values
V_{DS}	Drain to source voltage	2.5 V
V_{GS}	Gate to source voltage	−0.1 V
I_{DS}	Drain to source current	19 mA

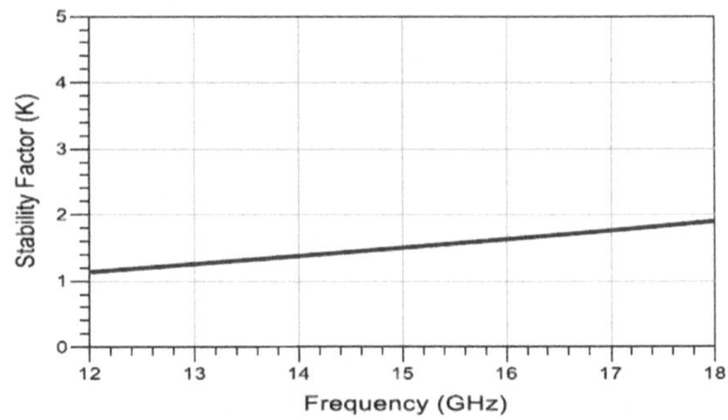

Fig. 4 Transistor stability simulation result after adding a resistor

3.2 *Stabilization Network*

An alternative approach to stabilize the amplifier is to load it with an additional shunt or series resistor on either the source or load side. If unconditional stability for this extended transistor (where resistor is connected) model is achieved, then the rest of the elements of the circuit can be optimized and performed to get the required bandwidth and gain. The approach we used in our design for stabilization is to introduce a resistor at the input in parallel with transistor and by calculating the value of resistor we have stabilized the transistor.

3.3 *DC Filtering and RF Choke*

The inadequate filtering choice of the DC biasing leads into instability. In active biasing network, the instability generally occurs at low frequencies because bias transistor is a low-frequency device and having no gain at RF. The DC and RF signals are co-existed. For designing of amplifier, there is no interference of these two signals. A RF choke is a filter which provides us high impedance but at the same time provides very low reactance. Microstrip band reject filters are used to make the choke because of their ability to maintain a nice RF isolation at high-frequency circuits. The gate and drain biasing network had been implemented by using the concept of stepped-impedance resonator (SIR) [11]. A wideband bias network is shown in Fig. 5 with dimensions of network and characteristic impedances as shown in Table 2 and Fig. 6 shown its s-parameters.

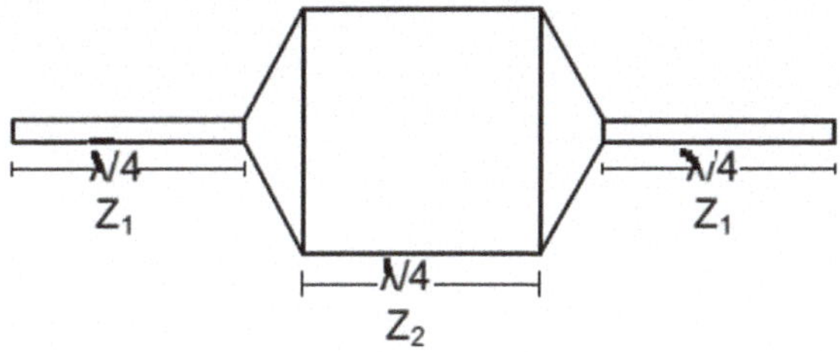

Fig. 5 Wideband bias network

Table 2 Dimensions of biasing network

Impedance	Length (mm)	Width (mm)
$Z_1 = 20\ \Omega$	2.799730	6.72373
$Z_2 = 120\ \Omega$	3.142240	0.310658

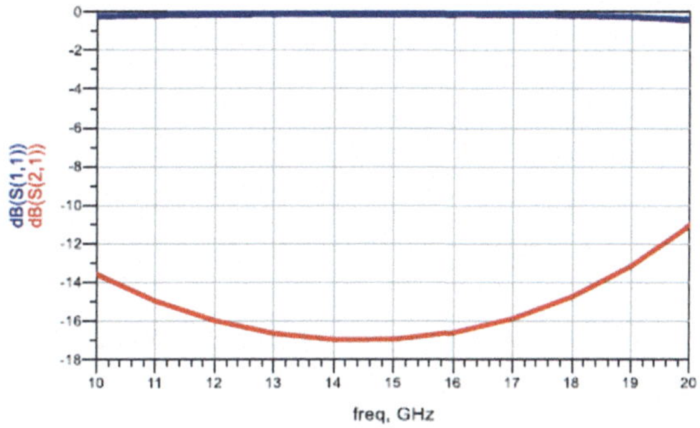

Fig. 6 S-Parameter of wideband bias network

3.4 Constant Gain and NF Circles

In design of LNA, minimize and maximize the noise and gain of the input signal, respectively, are important parameters. For this approach, some optimum values of source reflection coefficient (Γ_S) and load reflection coefficient (Γ_L) are taken to design the matching network for which the NF is up to a tolerable level and the gain is set to the required value. The intersection of gain and NF circle is located as shown in Fig. 7. From this intersection, the value of Γ_S is chosen to design the matching network and after finding value of Γ_S we can determine output reflection coefficient (Γ_{OUT}) using the following equation:

$$\Gamma_L^* = S_{22} + \frac{S_{12}S_{21}\Gamma_S}{1 - S_{11}\Gamma_S} \quad (6)$$

Γ_{OUT} is conjugate matched with the Γ_L value. After finding optimum values of Γ_S and Γ_L, the input and output matching networks have to be designed by using ADS Smith Chart utility (Table 3).

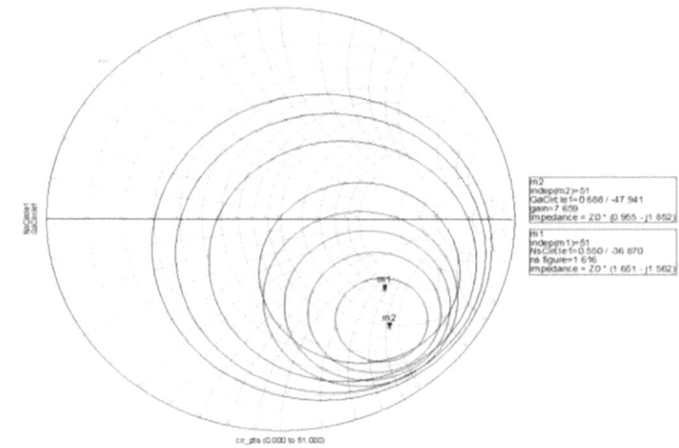

Fig. 7 Constant Gain and NF Circle

Table 3 Optimum reflection coefficient value

Impedance matching	Reflection coefficient value
Source (Γ_S)	0.440 – j0.33
Load (Γ_L)	0.4707 + j0.0132

3.5 *Input and Output Matching Networks*

The design of the matching network for the load and source side has to be done. The function of the matching network is to transfer high power or very low reflection coefficient which is done by converting the impedance looking into source and load side into 50 Ω which were selected after drawing constant NF and constant gain circles at the desired frequency, i.e., 15 GHz. The impedances were chosen in such a way that the NF of the LNA is low and gain is high. In the matching network, the first transmission line (L1) is in series with the load impedance which is used to convert the real part of the complex impedance to 50 Ω and second transmission line (L2) is an open circuit stub which is used to cancel out the imaginary part of the complex impedance which finally leads to match the complex impedance to 50 Ω. After getting the impedances, the dimensions of the matching network had been calculated (Table 4).

Table 4 Dimensions of matching network

Impedance matching	Width (mm)	L_1 (mm)	L_2 (mm)
Source	1.9280	1.8856	3.4667
Load	1.9280	2.1150	1.4977

3.6 Simulation Results

The simulated results include maximum gain, forward voltage gain, NF, input reflection coefficient and output reflection coefficient. The simulated results are shown in figures:

A maximum gain of 17.887 dB is obtained at 15 GHz and also gain is almost flat for the bandwidth of 1.5 GHz around the center frequency of 15 GHz. In Figs. 8 and 9 maximum available gain and forward voltage gain are plotted respectively. The forward voltage gain is 7.3649 dB at the center frequency 15 GHz but if the input matching network and output matching network have changed to increase forward voltage gain, then NF also increased. In designing of LNA, our main focus is to get the acceptable value of NF and forward voltage gain.

The noise figure of 3.869 dB is obtained at 15 GHz as shown in Fig. 10.

The input reflection coefficient and output reflection coefficient are found −27.010 dB and −15.176 dB at center frequency 15 GHz respectively as shown in Figs. 11 and 12 (Figs. 13 and 14).

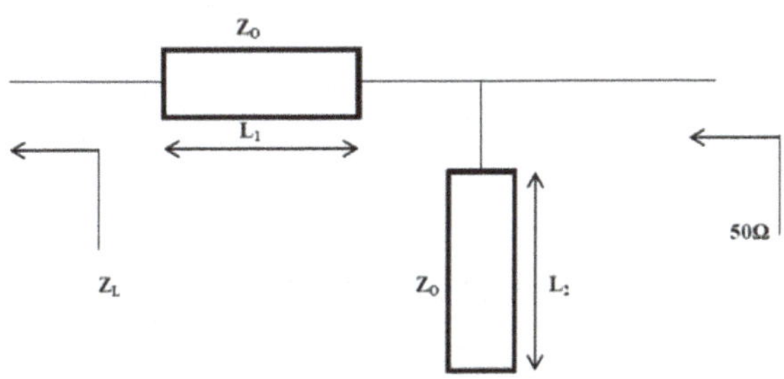

Fig. 8 Matching network

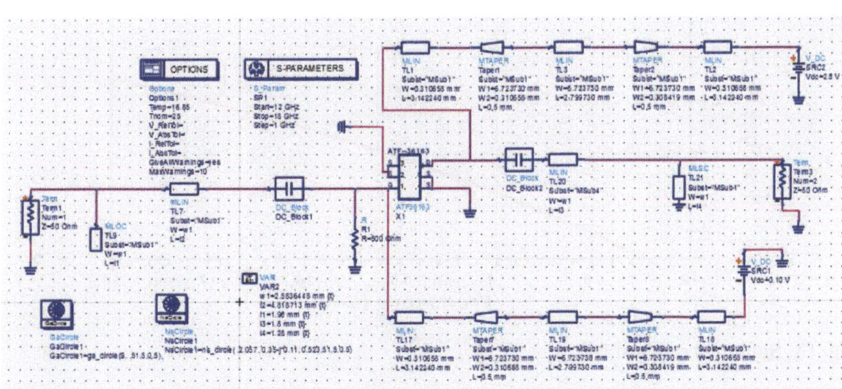

Fig. 9 Schematic (Agilent's-ADS) view of Ku-band LNA design

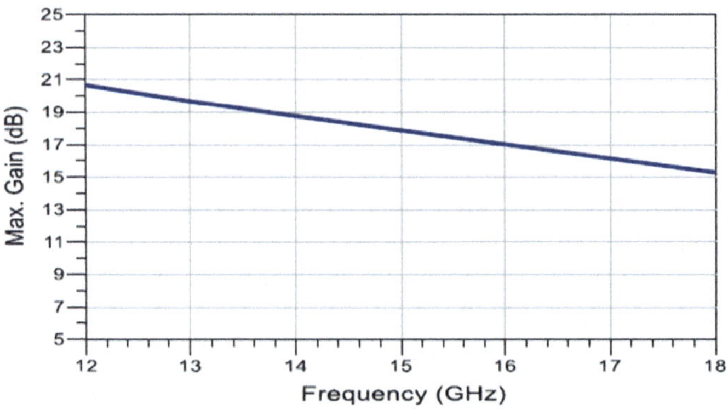

Fig. 10 Simulated result of maximum available gain

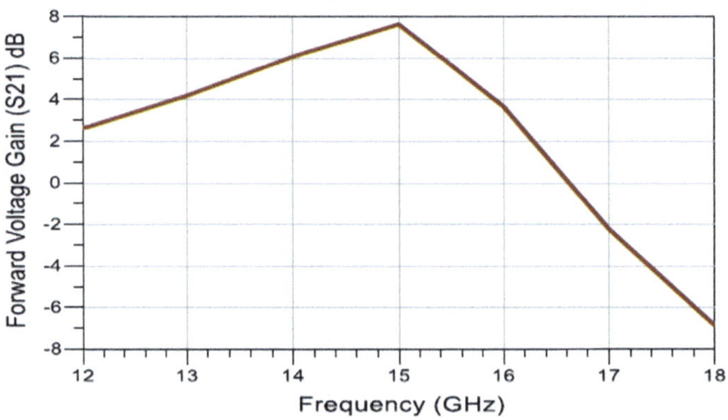

Fig. 11 Simulated result of forward voltage gain

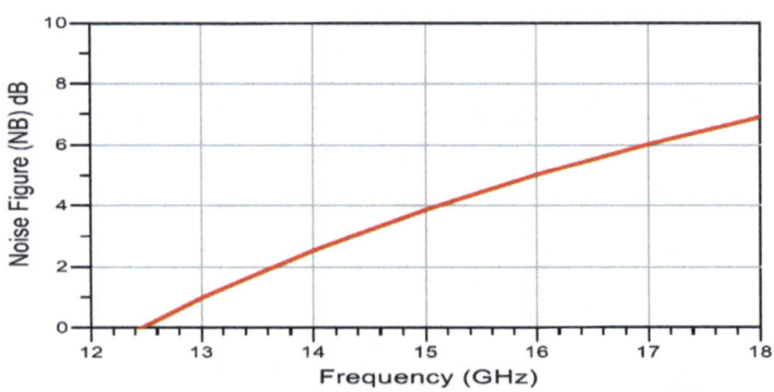

Fig. 12 Simulated result of NF

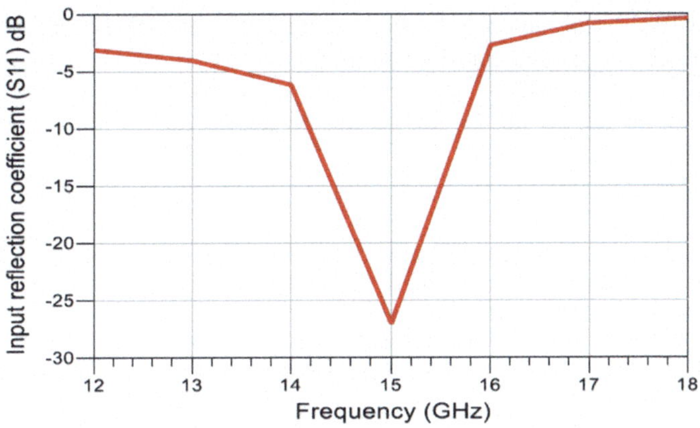

Fig. 13 Simulated result of input refection coefficient

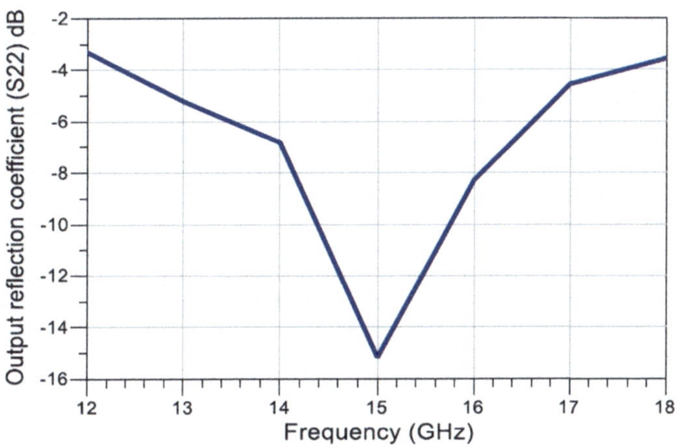

Fig. 14 Simulated result of output reflection coefficient

4 Conclusion

In this paper, a LNA is designed for Ku-band application by using ADS. The parameters related to microwave amplifiers explained including gain and NF circles. Selection of transistor was done after of comparisons between different transistors and the decision was based on the NF and gain parameters of the ATF 36163 device. The investigation on stability of the transistor is made which led us to a better understanding of the transistor characteristics and its performance. The most important part in microwave circuits is matching network design and matching techniques which are mentioned along with their design procedures. The simulated max gain value is of 17.887 dB, forward voltage gain is 7.6549 dB Input reflection coefficient is of −27.010 dB and Output reflection coefficient is of −15.176 dB. The noise figure 3.869 dB was obtained in simulation.

References

1. Aspemyr L, Jacobsson H, Bao M, Sjöland H, Ferndal M, Carchon G (2006) A 15 GHz and a 20 GHz low noise amplifier in 90 nm RF CMOS. In: Topical meeting on silicon monolithic integrated circuits in RF systems
2. Ehrampoosh Sh, Hakimi A (2010) High gain CMOS low noise amplifier with 2.6 GHz bandwidth. In: 1st International conference on communications engineering, University of Sistan & Balochestan, Iran
3. Challal M, Azrar A, Bentarzi H, Ecioui R, Dehmas M, Vanhoen Acker Janvier D (2008) On low noise amplifier design for wireless communication systems. In: 3rd International conference on information and communication technologies: From theory to applications (ICTTA), pp 1–5
4. Afshar B, Niknejad AM (2006) X/Ku band CMOS LNA design techniques. In: IEEE custom integrated circuits conference (CICC), pp 389–392
5. Wong S-K, Kung F, Maisurah S, Osman MNB, Hui SJ (2009) Design of 3 to 5 GHz CMOS low noise amplifier for ultra-wideband (UWB) system. Prog Electromagn Res C 9:25–34
6. Yeh H-C (2012) Analysis and design of millimeter wave low-voltage CMOS cascode LNA with magnetic coupled technique. IEEE Trans Microwave Theory Tech 60:4066–4079
7. Pozar DM (2000) Microwave and RF wireless system (Chapter 10), 3rd edn. Wiley, New York
8. Razavi B (2011) RF microelectronics, 2nd edn. Prentice Hall, Upper Saddle River, NJ
9. Ludwig R, Bretchko P (2000) RF circuit design (Chapters 2 and 4). Prentice-Hall Inc., New Jersey. ISBN: 0-13-095323-7
10. http://www.avagotechnologies.com/pages/en/rf_microwave/transistors/ (online)
11. Djoumessi EE, Wu K (2009) Dual-band low-noise amplifier using Step-Impedance Resonator (SIR) technique for wireless system applications. In: European microwave conference (EuMC), Oct 2009. https://doi.org/10.23919/eumc.2009.5296387